# INTEGRATED COST AND SCHEDULE CONTROL FOR CONSTRUCTION PROJECTS

# INTEGRATED COST AND SCHEDULE CONTROL FOR CONSTRUCTION PROJECTS

## Frederick Wm. Mueller, CCE

Senior Vice President and
Chief Administrative Officer
The Oppel Company
Baltimore

VNR VAN NOSTRAND REINHOLD COMPANY
—————————————————————— New York

Copyright © 1986 by Van Nostrand Reinhold Company Inc.

Library of Congress Catalog Card Number: 85-26502
ISBN: 0-442-26169-1

Manufactured in the United States of America

Published by Van Nostrand Reinhold Company Inc.
115 Fifth Avenue
New York, New York 10003

Van Nostrand Reinhold Company Limited
Molly Millars Lane
Wokingham, Berkshire RG11 2PY, England

Van Nostrand Reinhold
480 Latrobe Street
Melbourne, Victoria 3000, Australia

Macmillan of Canada
Division of Gage Publishing Limited
164 Commander Boulevard
Agincourt, Ontario M1S 3C7, Canada

15 14 13 12 11 10 9 8 7 6 5 4 3 2 1

Library of Congress Cataloging-in-Publication Data

Mueller, Frederick William.
    Integrated cost and schedule control for construction projects.

    Includes index.
    1. Building—Estimates.   2. Construction industry—
Management.   3. Scheduling (Management)   I. Title.
TH435.M84   1986      692'.5      85-26502
ISBN 0-442-26169-1

*To my wife, Jane, my constant helper throughout the entire writing, who through the several drafts listened to every word on tape at least twice and typed every word in the manuscript several times. As her understanding of the content and material grew, she served additionally as an editorial advisor, contributing immeasurably to getting the work into printed form. With all my love, Fred.*

# PREFACE

Management and administrative processes within the construction industry have been undergoing major changes in the last several decades. These changes have involved significant adjustments in management science and management techniques, brought about by the need for contemporary valid information with which to manage the construction process. In short, management in the construction industry is changing significantly; change will continue at an accelerated pace at least through the next decade. The responses required of construction industry management are now resulting in a movement away from an entrepreneurial management style to professional management techniques and procedures.

## THE COMPELLING ECONOMIC ISSUES

The issues forcing these changes are economic. The rising costs of construction and of money are forcing the buyers of construction services to be more demanding. Their demands are for more construction economies, more production, and more productivity than at any time in the past. Nowhere has this been more evident than in the Business Roundtable on construction and in the response of the construction industry to it.* To be successfully responsive, management in the construction industry will be required to use the best project management methods available for cost control, schedule control, and for financial and accounting controls. But responsive professional management can survive and will flourish within this more demanding economic environment.

The need (or rather, the demand by those who buy and pay for the products of construction) for better management information systems and the expectation that computer technology will make these available have been, perhaps, the two most significant factors in those bringing about changes. There have been a variety of new developments in these areas, some of which have been

---

* *More Construction for the Money: Summary Report of the Construction Industry Cost Effectiveness Project* (January 1983), The Business Roundtable, 1200 Park Ave., New York, NY 10166.

interesting and exciting, such as the advent of the critical path method for construction planning and scheduling and good contemporary job cost systems, the integration of which holds great promise. A number of attempts have been made to integrate cost and schedule controls and to make these a routine part of the construction management process. The beneficial results of integrated cost and schedule control will be quite extraordinary to the construction industry.

## USING THIS BOOK

The purpose of this professional reference on integrated cost and schedule control is to place in a single volume the current management science available for systems integration to facilitate cost and schedule control. While the context of the examples and the vernacular are of the construction industry, it will be clear to the project management professional in any industry that the procedures for integrated cost and schedule control presented here are universally applicable to any project management undertaking without regard to the industry context.

Other references have dealt with limited aspects of planning and scheduling and cost control. One of the primary objectives of this reference is to present a complete overview of the principal issues affecting integrated cost and schedule control for any project, beginning with the preparation of a project for construction and proceeding through its completion and closeout. The views presented take quite literally the concept that "control" means the ability to influence outcome. Thus significant material is presented on monitoring a project in work to expose poor performance and to correct that performance, controlling the ultimate outcome of the project from a cost and scheduling point of view.

This book has been prepared as a reference for the experienced management practioner. It, therefore, assumes a certain level of understanding and sophistication concerning the industry and its processes. It is the result of many years of experience in managing construction projects and in managing construction business enterprises. It fairly represents the author's utmost effort to utilize the best of contemporary management science.

## ORGANIZATION OF THE CONTENTS

It seemed convenient to divide the integrated cost and schedule control of a project into four major arenas of action. The first of these, **Part I, Preparing a Project for Construction** (Chapters 1 through 13), deals extensively with preparing a project for construction. A dominant view in this reference is that one of the vital keys to integrated cost and schedule control is formalizing the preparatory processes and dealing with these tasks in the context of their deserved importance.

In addition to the basic preparatory tasks and to physical construction in the field, a variety of administrative activities must be routinely addressed in order for the project to be managed and controlled in terms of cost and schedule. **Part II, Administrative Project Activity Flow** (Chapters 14 through 19), deals with administrative activity flow.

**Part III, Maximizing Construction and Production** (Chapters 20 through 25), deals primarily with the physical management of the project in the field environment utilizing the informational and administrative processes which were earlier described. The crucial statement of expectations for production management and the closely associated data capture issues are presented in detail. The theme of Part III is that the proper planning and scheduling of a job, with a well articulated, communicated, and managed statement of expectations, lead to good field production.

Even with the most intensive and thorough planning, scheduling, and budgeting, however, not *all* activities involved in the project plan will be performed as expected. **Part IV, Project Monitoring and Control** (Chapters 26–31) therefore covers the general area of project monitoring and control. Techniques are developed for measuring performance and for generating contemporary and valid management information on that performance. Management by exception to minimize the management work load is presented in detail. It is in Part IV that the true power of strong cost and schedule control to influence outcomes in construction becomes the most apparent.

## ALL WORK ENVIRONMENTS CONSIDERED

The procedures recommended in this reference fully integrate all major work environments in construction. The purpose of this integration is to maximize the quality and timeliness of the information provided for management purposes; it extends through all financial and accounting areas as well. The full and complete integration advocated by this reference also takes into account appropriate computer applications in all areas of control when this resource is available.

There can be little doubt in the minds of project managers and construction professionals of the value of integrated cost and schedule control. It is therefore hoped that the utilization of the processes and procedures described in this reference, which represent extensive research, development and testing over many years, will result in increased production and improved productivity in construction, assuring consistent profitability.

The efficiencies resulting from the full integration of all management disciplines will be essential to construction industry management in the future.

The contemporary management information processes presented in this reference allow management to act responsively and responsibly in the control of costs, schedules, and cash flows of construction projects. Active management is accomplished with minimum effort resulting from the efficiencies of integration; and, as a consequence, the demand for "more construction for the money" by those who buy and pay for the products of construction may be more nearly met.

FREDERICK WM. MUELLER, CCE

# ACKNOWLEDGMENTS

*Integrated Cost and Schedule Control for Construction Projects* represents an intensive effort on the part of the author covering a period of ten years. During this time, many people have been involved in many different ways. It would be impossible to give appropriate recognition to all who have contributed significantly to the development, understandings and content of the work itself. These include employers, mentors, and professional colleagues in the American Association of Cost Engineers, the American Society of Civil Engineers, the Project Management Institute, and the Construction Management Association of America.

One of my discoveries in working on this book was that a total commitment of time was necessary to reduce to writing many of the things being done routinely in the business environment in order to present them in a professional reference. In this regard, I am especially indebted to my colleagues at Norwood Industrial Construction Company, Inc., for allowing me the professional sabbatical which was required to complete the work, and to Charles M. Soltis for his encouragement to undertake the work in the first place. Without this assistance and without the innovative and creative environment provided by Norwood Industrial Construction Company, the task of compiling this reference would have been almost insurmountably more difficult.

To each of those who have been of help to me in so many different ways, including Gerald Galbo of Van Nostrand Reinhold, whose commentary on the early drafts of the manuscript was most helpful, and Walter Brownfield, also of Van Nostrand Reinhold, who labored with me over many months in improving the work's presentation, please accept my deepest appreciation. My expectation is that the material presented will be helpful in creating a better understanding of the processes of integrated cost and schedule control. I believe that integrated cost and schedule control holds the future of management science for the construction industry and trust that this reference will aid its implementation.

F. W. M.

# CONTENTS

# EXHIBITS

# INTEGRATED COST AND SCHEDULE CONTROL FOR CONSTRUCTION PROJECTS

# I
# Preparing a Project for Construction

The first 13 chapters of this reference deal extensively with the processes and procedures which are necessary for preparing a project for construction utilizing integrated cost and schedule control techniques. Chapters 1 through 6 develop the basic concepts which are essential to complete systems integration.

Since systems integration is inseparably bound to effective communications among the work environments and disciplines of construction, these concepts are developed in deatail in Chapter 1. Chapter 2 presents the result of basic research and reveals the discovery of the project activity as the lowest common unit of work which may be used for communications within all work environments and disciplines and which serves as the primary factor for systems integration.

The need for adequate job planning, scheduling, and budgeting is covered in detail in Chapter 3. The importance of setting out in advance the project statement of expectations to provide a basis for measuring performance and for evaluating that performance is developed in detail.

One of the major concerns in integrated cost and schedule control is providing contemporary and valid performance information. These techniques are greatly facilitated by single data capture processes as a part of integrated systems. These are more fully developed in Chapter 4.

Management information as clearly distinguished from accounting information in terms of time schedules and priorities is examined in Chapter 5. Techniques for quick and convenient generation of management information from performance data are fully developed.

No contemporary reference on integrated cost and schedule control would be complete without considering computer applications, an overview of which is provided in Chapter 6.

Chapter 7 through 13 more fully develop the idea of a project statement of expectations and, more specifically, the creation of a series of primary and derived performance models against which to measure actual performance. Chapter 7 deals with the project statement of expectations in the context of its use as a performance model. Chapters 8 and 9 deal with the two primary performance models which serve as the foundation for the remaining performance models and which, in fact, fix virtually all other performance parameters. Chapter 8 presents estimates and budgets as performance models and Chapter 9, the planned schedule as a performance model.

Five separate performance models are derived from the two primary performance models and these are discussed through Chapters 10, 11 and 12. The specific graphic techniques and tabular data associated with the graphic techniques are fully developed in these chapters.

Chapter 13, which turns out to be this book's longest and most complex discussion (because of the extra care its presentation requires), covers the analytical processes of intense interest from both a project management and a business management point of view. Production forecasting and analysis techniques are developed in great detail as are methods of cash analyses. Rational methods of developing the kinds of information required where specific performance data are not otherwise available are fully developed in Chapter 13.

# 1
# WORK ENVIRONMENTS WITHIN THE CONSTRUCTION INDUSTRY

The construction industry is characterized by its size, diversity and complexity. It is also characterized by its major work environments and disciplines. Unlike other industries, which tend to have a single focus, the construction industry is made up of a number of separate work environments and separate disciplines. Each of these tends to have its own particular characteristics while, at the same time, forming an integral part of the construction industry.

Any professional reference on integrated cost and schedule control must take into account those unique characteristics of each work environment and discipline. Systems integration presupposes effective interaction and communication within the construction business environment. In establishing an appropriate framework of reference, it will be useful to review the primary work environments and disciplines which make up the construction industry and to examine how data and information transfer between these environments. Effective communication of data and information is essential to the kind of integrated control systems which have the ability to examine performance trends, to influence outcome, and to modify adverse trends in performance.

## WORK ENVIRONMENTS IN CONSTRUCTION

The control processes for systems integration require effective working relationships as well as mutual understandings among divergent work environments and disciplines, each of which has its own internal technical requirements, vocabulary, and processes. Each discipline must effectively communicate the results of its operation and processes to those other dependent disciplines for information each needs to carry out its responsibilities. Each discipline and work environment also has its own internal processes which are carried out with a high degree of independence, and, further, each has external processes which are vital to others if the construction enterprise is to be carried out successfully and profitably. The internal and external processes of nine work environments and disciplines are presented in a functional context to examine informational requirements for systems integration. Since these are functional in nature, they are not intended to be a complete listing or to represent a framework for any organizational model.

The following nine different functional work environments are necessary to carry out the responsibilities and job tasks of the construction industry enterprise. These are listed generally in the way responsibility flows during the development of a construction project.

- Business management environment;
- Project management environment;
- Sales and marketing environment;
- Design and engineering environment;
- Estimating environment;
- Purchasing environment;
- Contract administration environment;
- Field construction environment;
- Financial management and accounting environment.

Each of these will be discussed in the context of its primary purpose or function and its general relationships with other areas. The communications issues including such items as languages and processes, internal communications, and both primary and secondary external communications will also be examined. A key discussion for each environment will be potential difficulties in communications.

## THE BUSINESS MANAGEMENT ENVIRONMENT

The construction business enterprise has at its center effective business management. Effective business management touches all other areas and work environments in direct and significant ways. It provides overall guidance and direction to the other work environments. It serves as a coordinating agency and synthesizes the efforts of all work environments into singular and purposeful management of the construction enterprise.

Exhibit 1-1 provides a simple diagram of the construction business environment, which clearly indicates business management at the center of the construction business enterprise. While it may be argued that construction should be at the center, and while the company management may have as its primary area of expertise some particular kind of construction, business management nonetheless provides goals, direction, purposes, and the overall management of the resources required to address each area of responsibility.

Business management universally touches all other work environments. In subsequent diagrams, business management is distinguished and set apart

**Exhibit 1-1. Construction Business Environment**

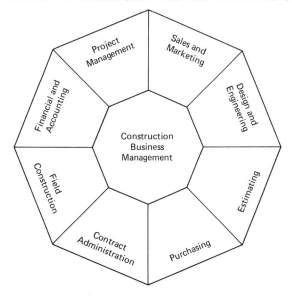

from other work environments because of its universal influence on all other areas. In cooperation with other work environments, it establishes sales and marketing goals and objectives, design criteria, estimating procedures, purchasing guidelines, contract administrative processes, and organization for field construction. It also relates to financial management and accounting in providing guidelines for cash management and in establishing working capital requirements.

The language of business management tends to be general in nature with its processes and procedures directed toward overall goal setting, long range planning, along with some monitoring and responding to general business trends. Its major focus may be on long range planning with a three to five year horizon and the development of more specific objectives in the one to three year range. These may include organizational growth objectives, business volume, and profitability. Short, mid, and long term planning is usually set out in the form of measurable objectives so that actual performance may be compared to planned performance. Internal communications within business management, being general in nature, tend to be clearly distinctive from those of the various technical specialties with which business management must communicate.

Business management must maintain primary communication channels with all other work environments. Within these interdisciplinary communications, business management must provide information in sufficient detail to be useful to the other work environments. Conversely, the other work environments have a similar responsibility to communicate with business management at the appropriate summary levels so that time and energy will not be dissipated on unnecessary detail which may be of great interest to a technical specialty but not useful from an overall business management point of view. Some degree of skill and organizational maturity is required to establish the appropriate level of communication.

Difficulties in communications between business management and other work environments are most likely to occur when business management fails to provide clear direction in terms of measurable goals and objectives and a clear statement of expectations, that is, of what it requires. Overall goals and objectives should be clear, in writing, and presented in ways which are useful and meaningful to the other work environments. Systems integration must take into account the communications requirements within the *construction* business environment and provide the vehicle for effective communication of information and data throughout the construction business enterprise. Communications to business management should similarly be provided in the same context as goals and objectives for easy, accurate, and convenient comparisons of planned performance to actual performance. These basic guidelines will be repeated throughout this reference.

## THE PROJECT MANAGEMENT ENVIRONMENT

Functionally, project management may be defined as the unifying element which results in construction itself. Project management is involved with all other work environments but in a significantly different way from business management. Project management is a resource coordination function which marshals the efforts of all other work environments specifically toward producing a construction project at or below budget, within the construction schedule, and at a quality appropriate to the contract documents.

Exhibit 1-2 shows the interrelationship between project management and the other work environments. Note that the business management environment is separately shown to indicate its unique and universal relationship to all other work environments in an overall control or management sense.

Project management must maintain continuing working relationships which are primary with all other work environments. It provides the appropriate technical support to the sales and marketing effort of the organization. It provides input into the design and engineering functions which are concerned with converting an owner's needs into technically accurate drawings and specifications. Project management is frequently called upon to provide cost engineering analyses and value engineering input for the design and engineer-

**Exhibit 1-2. Project Management Environment**

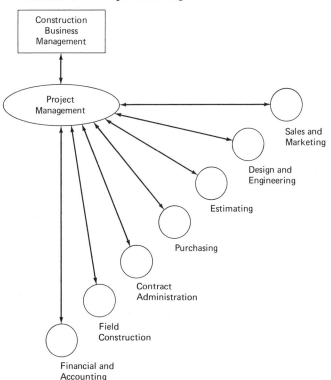

ing processes. Project management is inseparably bound to the estimating function, frequently maintaining the primary responsibility for the accuracy and validity of estimates even when a separate estimating work environment may be a part of the organizational structure.

The work of the purchasing environment is also of major concern to project management since this is the process by which the team is selected that will be required to work together in the construction of a project. Project management must also maintain close liaison with contract administration, which deals with the issues of contract terms and conditions, claims, extensions of time, and the protection of the rights of the parties to the contract. Project management must also maintain close coordination with field construction and provide assistance in expediting the flow of the work. Finally, project management must be inseparably bound to financial management and accounting, particularly in the areas of cash management and cash flow; project management must provide input for the generation of accounts receivable, assistance in collecting receivables, and review and approval of accounts payable.

Project management has languages and processes uniquely its own. Most of these are designed to facilitate communication and coordination with other work environments. These language and processes may include the use of planning and scheduling techniques such as the critical path method which may be second nature to project management but may not be very useful for field construction. The conversion of a logic network diagram to time scale for production control to be used by field construction may be required to facilitate communication with field construction. In the budgeting process, however, the budget must be responsive not only to the requirements of field construction but also to the requirements of financial and accounting.

All the external communications between project management and the other work environments making up the construction enterprise are of primary importance. None is more or less important than the others. In summary, the project management function must be the primary communications coordinator for all other work environments.

Difficulties may develop in communications where the assumption is made that certain information may or may not be needed by another work environment. One of the primary purposes of integrated cost and schedule control systems is to make the information and data generated by project management universally available to all other work environments which require such data and information and to provide it to them in the form required.

## THE SALES AND MARKETING ENVIRONMENT

The sales and marketing environment tends to be more specialized and more focused than either the business management environment or the project management environment. Sales and marketing's primary concern is that of generating the total business volume in keeping with goals established by business management. This includes establishing the proper mix of work between different types of construction, maintaining an appropriate client base and acquiring work through the various channels designated by business management. It is worth noting that in some construction business environments the estimating function may comprise the only sales and marketing effort on the part of the company. Even in these cases, the function of sales and marketing through estimating applies nonetheless. Sales and marketing is the key to establishing and maintaining the business volume of the kinds of work appropriate to the organization and at the levels of profitability established by business management.

Exhibit 1-3 shows diagrammatically the communications interface between the sales and marketing environment and other work environments.

**Exhibit 1-3. Sales and Marketing Environment**

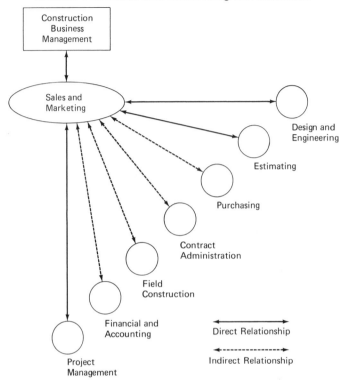

It is worth noting that in addition to the primary communications between business management and sales and marketing, other primary communications channels exist between sales and marketing and project management, sales and marketing and estimating, and sales and marketing and design and engineering. These primary communictions channels provide the sources of information used by sales and marketing in pursuing its responsibilities. Secondary communications channels informational in nature and helpful to sales and marketing and client maintenance include purchasing, contract administration, field construction, and financial and accounting.

The general relationships between sales and marketing and the other work environments must be such as to allow sales and marketing to fairly represent the qualities and strengths of the organization to potential clients. Of particular importance are the relationships between sales and marketing and project management and sales and marketing and field construction since these organizational resources represent the real product of sales and marketing.

Sales and marketing is confronted with several key communications issues.

First, it must have the ability to understand the languages and processes of project management, estimating, and field construction. Second, it must also have the ability to translate the highly technical language of these disciplines into the nontechnical language of the owner and the designer in order to properly communicate what is being proposed. It is equally important that sales and marketing communicate fairly the performance expectations being proposed by the organization. This requires the ability to communicate with the technical staff and to make certain that performance expectations are fairly presented to the client. Limitations in the organization's ability to perform, which could have an adverse effect on subsequent client relations if improperly handled, must be communicated by sales and marketing in a way which will not adversely affect those relationships.

The *internal* communications processes of sales and marketing are concerned with developing positive images which can be faithfully presented and used, converting sales and marketing goals and objectives established by business management into specific marketing strategies. These processes must support the organization's specific sales strategies which result in signed contracts for proposed projects through the negotiating or bidding process.

The *external* communications responsibilities of sales and marketing are concerned primarily with creating the interface between organization and client which fosters confidence and a desire to work together. The primary difficulties in communications experienced by sales and marketing arise in the conflict between the desire to make the deal and the realistic interest of the company.

## THE DESIGN AND ENGINEERING ENVIRONMENT

The design and engineering work environment in construction is concerned with converting the owner's needs and resource limitations into technically competent drawings and specifications which fairly and accurately reflect the scope of the work. The design and engineering function may be a part of the construction organization. More frequently, the design and engineering function will be a separate professional organization with its own separate interest, goals, and objectives. Design and engineering is nonetheless an essential and integral part of the construction business enterprise. The quality and accuracy of the work of the design and engineering function is frequently a major factor in how well the project is executed.

Exhibit 1-4 shows the primary and secondary relationships which exist between design and engineering and the other work environments of construction. The relationships, both primary and secondary, will vary, depending on organizational relationships between the design and engineering organization and the other entities within the construction enterprise.

**Exhibit 1-4. Design and Engineering Environment**

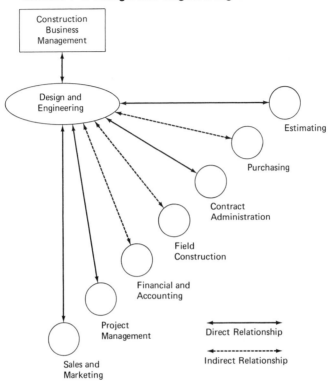

The general relationships of the design and engineering work environment to other work environments in construction are controlled by organizational and contract considerations. Where these functions are housed in separate organizations, special communications issues are involved. Contractural relationships are usually drawn between the owner and the design and engineering function and separately between the owner and the construction function. These create the basis for potential disagreements because of different perspectives and perceptions of the owner's needs.

The design and engineering functions have their own specialized language and processes. These must be articulated in the form of drawings and specifications which can be clearly understood and interpreted by such widely differing groups as the owner, the contractor, and the craftsperson in the field environment.

Internal communications issues also exist between the various disciplines in the design and engineering work environment. Architectural considerations may differ significantly from engineering considerations. Part of the problem of communication is resolving these differences, requiring special consideration. Similar differences may exist with the construction work environments which are required to produce the project shown in the drawings and specifications.

In addition to the primary communications between the owner and the design and engineering work environment, other primary communications channels will include the estimating, contract administration, project management and sales and marketing work environments. Field construction is considered a secondary communications channel since, by the time the drawings and specifications for a project reach the field construction stage, all the technical issues with respect to the project should have been resolved. Other secondary design and engineering communications channels include financial and accounting and purchasing.

## THE ESTIMATING ENVIRONMENT

The estimating work environment is responsible for preparing accurate and valid quantifications of projects to be bid in order to establish both the market value and fair value of each project based on historic cost and on the basis of estimating's data, and to prepare estimates for competitive bids and for negotiating. The estimating environment in many construction organizations is the primary sales and marketing function. In pursuing these tasks, estimating is responsible for obtaining bid documents for quantity surveying and pricing; for soliciting, collecting, and analyzing subcontractor and vendor prices; for preparing estimates and summaries; and, in cooperation with sales and marketing and business management, to determine bid prices.

Exhibit 1-5 shows the estimating work environment and indicates the primary and secondary communications channels appropriate for estimating. The business management environment continues to be a prime mover in the entire construction work environment; thus the primary communications channels between business management and estimating are separately depicted.

Since most companies are not successful on every estimate prepared for bid purposes, estimating needs to work with a high degree of independence. It is necessary to draw on other work environments such as field construction and project management for input and support in the estimating processes. Prebid analyses and market analyses are important tasks of estimating so that recommendations may be made to sales and marketing and to business management concerning the appropriate bid strategies for various types and classes of work.

Estimating, like other work environments, has certain languages and processes which are uniquely its own. Estimating must, however, be aware that

## Exhibit 1-5. Estimating Environment

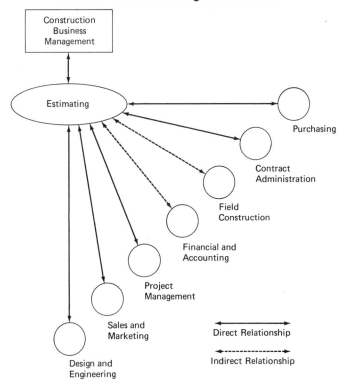

its work creates the foundation for all other informational processes which follow when the organization is successful in acquiring work based on the estimates prepared. In integrated systems, estimating should be done by activity and summarized by trade groups in order to facilitate subsequent budgeting, planning and scheduling tasks, and preparing a project for construction. Where quantity surveying is done separately from pricing, internal communications become particularly critical and some organized system of checks, balances and verification must be constantly employed to avoid errors and omissions in estimating.

Externally, estimating's primary channels of communication will be with sales and marketing and with business management. Validity in the estimating process is a crucial part of the overall interdepartmental or work environment relationships. Estimating will maintain primary working relationships and communications with project management in transferring the lead responsibility to project management as each new project is acquired. Estimating also maintains primary communications with design and engineering, frequently to assist project management in performing various cost engineering and value engineering tasks. The relationship between estimating and purchasing is particularly close since purchasing must be certain to buy from subcontractors and vendors exactly what has been sold to the client by sales and marketing based on the work of estimating.

Another primary communications responsibility of estimating is with contract administration. It may be reasonably assumed that there will be changes or disputes, or both, in any construction contract. Effects of changes in terms of cost, time, conditions of work, or conditions of the contract must be documented and evaluated. Certainly, estimating has a primary responsibility to determine the cost effect of the changes.

Secondary communications channels are maintained with field construction. It is assumed that by the time a project has been assigned to field construction, a project budget superseding the estimate will have been established and the production plan and schedule prepared by project management.

Secondary communications are also maintained with financial and accounting since other more formal documentation such as the production budget and schedule of values will have intervened between the estimate and the formal financial and accounting processes. Purchasing will also have intervened between estimating and financial and accounting with the more formal commitments of purchase orders and subcontracts which take the place of estimates and budgets.

It is necessary for estimating to work closely with sales and marketing and business management in selecting the kinds of work for bidding. The projects selected for bidding should bear a reasonable relationship to the organization's skills, expertise and ability to perform the work. In addition, estimates must be prepared with some understanding concerning how resources will be applied in the performance of the work when the bids are successful.

Few companies are successful on every bid they propose. Each estimate must, however, be prepared by the estimating environment on the assumption that the company will find it necessary to build every job which is bid.

## THE PURCHASING ENVIRONMENT

The purchasing environment is assigned the major task of simultaneously selecting the subcontractors and principal vendors who can perform the work required by the contract documents and who have the appropriate levels of skills and resources to become members of the project construction team. The selection criteria between price and performance are frequently in conflict; thus the purchasing function needs the input and expertise of several other work environments in order to effectively address its responsibilities. The major functional guideline for purchasing is to buy from subcontractors and principal vendors exactly what has been sold to the owner as represented by the contract documents.

Exhibit 1-6 shows the primary and secondary relationships between the purchasing environment and the other work environments in the construction enterprise. Business management also maintains its primary relationship with purchasing as it does with the other work environments.

Purchasing uses as its major control criteria (1) the plan and schedule for the project, usually prepared by project management, and (2) the budget, which may be established by project management or by business management. Budget issues should be dealt with prior to tendering a bid for the project so that the budgeting criteria become an integral part of the bid strategy. Purchasing must maintain very close primary working relationships with several other work environments. It also has language and processes all its own. It is especially concerned with the legal implication of agreements. Terms and conditions of contracts and purchase orders are its crucial consider-

## Exhibit 1-6. Purchasing Environment

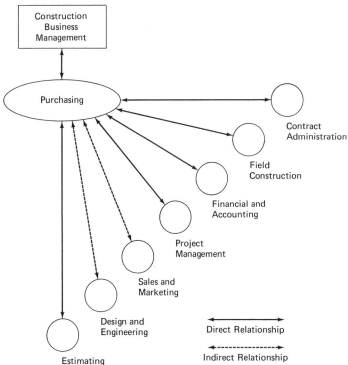

ations. The responsibilities of purchasing include the resolution of cost, schedule, conditions of the work, and terms and conditions of the subcontract and purchase documents.

Purchasing must maintain primary working relationships with estimating and draw upon estimating's knowledge when the contract documents may not fairly or completely describe the scope of the work. Some purchasing is frequently performed directly by project management; whatever the company's organization, the relationships between the purchasing and project management functions are, therefore, especially sensitive. Since purchasing establishes many of the controlling guidelines for financial and accounting, especially in dealing with accounts payable, the relationship between purchasing and financial and accounting must be especially close. The best written subcontracts and purchase orders to principal vendors will not eliminate the potential for disagreements in the field environment concerning the several responsibilities of the subcontractors and vendors.

Similarly, contract administration must rely heavily on communications from purchasing in order to resolve questions or disagreements concerning the scope of the work, especially inclusions and exclusions which result in claims, charges, and back charges.

Of less concern to purchasing are the design and engineering functions. By the time a project reaches the purchasing stage, the design and engineering issues should be fully resolved. Purchasing's responsibility is to buy what the contract documents require. There may be some input from purchasing into the cost engineering and value engineering processes. Such input usually occurs in a consulting or advisory role much earlier than the actual purchasing processes of the project. Similarly, the relationship between sales and marketing and purchasing is secondary in nature.

The purchasing environment is a service function charged with the distribution of technical data and pricing requirements to the subcontractor and vendor marketplace. It is responsible for collection and verification of competitive and comparative prices for the components of the project and in analyzing the data. It is also responsible for negotiating subcontracts and issuing purchase orders within the limitations of budget and schedule. Some management exception process is usually operative between purchasing and the business management environment when subcontracts cannot be let within the limitations of budget and schedule. The purchasing function must be carried out efficiently in order to maintain budget control of a project and it must be accomplished expeditiously so that materials, goods, and services will be available at the job site when these are required for field construction.

The difficulties encountered with purchasing and communications occur when subcontracts and purchase orders are not clearly written in the context of the general contract. The purchasing environment involves a high degree of discipline and organization. It deals with highly technical issues of legal consequence. These tend to be formally and efficiently handled when external legal counsel is used.

## CONTRACT ADMINISTRATION ENVIRONMENT

Contract administration in a construction organization is primarily a service function. It is responsible for seeing that the general contract and all subcontracts and purchase orders are faithfully performed and administered within the terms and conditions set forth for each. Contract administration is responsible for seeing that some master listing of claims and changes is maintained for the timely documentation of all claims and for assertion of these claims with the appropriate parties. In its truest sense, contract administration calls upon all parties to see that the contract is faithfully carried out. It is responsible for documenting all deviations from the contract. The *negotiations* required to resolve open disputes is not a normal contract administration function. Such negotiations are normally carried out by some other responsible work environment such as business management, project management, or sales and marketing.

In Exhibit 1-7, only sales and marketing is indicated as a secondary communications channel for contract administration. The relationship between sales and marketing and contract administration will deal primarily with issues of customer or client maintenance. Sales and marketing may develop a primary relationship with contracts administration in assisting in resolving claims and disputes with a client.

Contract administration, like purchasing, has its own language and processes. Its work is legalistic in nature and requires a high degree of discipline in following the many details of the contract administrative process. Effective contract administration requires a highly organized and systematic set of procedures which allow all contract administration issues to be identified upon occurrence in order to place these within an administrative system which, through its normal course and processes, result in the final disposition of all issues on a timely basis. Contract administration must be organized internally in a way that allows the efficient collecting of data from a wide variety of sources; it is dependent on routine flow of information and data from all the other work environments.

Since contract administration is most active when a project is in actual construction and since most problems come to light in the field environment, the working relationship between field construction and contract administration must be especially close. Feedback to contract administration from field construction must be within a systematic routine which allows for communication of data with the least amount of special conditions.

**Exhibit 1-7. Contract Administration Environment**

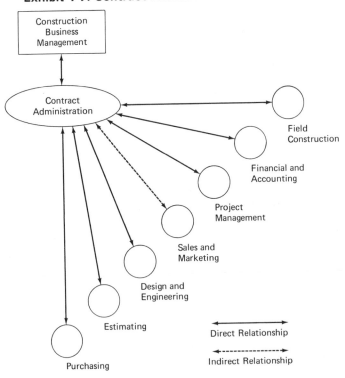

Contract administration is inseparably involved with financial and accounting. This involvement is intense both in the accounts receivable and the accounts payable areas. For accounts receivable, it processes claims to the owner for things which affect the project from a time, cost, terms and conditions, and conditions of work point of view with respect to field construction. Claims must be asserted and promptly brought to a conclusion so that financial and accounting may properly include these in accounts receivable for timely collection. Similar circumstances relate to accounts payable, where claims will develop with subcontractors and principal vendors out of the field environment. These must be similarly resolved on a timely basis so that the accounts payable process will not be impeded. The relationship between accounts receivable and accounts payable in the claims area is particularly important to maintain the buy/sell position which is a fundamental criteria of purchasing.

Contract administration must also maintain very close communications with project management as a primary source of information concerning proposed changes in the work which will come through the project management liaison with the client or with design and engineering.

Contract administration is especially involved with design and engineering since many claims issues will involve the quality of the contract documents including the drawings and specifications. Design and engineering interpretation of claims will frequently be required. Where claims or contract administrative issues involve known or deliberate changes in the scope of the work, contract administration will require design and engineering input to properly define the scope of the work. Contract administration is especially dependent on estimating to provide the formal documentation with respect to pricing to establish the cost or value of the work. In the same way, purchasing and contract administration must maintain close working relationships since, in many respects, their responsibilities are similar. Purchasing deals with establishing the working relationship with the subcontractors and principal vendors. Contract administration deals with maintaining that relationship so it remains orderly and productive.

Contract administration deals with the review and preparation of contract documents, review of terms and conditions of contract, and all of the administrative processes for handling changes in the contract. It must receive current and valid information on the production processes from the field environment and other areas to determine what changes in the work are under consideration and which require administrative processing.

## FIELD CONSTRUCTION ENVIRONMENT

Field construction is at the very center of the construction enterprise. It is where production is accomplished and where product is placed. It is the major arena (or battlefield) where issues of cost and schedule control are determined. The integration of cost and schedule control must become operative in the field environment if they are to be effective at all. The responsibility of the other work environments has to do with setting up a project for efficient production. The effectiveness of the planning and scheduling and budgeting processes for a job has significant impact on the ability of field construction to perform its responsibilities.

Field construction maintains its primary relationship with business management. Exhibit 1-8 shows the field construction environment with its primary and secondary relationships.

The field environment deals with the day to day assignment of resources, the productivity as well as the production of those resources, the scheduling of those resources, and the completion of the work. One of the major keys to integrated cost and schedule control has been discovered in carefully observing the languages and processes of field construction. Field construction usually assigns resources on a daily basis and by activities. It talks in terms of exact definable tasks which are to be started, worked on, and completed

**Exhibit 1-8. Field Construction Environment**

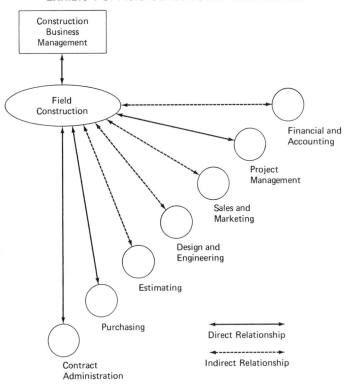

within a definitive time period by a specific resource. The language of field construction and its processs are *specific, direct,* and *measurable.* Field construction deals with measurements of feet and inches, cubic yards of concrete, specific equipment required, specific work force requirements, and similar specific items. Internal communications within the field construction environment must constantly be detailed and specific if the project is to be accomplished with the highest degree of productivity attainable.

Since field construction is ordered by activities and the specific resources to perform those activities, it is reasonable to base the project budget and schedule, otherwise known as the *project strategy,* on the specific informational needs and processes of the field environment.

Field construction maintains a primary working relationship with project management. Project management and field construction are responsible for preparing the game plan and for establishing and maintaining the appropriate working relationships within the total construction team. Purchasing is especially important to field construction since the ability of field construction to perform its work depends on the timely purchasing of all goods, services, and materials required for the project and in expediting their availability at the job site when these are required by the construction schedule. Contract administration is also crucial to field construction since it provides for the resolution of the minor disputes and disagreements which may stand in the way of effective and efficient production.

Estimating stands as a secondary consulting resource to field construction since other, more formal processes, including budgeting and purchasing along with planning and scheduling, will have intervened between field construction and estimating. Similarly, design and engineering is available in the secondary relationship to field construction to resolve unanswered design and engineering questions which may occur during the construction process. A test of the efficiency of the design and engineering process is the absence of the need for field construction to be routinely and regularly in contact with design and engineering. Field construction's relationship with sales and marketing is secondary in nature and relates primarily to issues of customer satisfaction on the quality of work and quantity of production.

Relationships between financial and accounting and field construction deal principally with informational flow processes from field construction to financial and accounting. These relationships are considered to be secondary since they should be routine and factual in nature.

Field construction receives information on the product to be produced, changes which may occur in the product requirements, assignment of subcontractor and vendor resources to perform certain portions of the work, scheduling requirements, and budget constraints from the other work environments. Field construction provides information on its performance by reporting activities started and completed (and/or percent completed), by reporting labor as used in completing activities, and by reporting on subcontractor and vendor performance.

The methods used by field construction in assigning resources to activities and in capturing data on performances are a significant key in creating effective and usefully integrated control systems.

## FINANCIAL AND ACCOUNTING ENVIRONMENT

Financial and accounting is responsible for the overall management of the financial resources of the organization. It includes cash management, the management of accounts payable and accounts receivable, and all other formal accounting functions such as payroll, general ledger, and similar items. It is responsible for verification of project productivity and production in a historical and job cost context.

Internally, financial and accounting is a primary service function, maintaining the financial and accounting records of the organization. It is responsible for providing financial and accounting information to business management and other areas. Cash flow and cash management are critical accounting functions requiring current valid data.

Exhibit 1-9 shows the primary and secondary communications relationships which are part of the financial and accounting responsibilities.

Financial and accounting has its own special languages and processes which tend to be outside of the mainstream of construction and which are generally not understood by the technical work environments. Financial and accounting, therefore, has special communications problems. Internally it is dependent upon receiving information from all the other work environments and for interpreting or translating this information into useful accounting data.

Since financial and accounting concerns itself primarily with areas involving historic costs and expenditures, its principal relationships tend to focus on the other work environments with similar interests and responsibilities. Of

**Exhibit 1-9. Financial and Accounting Environment**

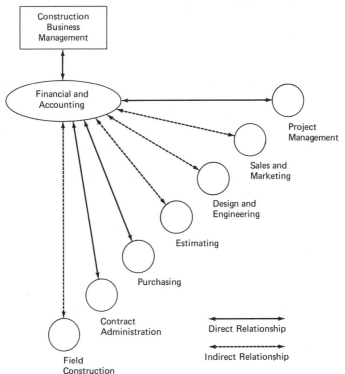

course, its relationship with business management is similar to those of all other work environments. Since project management may be identified as having a primary cost control responsibility, financial and accounting has a primary relationship with project management, particularly regarding the proper establishment of budgets for cost control. Here accurate information is crucial.

Purchasing involves a major commitment of corporate resources and the establishment of financial obligations. Purchasing is, therefore, another of the primary relationships for financial and accounting. Proper documentation of all commitments in the form of subcontracts and of purchase orders to principal vendors is especially important to financial and accounting; the terms and conditions of these agreements and orders control major corporate financial functions such as cash flow and cash management.

Contract administration is another primary relationship with financial and accounting since the former represents one of the main sources of cost accounting data. Contract administration deals with asserting and perfecting claims affecting cost and cost recovery, always important.

Financial and accounting's relationship with field construction is primarily concerned with the flow of performance information from field construction to financial and accounting. The principal concerns are raw data such as payroll hours and percentages of completion. Financial and accounting must be assured of the accuracy of the information reported so that it may be used for generation of payroll, for generation of requisitions, and for processing of accounts payable.

Estimating, design and engineering, and sales and marketing are similar secondary relationships for financial and accounting. These are somewhat removed from the actual production and cost accounting issues concerning financial and accounting.

## SUMMARY

The differences among the work environments which have been discussed are functional. Construction organizations will group and structure these in various ways depending on their individual organizational designs. The purpose in reviewing the disciplines and work environments is to focus on the data and informational interdependency and the consequent need for integrated systems of cost and schedule control.

The communication issues appear rather staggering when viewed in the context of these complex relationships between work environments having different special languages and processes. Effective communication can be established by the use of common systems and by integrating data and information flow. These procedures thus become important issues. While the primary focus of developing integrated systems is on the interfaces between work environments, the internal processes within each work environment may also be affected in significant ways. The principal impact on internal processes and systems within each discipline involves the anticipation of a need for information *by others*. The preparation of raw data and information must be in a form that can be conveniently transferred and efficiently utilized without the recasting of data or information by the recipient discipline. The successful integration of cost and schedule control is singularly dependent on both anticipating the need for information and providing it in the most useful, efficient, and timely way.

A major factor in developing procedures for transfer of information and data between work environments is limiting data and informational transfer to that which is useful or required. Considerable cooperation and dialogue is required to determine what information is useful and required, and to eliminate the transfer of information which may be of intense interest to one discipline but of no interest to another. Transferring unnecessary information always consumes organizational resources without benefit.

The processes for integrated cost and schedule control which will be presented are directed to the significant improvement of communications as a part of providing informational transfer in the most efficient way. Adjustments in the traditional methods of handling data and information will be required to gain the maximum effectiveness of integrated systems.

# 2
# COMMON DESCRIPTIONS OF WORK

Any serious attempt at systems integration must recognize that significant differences in both vocabulary and processes exist among the work environments of construction. These differences, which were reviewed in some detail in Chapter 1, increase the difficulty of systems integration but by no means make the task impossible. Since all work environments and disciplines involved in the construction enterprise must deal in one way or another with the products of construction and the processes by which construction projects are built, the task, though difficult, can be successfully accomplished. While the focus of responsibility and activity varies among the disciplines (sales and marketing, estimating, purchasing, field construction, project management, design and engineering, contracts administration, and overall business management), each will have as its focus the *product of construction.*

Since all work environments and disciplines deal in one way or another with the products of construction, it is not unreasonable to expect some commonality within all work environments and disciplines which may be used as the primary vehicle of communications and integration. The discovery of this commonality is the principal challenge in systems integration. With its discovery, the descriptors and processes may be consistent among the disciplines and work environments and will allow for the transfer of both data and information between the disciplines, eliminating the present practice of recasting both raw data and information preparatory to its use upon each transfer from discipline to discipline. The significant challenge is to eliminate the artificial or contrived differences and distinctions among the disciplines and work environments of construction which tend to polarize processes and procedures unnecessarily. *Appropriate* distinctions within each work environment must be recognized. What must be eliminated are the differences *which serve no useful purpose* and hence encumber the informational processes.

## JOB COST CHARTS OF ACCOUNTS

In the search for some commonality to serve as a vehicle for interdisciplinary communications, some initial help is derived from present practices in the industry. The generally accepted use of charts of accounts for both job costing and accounting may provide an initial basis for examining informational processes. The job cost chart of accounts may therefore be especially useful as a beginning point for examining commonalities for systems integration.

While job cost charts of accounts are in common usage for budgeting and job costing, the makeup of these charts of accounts may very widely. In some cases, these are specifically tailored to the needs of the particular construction project or company. In other cases, charts of accounts are more generic or general in nature. The construction industry has been working on a set standard descriptions of work for many years. The standard descriptions of work used in drawings and specifications provide a basis for more uniform descriptions of the work involved in construction. Where such descriptors and charts of accounts already exist, these may be conveniently utilized in developing job cost charts of accounts.

Systems integration requires a common job cost chart of accounts. The job cost chart of accounts, however, needs sufficient flexibility in the way it is developed to allow for an adequate level of cost control and to simultaneously allow for an added level of detail in planning and scheduling. The idea of using the job cost chart of accounts for planning and scheduling purposes has a certain degree of novelty associated with it. This concept will be developed more fully since it serves as the basis for integration of cost and schedule in the least complicated way. The chart of accounts must, therefore, be designed to accommodate the maximum level of detail required for either budget preparation or for the preparation of the plan and schedule for a project. The *common* listings for budget and schedule are important considerations.

## THE PROJECT STRATEGY AND STATEMENT OF EXPECTATIONS

The *project strategy* may be defined as the plan for building a project. In its simplest form, it defines how a project is to be physically constructed. When scheduling information is added to the plan, it is concerned with how long that construction is expected to take. Budget information states what costs are expected to be incurred in the performance of the project. More specifically, what is required to express a project strategy and statement of expectations is (1) a plan for its construction, (2) a budget, and (3) a schedule.

Depending on the procedures and practices within any construction enterprise, these planning, scheduling, and budgeting processes may be informal to the extent that they are seldom reduced to writing, or they may be rather elaborate. In either case, they are not normally integrated in the sense that any item found in the project plan and schedule will also be found and described in precisely the same way in the project budget. In order to integrate cost and schedule, however, one of two alternatives must be considered. The first, and perhaps the most complex, is some means of translating budget items into schedule items and schedule items into budget items. These translation and correlation processes tend to become so involved that they are simply ignored as a practical issue. As a result, the planning and scheduling of most projects is carried out with one criteria and one set of information while the budgeting processes are carried out with an entirely different set of information. As a result, the cost control functions and the scheduling control functions are carried out independent of each other with little or no attempt for integration other than at the crudest summary levels.

The second alternative is to attempt to devise a scheme which allows a common description of the items in budget and schedule with both budgeted cost and time assigned to each. The ability to look at a single item and deal with its budgeted cost and budgeted time has many interesting possibilities. For example, *performance* information on an activity could provide the basis for both cost and schedule control. It could provide the basis beyond the development of a complete project strategy to the implementation of that strategy to data capture on actual performance and would further provide for comparing actual performance to planned performance. Such data and informational processes would, then, provide sufficient current and valid information to design and implement corrective measures where these are required.

## MAJOR IMPEDIMENTS TO SYSTEMS INTEGRATION

What begins to emerge as a major criteria for simple integrated cost and schedule control is a common informational set for both costing and scheduling a project. As we have seen, a major impediment to the use of a common informational set is that each discipline uses data available in significantly different ways. The kind and quantity of data required also varies significantly among the disciplines. These differences must be recognized and resolved in order to establish the common informational set required for systems integration.

The difference in the kinds of information and data required by the separate disciplines is worth examining. Business management is interested in general goals and objectives, summary performance information, and the ability to generally forecast the outcome of each and all projects on a timely basis. Field construction, on the other hand, must be concerned with individual activities assigned to the work on a day by day basis and with specific feet and inches kinds of detail crucial to the proper performance of each project. Field construction interest in each activity involves both budgeted cost and budgeted time so that actual performance may be measured against expected performance.

In the context of summary and detail, the remaining work environments in construction fall somewhere in between. Sales and marketing is interested in the general composition of the bid package and general descriptions of the scope of the work. Design and engineering must concern itself with specific and precise detail so that these may be articulated in the form of drawings and specifications. The estimating discipline deals primarily with trade groupings along the general lines of the way the technical information is presented in the drawings and specifications. This activity usually involves dealing with the quantitative and qualitative aspects of the project in a way that allows for establishing a cost or value for each item. The purchasing environment focuses more directly on the contract scope of the work, proper description of the work, the cost to be paid for the work, the time allowed for its performance, the conditions under which it will be performed, and the terms and conditions associated with the performance of the subcontract.

Contract administration monitors carrying out the contracts, subcontracts, and purchase orders according to the terms and conditions of those documents. Contract administration also is concerned with changes in the work and the effect of those changes upon the contract. Financial and Accounting concerns itself with cash flow and cash management, direct job costs, indirect job costs, overhead and profitability.

Project management, with its general responsibility for the project, establishes the project strategy and manages the construction processes within the strategy for maximum profitability.

These very wide variances present significant impediments which must be overcome in order to develop a common informational set which may be used for systems integration. The different areas of *interest* among the work environments of construction present a major challenge in designing an integrated cost and schedule control system. What must be discovered is some commonality capable of providing maximum detail for production control while at the same time providing a variety of summarizing capabilities to meet the requirement of all other disciplines in construction.

On the other hand, while the motivation for systems integration is very strong, what must be avoided is unnecessarily complicating the work of any particular discipline for the sake of convenience or systems integration based on the requirements of one or more of the other disciplines.

## THE ACTIVITY AS THE COMMON DENOMINATOR

Two separate developments in recent years have led to identifying that commonality in construction which may be used as a major communications link between all disciplines of construction. First of these was the development in the late 1950s of a logic network technique for project analysis, planning, and scheduling. This technique was originally developed for use in the space program and for the development of the Polaris submarine project. Two versions of the logic network technique emerged, each with a different focus of attention.

Program evaluation and review technique (commonly known as "PERT") was primarily a probability analysis focusing on the likelihood of significant events, or milestones, occuring at desired times. PERT was less concerned with the activities required to achieve the milestones and focused its attention on the occurrence of the *milestones* themselves. PERT was particularly useful in the long term undertakings related to the space program and other similarly very large projects.

## The Critical Path Method and the Activity

The critical path method (commonly known as CPM) was developed with the focus of attention on the *activities* required to complete the project and the relationship between them. The duration of those activities and the resources required for the performance of the activities were used to arithmetically calculate the scheduling consequence of the project. The critical path method with its focus on the activity or things to be done held great promise for the construction industry since the construction industry was similarly activity oriented. As a result, a variety of critical path method programs were developed for construction application.

As it subsequently developed, there were two significant problems with CPM applications for construction. The first of these was the assumption that the detailed planning of a project would automatically result in project performance consistent with the project plan and schedule. The second came from the fact that the CPM techniques as powerful analytical tools created significantly more information than was needed or useful in production control in the field environment. The result of these two circumstances was for the critical path method to fall into disfavor in construction as a practical production control tool. Subsequent experience with the critical path method, however, confirms its validity as a powerful analytical tool and confirms that using the portions of the analytical output which are of value while discarding the remaining data provides worthwhile production control capabilities. These differences and distinctions will be more fully developed.

The second, more recent, major development is the clear recognition that job site construction is activity oriented and can be conveniently and effectively designed and described in the context of a roster of activities required to perform the project. In the search for a commonality to be used for work descriptions throughout all work environments and disciplines, the activity of job site construction emerges as the most likely choice for common descriptions. The technical description of an activity as a *time consuming task* was developed primarily out of the critical path method and other logic network techniques used for the planning and scheduling of projects. The use of activities as the basis for assigning the work force and resources was, however, more a matter of confirming how construction projects were built rather than as a result of any great new discovery. The use of activities on virtually all construction projects can be confirmed, even when these are not formally recognized or articulated in the form of a CPM schedule.

The activity basis for assigning the work force in the field envirnoment follows the commonly accepted practice in the industry. The activity may be conveniently utilized in developing formal production plans and schedules. Activities may also be grouped into work packages as required for purchasing. They may be summarized into trade groups for estimating purposes, and they may also be summarized into other categories appropriate for both accounts payable and accounts receivable.

The value of the activity as the commonality, or lowest common unit of work, is that, while the need for detail required for both cost budgets and scheduling is being addressed, activities may be summarized at whatever other levels may be appropriate to each work environment and discipline. A listing, or roster, of activities for a project may be prepared with the maximum detail required either for cost budgeting or for scheduling purposes. Once these descriptors and the roster of activities are established, groupings and summaries may be used for all the communication between work environments and disciplines. It would appear that the activity can indeed be used to create a mainstream of communication through the overall construction environment.

## The Technical Characteristics of the Activity

Further confirmation of the activity as the commonality throughout all work environments and disciplines occurs via an examination of its technical characteristics. A project may be defined as an undertaking involving a group of interrelated activities having a beginning, or origin, and an objective. The objective is accomplished as a result of completing the activities required to meet that objective. The critical path methodology is built on the desire to clearly articulate the interrelationship between activities and, as a consequence, to clearly and concisely define how the project is to be performed. Exhibit 2-1 shows three of the four relationships which may exist among activities. In the fourth relationship, the activities are unrelated.

An activity may be defined both as a time and resource consuming task and as an indivisible portion of the work. If one activity is said to be *precedent* to another one, it must be completed before the subsequent activity can commence. An activity is *subsequent* when it cannot start until the previous

**Exhibit 2-1. Activity Relationships: A Logic Network Diagram**

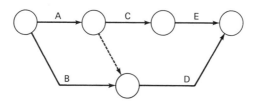

1. Precedent: A is precedent to C.
2. Subsequent: E is subsequent to C.
3. Concurrent: A and B are concurrent.
   D and E are concurrent.
4. Unrelated: B is unrelated to E.

activity is completed. Activities are *concurrent* when (1) they may be performed in the same time frame and (2) share a common event at either the head or tail. Activities are unrelated if they do not share a common event. All activities begin and end in an event; an *event* is a point in time between the completion of one activity, or group of activities, and the commencement of another activity, or group of activities.

Within the critical path methodology, each activity represents a *specific and definitive piece of work*. Each activity can, therefore, be quantified or measured and its value established by using the standard quantity times unit price equals cost algorithm ($Q \times U = C$). The available resource may be used to calculate the duration of the activity. Through simple computation routines, both the cost (budget) and the duration (time) may be established for each activity. When these two are combined within the context of a logic network diagram, the schedule may be calculated and the rate of expenditures may be determined based upon the schedule for the work. For cost and schedule control purposes, these processes assign to each and every activity making up the budget and the schedule, a planned cost and a planned duration.

Within the context of the logic network or arrow diagram, the overall schedule is also established. All other work environments and disciplines in construction may deal with each activity, groups of activities, or the entire project from the point of view of budgeted costs, budgeted time, quantities required, or quality of the work to be performed. As such, the activity (within the context of the critical path method) provides for precise communications throughout the construction environment where equally competent people will come to the same conclusion concerning the attributes of each activity. This is indeed an extraordinary communications capability.

## The Use of the Activity in Most Work Environments

To qualify an activity as the most likely candidate for commonality in all work environments and disciplines, it is necessary to examine how the activity may be utilized within each construction industry work environment and discipline. The business management environment will use activities summarized for business management, normally through multitiered summaries involving an increasing amount of detail which may be accessed to the extent business management requires.

In a multioffice environment, the first summary level may be the work of each individual office. To the extent of management's further interest in the composition of the work of each office, an individual office summary comprising single line performance information on each project might be examined. Where one or more projects may indicate some situation requires further examination, a trade summary of the individual job may be examined. Where the trade summary may indicate even further interest, the individual activities making up the trade summary may be explored.

It is worth noting that the validity of information at each summary level is dependent on the detail being valid and current at the activity level within each project. When the activity is used in this way for production control purposes, business management may be assured of the validity of any subsequent summary level. In this context, the general guideline is to plan and schedule by activity at the level of detail at which control is to be exerted and to use summaries to provide more concise information for business management purposes.

Project management has the overall responsibility for each project including the planning and scheduling tasks. In cooperation with estimating, project management must establish the roster of activities for the project as early in the processes as is practical. Once the roster of activities has been established, it may be used for all planning and scheduling tasks and for estimating tasks as well. As the project matures, the roster of activities is used also for budgeting purposes.

Estimating uses the roster of activities, developed by project management, to create the estimate and subsequently budget. When the estimate and budget for each project are established at the activity level, maximum information is generated for subsequent construction processes.

Purchasing may also use the roster of activities for integration of purchasing, accounts payable and accounts receivable with both the schedule and budget. The roster of activities within each work package may be used for creating a payments schedule for each subcontractor and principal vendor on the project.

Design and engineering is, perhaps, less concerned with the roster of activities than the other environments although some knowledge of the roster of activities which is to guide the construction of the project is very helpful in the design process. Design and engineering needs to take into account issues involving resources available for the construction of the project. This information is best communicated to design and engineering through the roster of activities containing both budget and schedule data.

Contract administration may use the roster of activities as a primary administrative control tool. Added activities which are candidates for claims which may require price adjustments with subcontractors and vendors come to light when inconsistent with the roster of activities. New activities may be assigned for contract administrative purposes as required to give visibility to claims and changes in the work.

Field construction, perhaps, requires the greatest amount of detail and should be the primary source of information concerning the detail to be included in the roster of activities. What is to be avoided, however, is the notion that more is better in terms of detail. What specifically is required by field construction is a roster of activities which can be used to present budget control information and schedule control information at the level at which control is to be exerted. Although a roster of activities which is too gross in nature does not provide sufficient control, details which are too minute unnecessarily burden the construction processes to no particular benefit.

Financial and accounting has the overall accounting and job costing responsibility. It s work can be accomplished most conveniently and when the information communicated to financial and accounting comes directly from field reports on the status of activities as these relate to budget. There is no compromise to the regular and formal accounting and financial management procedures by the use of the roster of activities to communicate data. In fact, both the timeliness and validity of data communicated is significantly improved when the roster of activities is used.

Sales and marketing, perhaps, is the least concerned with the detail involved in the use of activities since its primary concern is competitiveness in the marketplace and the validity of cost estimates for sales and marketing purposes. However, preparations may be made by the other work environments in support of sales and marketing, principally estimating and project management, but no compromise resulting from their use in the sales and marketing endeavor will hamper the use of the activities for estimating, for planning and scheduling, for budgeting and for purchasing purposes.

All factors considered, the activity seems to have all the attributes necessary to serve as the common unit throughout all work environments and disciplines. The activity has the flexibility required to address the common need for information and data and to provide an adequate interface among all work environments. It allows the transfer between work environments and disciplines of only that data which is important or useful while maintaining the integrity of language and process within each work environment and discipline.

The most common usage of the roster of activities for a construction project is in the area of planning and scheduling. This is especially true where the critical path method or other logic network techniques are utilized. It is essential to integrated systems that the roster of activities by descriptors and account numbers be an integral part of the job cost chart of accounts. This requirement is important because all planning, scheduling, and budgeting for each job will be done with its roster of activities and all performance data and information will be reported on the basis of the roster of activities.

## ACTIVITY SUMMARY LEVELS

The activity may now be clearly defined as the lowest common unit of work for integrated cost and schedule control. The resolution of summary processes can also be identified to allow a wide variety of summations of performance data from the roster of activities for all work environments and disciplines. With these conclusions, the process of integration resolves itself into the use of a roster of activities for budgeting, for scheduling, and for simple and direct summarization of data and information designed to meet the requirements of each work environment and discipline.

What remains are the clear definitions of the kinds of information required, the transfer processes between work environments and disciplines, and the ways in which data is to be provided. Also a matter of consideration are methods of capturing and maintaining data for maximum utilization. This involves not only the job cost chart of account descriptors but also a series of primary and subordinate numbering systems which allow for both quick and convenient summarization of data. These subordinate numbering systems provide the vehicle for summarization in whatever detail and format are required within a particular construction environment.

## THE MAINSTREAM OF COMMUNICATIONS

Whenever business management problems are discussed within any business environment, especially construction, one of the major ongoing problems, as we have seen, is communications. However communications may be defined, the use of a roster of activities to set out the project expectations and data capture procedures which record performance in the same way that the statement of expectations has been prepared will greatly facilitate communications. Further, the transfer of data and information between work environments can be greatly facilitated by integrated cost and schedule control procedures using the roster of activities.

Precision in communications is also greatly enhanced by these processes since all work environments and disciplines will be working on the same informational set.

## DIFFICULTIES IN ACCEPTANCE OF ACTIVITY BASED INTEGRATION

With the great diversity of work environments and disciplines within construction, some difficulty will be encountered in establishing the activity as the basis of communication. The technical mindset must be overcome in order to use the construction activity as the basis for all informational processing. Each work environment will have its own particular agenda relating to the use of the activity. The estimating and budgeting areas may experience the most difficulty.

Estimating is customarily done by trade groupings following some standard job cost chart of accounts. Indeed, imbeded in the industry is a mindset toward estimating by trade groupings and summarization by common work types. By comparison, estimating by activity requires advanced preparation of the roster of activities, which should be prepared basically in the same format which subsequently will be required by planning and scheduling. The roster of activities then becomes the basis for quantification and pricing. Summarizations may still be made by traditional estimating methods. The significant difference is that the pricing levels are smaller increments and, therefore, somewhat more tedious. The principal benefit for integration of cost and schedule control is that each activity can be separately budgeted and separately scheduled. The ability to separately budget and schedule every activity in the project is crucial to setting up a project for integrated cost and schedule control.

Since estimating, budgeting, planning and scheduling comprise the major tasks in setting a job up for production, the other major tasks to be performed are the beneficiaries of this work. Since the communications interface among other disciplines involve appropriate summarization of data, the difficulties to be encountered here are less significant.

## SUMMARY

The construction activity is indeed a viable candidate upon which to build integrated cost and schedule control. It accurately reflects the way projects are built in the field environment. There is no doubt that projects may be budgeted and scheduled by activities. All purchasing may be done by activities and all other construction industry related procedures may be built around an activity based approach to construction. The integration of cost and schedule control by means of the activity is primarily a matter of incorporating information concerning the way projects are built into the core for systems integration.

# 3
# THE PROJECT STATEMENT OF EXPECTATIONS

A major consideration in designing a system for integrated cost and schedule control is the process by which performance will be measured or evaluated. Since control implies the ability to influence outcome, the processes for measurement of performance must provide current valid information on actual performance. Current valid information on actual performance, however, is not of itself sufficient. What is also required is some clear, well articulated performance models or set of performance standards against which actual performance may be measured. It is not sufficient to know, for example, that the actual performance is 100 units, that the actual cost per unit is $10, and therefore the actual cost is $1,000. This actual performance data becomes relevant only when we know *in advance* that we expect to place 100 (or whatever) units and that we expect to expend $10 per unit, for a total cost of $1,000. With information on both *expected* performance to be and *actual* performance, we can generate measurements which compare (usually as ratios) actual performance to planned performance.

## PERFORMANCE CALCULATIONS

The example given of 100 units at $10 per unit equals a cost of $1,000 represents an example of the simple quantity times unit price equals cost algorithm common to both estimates and budgets. If, for example, 120 units were actually required to be placed rather than 100 units, it is possible to calculate quantity performance by comparing actual to planned:

$$\text{Actual quantity/planned quantity} = 120/100 = 1.2$$

The 1.2 represents a quantity performance ratio (QPR) indicating that 20% more product was actually required to be placed than planned. If the actual cost, however, per unit was $8 instead of $10 as planned, a unit price performance ratio (UPPR) could be calculated.

$$\text{Actual unit price/planned unit price} = \$8/\$10 = 0.80$$

The 0.80 represents a unit price performance ratio of .8 and reflects a unit price performance 20% below the planned performance ratio. A further comparison may be made by comparing the total cost of actual performance to the total cost of planned performance. The following formulas show these calculations:

Actual cost performance = 120 units × $8 per unit = $960
Planned performance = 100 units × $10 per unit = $1,000

The calculation of the cost performance ratio (CPR) for a total cost may be calculated:

Actual total cost/planned total cost = $960/$1,000 = 0.96

The solution, 0.96, represents the (total) cost performance ratio of the activity and indicates that while the quantity performance was 20% over planned and while the unit price performance was 20% below what was planned, the net cost of the activity is at 4% below planned or budget.

The purpose of these illustrations is to demonstrate that quantity and unit price issues are more interesting and revealing than total cost comparisons.*

In a similar way, performance with respect to time required to perform each activity may also be made using the same illustration. Assume a 10 unit per day placement rate:

Total quantity/quantity planned per duration unit
= 100 units/10 units per day = 10 day duration planned

Similar comparisons may now be made for time performance as have already been made for cost. The quantity performance ratio, of course, is unaffected when comparing quantity to quantity for time calculations. Quantity performance ratio remains at 1.2. For purposes of illustration, assume actual rate of placement equal to planned rate of placement. The rate of placement ratio is calculated as follows:

Actual rate of placement/planned rate of placement
= 10 units per day/10 units per day = 1.00, unit placement ratio

This unit placement ratio indicates that the rate of placement actually experienced equals the rate of placement planned. An interesting comparison occurs when the total times are compared:

---

* The cost performance ratio alone would indicate satisfactory performance while ignoring a significant quantity problem.

Actual quantity/rate of placement per day

$$= 120 \text{ units}/10 \text{ units per day} = 12 \text{ days}$$

Planned quantity/planned rate of placement

$$= 100 \text{ units}/10 \text{ units per day} = 10 \text{ days planned duration}$$

The time performance ratio for overall time may now be compared:

Actual activity duration/planned activity duration

$$= 12 \text{ days}/10 \text{ days} = 1.2, \text{ time performance ratio}$$

It is worth noting that the 1.2 time performance ratio may be explained either by a variance in quantity, actual compared to planned, or by a variance in the rate of placement, actual compared to planned. This kind of detail gives considerable visibility to actual performance and provides the kind and quality of information necessary for effective cost and schedule control.

## THE IMPORTANCE OF THOROUGH PROJECT PLANNING

The examples given serve to illustrate that more is required than high quality contemporary actual performance data. If, for example, it is known that the actual cost was 120 units placed at $8 per unit for a total cost of $960, the obvious question is, "So what?" Without the knowledge that planning calls for placing 100 units at $10 per unit for a total cost of $1,000, the actual performance data for integrated cost and schedule control is meaningless. While 120 units at $8 per unit, or $960 in all, represents high quality job cost information, only in the context of how the actual activity performance compares to the planned performance can relevant and meaningful information may be generated.

### Master Planning Checklist

Most construction projects will undergo some kind of planning prior to construction. Basic planning, no matter how crudely or informally carried out, involves four basic areas of consideration and four related questions to be asked. Exhibit 3-1 provides a summary planning checklist which may serve as a basis for the major considerations of project planning.

Each of these questions is answered either formally or informally during the course of the completion of every project. On the assumption that time and energy must be expended to plan how the project will be built, to analyze the resources available for its construction, to determine the schedule for the construction of the project, and to establish the expected cost of the project, the question which should be raised is:

## Exhibit 3-1. Master Planning Checklist

1. **Planning.** How will the project be built?
2. **Resources.** What resources are required and are available to complete all activities required to complete the project?
3. **Schedule.** Based on the way in which the project will be built, the resources available, and a specific start time, when will the project be completed?
4. **Budget.** Based on the quantities of product to be placed and the quality required by the specifications, what is the fair value cost of the project?

When is the most effective time to attend to these tasks so that the maximum benefit may be derived from the effort expended?

Clearly, the answer to the question is that the entire planning process should be completed as early as is practical. The latest acceptable time would occur prior to the commencement of construction and expenditure of resources. The next question is:

How detailed should planning be?

Thorough planning is obviously required for the expeditious and economical completion of the project. Thorough planning, however, implies *detailed* planning. While detailed planning is certainly required, the amount of detail to be included in the project plan requires further consideration. As noted earlier, the original critical path methodology holds that more detail is better; thus choosing the appropriate level of detail for the planning, scheduling and budgeting of a project, becomes a major consideration. Experience within the construction industry recently has led to a practical guideline for the amount of detail to be included in the project plan, the project schedule, and the project budget. The maximum amount of detail to be included in the planned schedule and the budget is the level of detail at which control is to be asserted and the project strategy implemented. In simpler terms, the project should be planned the way the contractor intends to build it with sufficient detail for controls, but no more detail than required to control the project.

## THE PROJECT STATEMENT OF EXPECTATIONS AS A MASTER PLANNING STRATEGY

The project statement of expectations including a plan, schedule, and budget provides a sound basis for building a master project planning strategy. When the plan for building the project is clearly articulated, comparisons to how the project is actually being constructed may be made conveniently. When the calendar-dated schedule is set for the project and for each activity within the project, comparisons to actual time performance may be made easily. The same is true with respect to the cost performance of the project. The project statement of expectations for the plan of building the project, for the schedule, and for the cost of the project provides the framework for asking the question essential to integrated cost and schedule control:

How is the project performing compared to how it was planned or expected to perform?

The master planning of a project must cover all aspects of the project. While the statement of expectations is usually directed toward controlling cost and schedule, other aspects of the project require planning with equal care.

## PLANNING FOR ADMINISTRATION

Planning for the proper administration of contracts, subcontracts, and purchases does not customarily receive the attention that it desires. Effective contract administration with respect to changes as well as routine matters such as collection of payments are significant issues in cost and schedule control. They are especially important when cost and schedule control are to be integrated. It is difficult to control the schedule of the project when there continue to be unresolved questions concerning the performance of the project. It is equally difficult to control the cost of the project when significant additional costs are being incurred without the basis for compensation being properly established. Planning for contract administration involves a thorough knowledge of contract requirements and the presence of administrative procedures which routinely and efficiently carry out contract administrative tasks. Procedures for effective contract administration may vary significantly among different organizations. What should *not* vary is the timely handling of all necessary administrative tasks required by the contract.

## PLANNING FOR FIELD CONSTRUCTION

Field construction is at the very center of the construction enterprise. This is where production is accomplished, the major battlefield where issues of productivity and of cost and schedule control are determined. Planning for field construction is of paramount importance because when properly done it provides the foundation for integrated cost and schedule control throughout the entire construction enterprise.

The way in which field construction is accomplished is worth noting. When recognized, this can form the basis for integrated cost and schedule control, both of which concern the ways in which resources are applied to the accomplishment of work in the field.

First, all resources (whether supplied by subcontractors or principal vendors, suppliers, or in-house capabilities) are applied in the field environment on an activity by activity basis. For example, when the plumbing workforce arrives at the job, they are not assigned to perform "the plumbing." They are assigned to specific activities *within* the plumbing work package such as rough-in toilet rooms, first floor; or set fixtures, third floor. Similarly, when the carpentry workforce is assigned to the project, the carpentry supervision is not told to perform the carpentry work. They are assigned to specific activities within the carpentry work package such as to complete the rough framing for the mansard roof or to hang the doors on the east wing, second floor. Since competent supervision will assign workforces in the field environment activity by activity whether or not a formal plan for the work by activities exists, the argument for planning this way becomes very compelling.

The second characteristic of the application of resources is that each resource is assigned on a specific time interval basis which will vary with the size and complexity of the job—from a daily to perhaps a weekly time frame as determined by the general supervision on the job. The actual activity work assignments, however, are made by immediate supervision on a daily basis.

With recognition of how work is accomplished in the field environment, it seems logical that planning for field construction should anticipate actual construction for maximum validity in the planning process. This means that the planning for field construction should be accomplished on an activity by activity basis with resources assigned on a short interval basis (usually daily).

The recognition of these two factors is a major step forward leading to integrated cost and schedule control They meet the basic criteria of planning the way the project is intended to be built and to build the way the project is planned. It is also noteworthy that planning by activity and assigning resources daily is a very close match to the way in which the critical path methodology is developed for planning, resource analysis, and scheduling of construction projects. While it is possible to use any planning and scheduling method for integrated cost and schedule control which deals with the activities of the project and the duration of each activity based on available resources, the critical path methodology *anticipates* these requirements and can be, therefore, very effective in planning for field construction.

In using the critical path method for planning and scheduling a construction project, the major problem is selecting the appropriate level of detail to be included in the critical path network or arrow diagram. The key to this problem is to select activities which are generally sufficient in detail and sufficiently short in duration to be meaningful while, at the same time, not encumbering the schedule with greater detail than will be used for production control. Clearly, the key to planning for field construction is the creation of a roster of activities with activity descriptions of sufficient detail (but not too detailed) which relate to the way in which the resource will be assigned in the field environment.

Planning and scheduling by activity for field construction requires the formal articulation of the project plan. This requires considerable front end planning of the project, not commonly the practice in the industry. In order to deal appropriately with preplanning, the project plan or strategy is represented by a logic network diagram and must have sufficient flexibility to allow for normal field variances while at the same time forming the basis for not only schedule control but cost control as well.

## PLANNING FOR RESOURCE REQUIREMENTS

The critical path methodology allows for the independent planning of a project once the roster of activities for the project has been generated, without regard to the resources available to perform the work, the consequent duration of each activity, or the actual schedule for the project. This clear and definitive separation of planning tasks from resource analysis and scheduling calculations is one of the particular niceties of the critical path methodology not conveniently found in other planning and scheduling techniques.

Once the plan for a project has been completed and articulated in the form of a logic network or arrow diagram, two kinds of resource analyses may be made. The first of these concerns a resource and the rate of application of that resource to the completion of each activity. The concept of usual practical resource is used to determine workforce size for labor activities. With each activity budgeted separately, the calculations to determine duration are relatively straightforward. An example of these calculations follows:

Assume a labor activity is budgeted at $12,000; this consists of 120 units at $100 each = $12,000. Assume that the workforce required is 10 and the average wage rate is also $10. The cost per workforce hour = 10 × 10 or $100 per workforce hour. At 8 hours per day, the cost per daily duration unit = 100 × 8 or $800 per day.

Duration may be calculated as follows:

Total budgeted cost/cost per duration unit
$$= \$12,000/(10 \text{ workers} \times \$10 \text{ per hour} \times 8 \text{ hours per day})$$
$$= \$12,000/\$800 \text{ per day} = 15 \text{ days}.$$

In the case of those trades where cost estimating is manhour based, the estimated manhour data may be used directly to calculate duration. It is clear, however, that the duration of any activity is a function of the resource available for its performance. Two criteria determine the rate of application of that resource: (1) the customary or usual resource, and (2) the practical limitation of the work environment. In calculating duration, each activity is considered independent of all other activities.

In a construction project of any complexity or duration, calculation of duration for individual activities does not take into account the demand that may be placed on the same resource for performing two or more concurrent activities. When this occurs, the question to be resolved is whether or not additional resources are available or whether resource leveling will be required. In either case, the issue of resource analysis becomes an important one in project planning even though the initial concept of planning may assume unlimited resources. Resource analysis is required in order to give validity to the overall planning for the project.

## PLANNING THE BUDGET FROM THE ESTIMATE

The budget is the project cost statement of expectations. The budget is derived from the estimate, with the level of detail included in the budget taken from either the planning and scheduling requirements or from the cost control requirements, whichever are greater. In either case, the roster of activities used for planning and scheduling and that used for budgeting must be indentical. It is through the common roster of activities for both budget and schedule that the foundation for systems integration is developed.

Preparatory to developing the budget from the estimate is creating the roster of activities for the project—the first priority. Once the roster of activities has been prepared, the estimate is then converted to the budget by assigning quantity, unit price, and cost to each and every activity on the activity roster. The standard estimating/budgeting algorithm is used:

$$\text{Quantity} \times \text{unit price (quality)} = \text{cost}$$

Beyond dealing with the roster of activities, two major issues which need to be taken into account when converting an estimate to a budget are the more detailed evaluation of the resource requirements for each activity and the provision of a basis for both cost and schedule control by activity. Several examples have already been given for the quantity unit price equals cost algorithm.

An additional consideration is the way in which activities in the budget will be grouped into work packages. When each activity is dealt with individually, the work package assigned to a subcontractor, principal vendor or supplier consists of some subset of the master roster of activities. This subset, when grouped and totaled, comprises the budget for that particular work package. The grouping of items or activities into work packages will depend on the common practices of subcontracting within a particular geographic area and the ways in which subcontracts and purchase orders to principal vendors are negotiated. This methodology provides for the budgeting of individual activities summarized to work package levels and also provides the basis for actual purchases to be compared with budgets not only at the work package level but also at the indivdual activity level.

## THE BUDGET AS A COST STATEMENT OF EXPECTATIONS

Planning of the budget from the estimate allows the opportunity for complete review of the costs of the project; where duplications have occurred in the estimating processes, these may be eliminated. Errors or omissions picked up in the original estimates should be corrected in the budgeting process. Risk factors which have been assumed in the estimating processes may be isolated as separate activities to allow for contingencies rather than built as contingencies into each budgeted item.

When the budget has been formalized and acknowledged by all interested parties, it should fairly represent a cost statement of expectations. The budget should be realistic and achievable. It should serve as the basis for all cost control issues relating to the project. For items to be purchased through subcontractors or principal vendors, the budget should serve as the purchasing guideline, the objective being to purchase at or below budget. For those activities performed using the organization's own resources, the budget should serve as the clear basis for measuring performance. In general, the budget should serve as the overall cost control vehicle, which in turn should be viewed as an independent function separate from job cost accounting. Cost control using the budget as the basis for comparison should routinely provide contemporary cost information and provide information on trends and the cost performance of the project.

## PLANNING THE EXPECTED SCHEDULE

One of the niceties of the critical path methodology is that it allows the complete separation of planning and budgeting from scheduling. Part of the information required for scheduling comes from the calculations dealing with the duration of each activity. The separate resource analyses should be addressed at the time these calculations are made to determine if it is necessary to utilize other available methods to balance resource requirements. Once these planning tasks have been accomplished and the scheduling calculations made to establish the duration of each activity, two additional steps are required to formally establish the planned or expected schedule for the project.

The first of these is to perform the classic critical path method forward and backward pass calculations. These establish the time earliest (TE) and time latest (TL) for each event in the logic network diagram. Chapter 9 provides an elementary overview of the critical path method.

These also establish the time boundaries for each activity including early start (ES), late start (LS), early finish (EF), and late finish (LF), and establish the critical path throughout the project. The critical path indentifies those activities which, based upon the project plan, will control the overall duration of the project.

The second step in setting the project schedule is calendar dating. It is worth noting that the calculation of time boundaries is expressed in *duration units* and is independent of calendar dating or the schedule of the project. This is a particular nicety of the critical path method because it allows establishing the time boundaries very early in the planning process, when the scheduled start date may not be known.

When the prerequisites to the commencement of field construction are accomplished and the actual start date is known, the schedule for the project is set by establishing the calendar date for the start of work. Establishing the calendar date for any activity or event throughout the network will automatically establish the calendar dates for all of the *remaining* events and activities. Typically, the start date is established and the calendar dates are carried forward through the completion of the project.

It is also possible to establish the probability of meeting any expected milestone using critical path analysis. For example, given a calendar completion date for a project, it is possible to validate that completion date by comparing the planned calendar date start with the current calendar date. If the planned start date is later, the project can be completed within the planned time.

## THE SCHEDULE AS A TIME STATEMENT OF EXPECTATION

The schedule for the project (especially one which has been logically derived) not only provides a very detailed time statement of expectations for the performance of the project but also provides specific calendar dating to establish the time performance boundaries of every event and activity throughout the project. The logic network diagram used in the critical path method, however, tends to be somewhat confusing since it does not contain the *calendar* dates for the start and completion of each activity. The preferred current practice is to complete the critical path analysis using the logic network diagram and then to convert the logic network diagram to a time scale network for production control purposes. The *time scale network* has the characteristics of the typical bar or Gantt chart schedule in that it shows calendar dated start and completion while displaying this information on an activity rather than a trade basis. The time scale network is the preferred vehicle for displaying the time statement of expectations for the project.

The time scale network, calendar dated, has many useful functions. Particular occurrences such as completion of foundations, completion of the weather envelope, and similar items may be highlighted in the form of milestones on the time scale network.

The time scale network also has other values as a project statement of expectations. For example, from a purchasing point of view, each event which marks the beginning of a significant product placement in the field, such as structural steel, indicates the need for that material being on the job site at that calendar date. This means that the calendar date for the delivery of every product required for the job is clearly established with the time scale network, certainly a significant time control vehicle for purchasing. With a reverse sequencing analysis from the delivery event, it is possible to determine the probability of the product being at the job site when required—an additional significant advantage of the schedule as a time statement of expectations. The schedule should clearly represent the overall expected performance of the project in terms of time. It should also provide the basis for comparing actual performance to planned performance.

## THE BUDGET AND SCHEDULE AS PERFORMANCE MODELS

The budget and schedule have been presented as statements of expectations. As such, they provide all interested parties with a clear indication of the expected performance of the project in terms of cost and time. In an integrated cost and schedule control system, the budget and schedule may also be considered as planned performance models against which actual performance may be measured. The use of performance models is an important consideration in any integrated cost and schedule control system. The standard for comparisons may be verbalized in the question, "How are we doing compared to how we planned to do?" This is expressed arithmetically as a formula for performance ratio calculations:

Actual performance/planned performance = performance ratio

When the performance models are planned on the basis of the way in which the project is expected to be constructed, they become effective production control tools. When data on actual performance is captured in the identical way and format as the planned data is presented, simple ratio calculations provide a quick and simple way of creating performance information on the project as performance ratios. The ratio calculations represent a powerful method of creating management information when actual performance is to be compared to the performance models.

## LEVEL OF DETAIL AND SUMMARY REQUIREMENTS

For continued usefulness as well as validity, performance models which are generated on an activity basis for both cost and schedule must be planned at the level at which the work is expected to be performed. The level of detail included in each activity continues to be a key issue in integrated cost and schedule control. Descriptions of work which are too gross will not provide vehicles for control; descriptions of work within activities which are too fine will provide more detail and information than is useful. To repeat, the level of detail for control purposes required by either budget control or schedule control is the most important consideration in establishing the roster of activities.

The technical nicety about using a roster of activities to set the level of detail at which the project will be controlled and at which the schedule will be implemented is that any summary level may be made conveniently for all informational requirements. Activities may be summarized into work packages relating to the way in which subcontracting and purchasing are accomplished. Summaries may also be made by trade groupings for convenient cross-referencing of the contract drawings and specifications. Further, summaries may be made of the specific resources to be used in the performance of the work, for example, labor, material, equipment, subcontracts and principal vendors. And time related summaries may be made. For example, all the activities to be performed during a particular time frame such as a billing period may be conveniently summarized, to provide very useful information on cash management requirements.

Experience repeatedly indicates that it is easier and more convenient to work with an appropriate level of detail, summarizing as required, than to begin with gross figures and to attempt to break these down accurately. The appropriate level of detail for summarizing is yet another key to integrated cost and schedule control.

## EVALUATING THE STATEMENT OF EXPECTATIONS FOR VALIDITY

The processes involved in preparing the statement of expectations for the cost and schedule control of a project are well defined. The analytical tools available for creating the statement of expectations are also well defined. A rather common error in these detailed analytical processes is the assumption that because the methods are intricate the output is valid. For example, an early assumption with regard to the critical path method was that planning the project using the critical path method would automatically result in control of the schedule. The analytical processes gave the appearance of validity and of control, which might have not been warranted. On the contrary any statement of expectations must be constantly evaluated and its validity continually reconfirmed. To the extent that validity is reconfirmed, the project statement of expectations becomes a useful working tool for production management. To the extent that validity appears to be weak or defective, the statement of expectations must be adjusted, amended, or revised to assure its validity.

Neither does control automatically result from detailed planning, scheduling, or budgeting. Control results from diligence in monitoring of actual performance processes by comparing these to planned performance. This diligence must be directed first to identifying adverse trends in performance; second, to designing corrective measures to change or control the adverse trends; and third, to implementing the corrective measures. Continued diligence is required to monitor the consequence of the corrective measures to see that the planned objectives are, in fact, the objectives being achieved.

Evaluating the validity of the statement of expectations also requires a continual update of schedule information based on actual performance. It is clear that the work which can be performed for a specific calendar week is less dependent on what may have been shown for that week in a schedule prepared months earlier than it is on actual current performance and on what was, in fact, completed the week previous. Continual updating to accurately reflect current performance is essential to maintaining the validity of the statement of expectations.

## SUMMARY

The project statement of expectations is prepared by a process which results in the schedule as a time statement of expectations and in the budget as a cost statement of expectations. As these are established and their validity confirmed, the budget and the schedule become planned performance models against which actual performance may be measured. The time statement of expectations and the cost statement of expectations as represented by the schedule and budget makes possible integrated cost and schedule control.

# 4
# SINGLE DATA CAPTURE AND
# INTEGRATED SYSTEMS

There are certain basic considerations involved in integrated cost and schedule control systems. The general procedures for creating useful management information require that certain minimum raw data be collected on a timely basis, and the information generated from the manipulation and processing of the raw data must be summarized and distributed consistent with the informational needs of management.

## BASIC CONSIDERATIONS

Several basic considerations of integrated systems similarly relate to single data capture procedures. Economies of clerical time and effort become important in systems integration. Data capture procedures must anticipate all raw data needs and requirements prior to systems design. This consideration anticipates that raw data may be quickly and conveniently distributed to all management information systems once the data has been captured.

## ORIGIN OF "SINGLE DATA CAPTURE"

The term "single data capture" was developed in data processing. It is especially useful in integrated cost and schedule control systems whether these are manual or automated. The two major factors are convenience and simplicity in collecting raw data and the distribution of that raw data to all sources requiring the data for processing.

## PROCEDURES FOR SYSTEMS DESIGN

Procedures for single data capture systems design require a thorough analysis of the sources of raw data and the kind of raw data to be captured. More critical attention must be given to all of the other work environments which require the raw data and the form in which that raw data must be presented to each. Time invested in analyzing raw data sources and the use of the data throughout integrated cost and schedule control systems will significantly improve the efficiency of such systems. The specific procedures for designing single data capture processes are:

1. Identify all raw data sources supplying data into the system.
2. Identify all uses which will ever be made of that data.
3. Design the data capture tools consistent with the needs for raw data.
4. Capture raw data as close to the source in time and physical proximity as possible.

Note that these procedures require the least amount of restructuring of data prior to its use. The suggested procedures are only a small part of the overall design of integrated cost and schedule control systems. They are, however, critical in maintaining a level of efficiency and usability in the systems design.

## THE NEED FOR PERFORMANCE INFORMATION

The need for performance information for management purposes involves two general areas. The first of these relates to the budget for cost control. The second relates to the schedule for time control. These two also represent the two primary performance models used in the planning, scheduling, and budgeting of a project against which all subsequent performance may be measured. A number of illustrations could be cited to demonstrate the value of single data capture of raw data. Imposing specific structure on the data capture processes, however, is required. A key issue in systems integration involves the capture of data in a form which is useful in all work environments; that is, the data must be in a form immediately usable by all work environments without extensive recasting or reworking of the data.

In every case, the common vehicle for generation of the statement of expectations on job performance is the activity. Activities relate in detail to the way the job is planned and the ways in which that plan is implemented, managed, and monitored. It is a convenient reference for the use of numerical cost codes, descriptors, and other references throughout all the work environments.

Research has clearly identified that the production activity with the appropriate numerical cost code references and descriptors as the most discreet informational unit within a construction enterprise. The activity is not only the key to any realistic approach to integrated cost and schedule control; it is equally important if the major clerical tasks associated with integrated systems using single data capture are to be carried out in the most efficient way.

Two kinds of performance information are critical to integrated cost and schedule control. The first of these is the performance information which allows production management to measure actual performance against planned performance. The second is the summary kind of management information required for business management purposes. To determine the kinds of raw data to be captured, it is necessary to have in place a budget for cost control and a schedule for time control for any job to be undertaken. These two provide the basis for designing single data capture procedures from which raw data then can be quickly and conveniently processed into the kinds of information required. The key in both cases is to plan for cost control and schedule control in the way the project will be managed. Essential to the process is capturing the data on performance in exactly the way the project has been planned.

## PLANNED VS. ACTUAL COMPARISONS

The most convenient way of generating production management information from raw data is through the comparison of actual data to planned data. These performance ratios allow production management information, the most critical to controlling cost and schedule, to be generated and used while there is opportunity to influence performance. Information for other work environments, including management information, is less critical from a *time* point of view, but is equally important to business management. Single data capture anticipates all needs for raw data; and even when there are differences relating to when the information is required, single data capture anticipates that certain raw data will be required at some time in the future.

## PERFORMANCE MODELS

The two primary performance models, specifically the budget and the schedule, are required for production management. The consequence of production management is required for management information. The primary purpose of the derived or secondary performance models is to determine the consequence of the budget and schedule. For example, the cost curve and the production curve show the effect of the budget and schedule in terms of cost accumulation and the value of that cost from a contract point of view. The Schedule of Values curve shows the value work in place. The income curve indicates cash income resulting from the budget and the schedule. The cash requirements curve indicates when cash will be required and how much. A comparison of the cash income curve to the cash requirements curve will show the planned cash position of the job at any time. Chapters 7-12 deal extensively with all aspects of these performance models.

The planned data provides information on what is expected to happen.

The actual data provides information on what is actually happening. The single data capture procedures allow the raw data to be used by any of the work environments when the data is required for processing.

## FIELD PERFORMANCE AS A PRIMARY RAW DATA SOURCE

The major focus of the construction business enterprise is field construction. Field performance is, therefore, a primary source of raw data. The single capture of this raw data for all systems requirements is especially important. The planning and scheduling processes will have resulted in establishing specific activity start completion dates. Data capture on field performance for activities will be the actual start date of an activity which may be compared to the planned start date and the actual completion date which may be compared to the planned date.

The capture of raw data on payroll hours by activity is equally important and illustrates the multiple need of raw data by several work environments. Payroll, of course, requires reporting of payroll hours for processing. Actual man hour requirements may be measured against planned man hour requirements. Job cost requires that payroll hours be converted to cost to compare actual performance to planned performance.

Another kind of raw data to be captured concerns material deliveries. The data required deals with complete deliveries as well as partial deliveries.

Field performance also relates to such items as additional work or changes in the work and actions required to remedy defective work. Further, field performance is the primary source of raw data on charges and backcharges affecting the performance of subcontractors and principal vendors.

All of these are essential sources of raw data needed for performance measurements.

## PURCHASING AND SUBCONTRACTING AS A PRIMARY RAW DATA SOURCE

Purchasing and subcontracting activities provide yet another significant source of raw data. Of specific concern are the activities grouped into specific work packages. The work package distribution as part of the resource for the job is also essential raw data. The value of work included in work packages, described by the value of each activity within the work package is also raw data to a number of work environments. The terms and conditions of subcontracts and puchase orders become essential raw data to the processes relating to cash analysis and cash requirements. The terms and conditions of the subcontracts and purchase orders control when payment is to be made and how payment is to be made, therefore become essential raw data to the business management environment.

## ACCOUNTS PAYABLE AND ACCOUNTS RECEIVABLE AS PRIMARY RAW DATA SOURCES

Accounts payable and accounts receivable invoices are generally processed by accounting. In the case of accounts payable, the raw data is received from external sources in the form of invoices to be paid. Accounts receivable, on the other hand, generates invoices in the form of accounts receivable to the owner. Both represent raw data essential to preparing management information on actual cash income and on cash requirements. This actual information may be compared to planned information for further analysis.

## DATA CAPTURE PROCEDURES

A number of data capture procedures and criteria tend to improve single data capture for integrated cost and schedule control requirements. The first of these is to anticipate all needs for raw data and to provide for the handling or storage of raw data until it is required.

Another procedure which greatly simplifies single data capture is to plan the job in the way that it will be built and managed. This procedure assures that the actual raw data captured from field performance will match the planned data. This matching of actual and planned performance data is especially helpful in generating production management and business management information. Data capture instruments should be carefully designed so that they anticipate all needs for data and are in keeping with the way the project or job has been planned. The data capture instruments may be time sheets, daily job reports, receiving and delivery tickets, and turnaround documents reporting activity start, percent complete, and completion.

It is especially important that the capture of data be carefully matched to the job plan. Each inconsistency will significantly add to the clerical effort required in manipulating raw data before it can be processed into useful information. In any event, every effort should be made to avoid the need for restructuring or recasting raw data prior to its use in generating management information.

## DISTRIBUTION OF RAW DATA FOR PROCESSING

The distribution of raw data for processing purposes is an important consideration in single data capture procedures. Some work environments tend to claim ownership of raw data for which they have the primary need and responsibility. Single data capture, however, requires that all raw data is shared and equally accessible to all disciplines. Data capture procedures structured in the data capture system represent the most effective way of providing information and accessibility to raw data when needed by any of the several disciplines.

Exhibit 4-1 shows the flow of raw data for processing. The primary sources of raw data, which include field performance, accounts receivable, accounts payable, purchasing and subcontracting, should feed into a data capture system. The data capture system should have the capability of making accessible the raw data from the primary sources to other informational areas.

For example, job cost should be able to access payroll hours directly for conversion of hours to job cost without having to wait on payroll to produce such information. Payroll hours are also a part of production control, giving further reason that this information should not be tied in or made dependent on the processing of payroll.

The control information system generally consists of payroll, production

## Exhibit 4-1. Flow of Raw Data for Processing

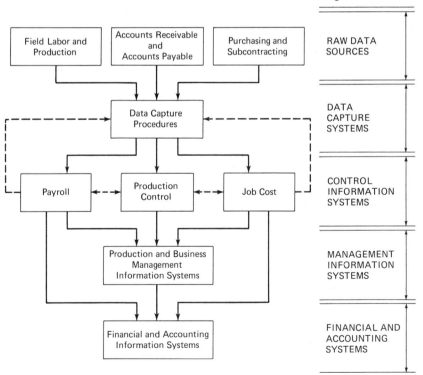

control and job cost, each of which makes significantly differing demands on the identical raw data in terms of time. Each also provides additional feedback into the data capture systems as its routine processing is accomplished. For example, since job cost requires payroll information on a daily basis so that the job cost amount is available on a daily basis, certain information is generated from payroll hours for job costing purposes prior to weekly payroll. Then when the labor portion of job costs is processed through the routine weekly payroll procedures, verification must be made that the formal labor cost information coming out of payroll is consistent with that processed through job cost.

The management information systems provide summary data from control systems. A variety of summarizations including varying levels of detail will be required in most business management environments.

## SUMMARY

Single data capture procedures are a key to integrated cost and schedule control systems. Properly designed, they allow for the most efficient way to collect raw data on performance and to make that raw data available when required to the variety of work environments and disciplines within the construction business enterprise. Accessibility to raw data for processing to information without unnecessary inter-discipline dependencies is especially important. Equally important is anticipating the need for raw data throughout the integrated system so that data may be captured on performance for all purposes while the performance is underway. Care in designing data capture procedures will result in efficient methods of collecting data for all informational requirements. Single data capture is dependent on the use of the activity for all job related descriptors throughout all work environments. In data processing environments, the utilization of single data capture procedures and the automatic updating of all necessary files with single data entry are relatively simple. Single data capture is equally valuable in manual systems since it eliminates the laborious task of recasting data to translate raw data from the language of one work environment to another.

# 5
# MANAGEMENT INFORMATION SYSTEMS

*Management information systems* (*MIS*) is one of those terms developing out of the data processing environment. Definitions of management information systems vary widely in different industries and in different work environments within specific industries. The timeliness and use of management information also varies widely. Within the context of integrated cost and schedule control, management information systems provide information to management contemporary in nature, valid within an acceptable variance, containing only the kind and depth of information required by management for management purposes.

## HISTORIC OVERVIEW

One of the more significant impediments to understanding contemporary management information systems comes from pre-data processing perceptions of management information. Historically, management and accounting information have been considered synonymous. Job costs have been considered a part of accounting information which provided management information. As a result, an inappropriate dependency for valid information on project performance has developed around the accounting functions. In examining most construction business environments, however, it quickly becomes apparent that the historic nature of most accounting and job cost information systems makes this kind of information unusable for contemporary management purposes.

Typically, accounting and job cost information is geared to monthly cycles. The output from these information systems may typically be two to four weeks outside the performance period. This means that job cost information coming through formal accounting procedures could present data to management 60 days out of phase with actual performance—simply not acceptable.

Such historic information, while essential from an accounting and financial management point of view, is of no real value in the day to day management of the business enterprise. It is of even less value to both field construction and project management, where demands for contemporary information on performance continues to be essential to identify adverse trends in both cost and production performance. Better management information systems methods can and should provide contemporary information while performance is underway, allowing adverse trends to be detected and corrective action implemented while the outcome of performance can be influenced.

## INDUSTRY DEMANDS

The urgent need for contemporary information in the construction industry pertaining to both cost and schedule performance has created the more specific demand for effective management information systems. These management information systems, while using the same raw data base as financial and accounting, are contemporary in nature. The information they provide differs materially in timeliness and content from the historic information provided by accounting information systems. For efficiency, MISs are designed to take advantage of single data capture of raw data. The raw data itself is distributed to financial and accounting for financial information systems and separately distributed to management information systems for immediate processing. MISs provide immediate (daily or weekly) cost and schedule performance information for field construction, project management, and business management.

Single data capture becomes crucial in maintaining the clerical efficiency of both the management information systems and accounting information systems. Sharing raw data is accomplished through properly designed data capture and distribution processes which clearly recognize the distinctly separate needs for (contemporary) management and (historic) accounting information.

Separate processing of raw data for management information makes possible the provision of daily and weekly data on both cost and schedule performances. Daily and weekly information on production performance can be provided similarly.

## CONVERTING RAW DATA INTO
## MANAGEMENT INFORMATION

As we have seen, management information in the context of integrated cost and schedule control must be contemporary, valid within an acceptable variance, and presented in the form required. The processes by which raw data is converted into management information are of particular importance in the design of a management information system.

It is not sufficient for field construction to know how much has been expended in performing certain activities. It is not sufficient to know which activities were started and were completed during the reporting period. Neither is it sufficient to know the value of product placed during the reporting period.

One of the fundamental functions of integrated cost and schedule control is the creation of performance models embodying expected performance against which actual performance may be measured. What is lacking is some *method* which allows comparisons to be made between actual performance and planned performance. In this context, three questions may be raised concerning the three measures of performance:

1. **Cost Control.** How much was spent compared to the amount planned to be spent during the reporting period?
2. **Schedule Control.** How do actual start and completion dates for a given group of activities compare with planned start and completion dates during the reporting period?
3. **Production Control.** What is the value of production product in place compared to the value of product planned to be put in place during the reporting period?

These three questions are keyed to certain elements of a management information system. They can be answered only if there is a clear statement of what is expected (or planned), and if there is a basis for measuring actual performance against planned performance.

## PERFORMANCE RATIOS

The foregoing questions focus attention on a significant and novel concept which allows the quick manipulation of raw data into useful management information—simple ratios comparing actual performance of various kinds to planned performance.

In terms of cost and schedule control and of production, these comparisons may be identified as the cost performance ratio (CPR), the time performance ratio (TPR), and the production performance ratio (PPR).

Since the idea of scheduling and budgeting by individual activities has already been presented, the framework is now in place to enable direct comparisons between actual performance and planned performance in these three control areas. The keys to a workable management information system in these three areas are planning by activity, assigning resources in the field by activity and capturing data on performance by activity (cost, time, and production) so that comparisons may be quickly and directly generated with the minimum manipulation of raw data.

An illustration at this point will prove helpful. For example, assume that the actual cost of performing an activity is $900 as reported by the data capture procedures. Assume also that the budget for that activity is $1,000. From these data, conclusions about cost performance of this activity may be drawn quickly and accurately simply by dividing the actual expenditures by the budget:

$$\text{Cost performance ratio (CPR)} = 900/1000 = 0.90$$

The result of this computation provides immediate information on the cost performance of this activity as a cost performance ratio (CPR). The CPR of 0.90 indicates that the activity has been performed at 10% below the budget.

Available, then, through this process is an immediate conversion of raw cost data into management information in the form of a cost performance ratio.

In such cases, the actual expenditure for the completion of an activity, a work package, a trade group, or any other summary group of activities is divided by the budgeted value for that activity or group of activities. The result of the calculation, a cost performance ratio, compares the actual performance to a standard, here cost vs. budget. Similarly, this same process may be applied to the activities assigned to a subcontractor on a job where the work package consists of a subset of activities in the master budget. Once

## Exhibit 5-1. Percentage Completed and Value of Product in Place through a Particular Period

### THE ABC SUBCONTRACTOR, INC.

| (1)<br>WORK PACKAGE<br>ACTIVITIES | (2)<br>PAYMENT<br>SCHEDULE | (3)<br>PERCENT (%)<br>COMPLETE | (4)<br>VALUE IN<br>PLACE |
|---|---|---|---|
| 1 | $10,000 | 100.0 | $10,000 |
| 2 | 15,000 | 66.7 | 10,000 |
| 3 | 20,000 | 50.0 | 10,000 |
| 4 | 30,000 | 25.0 | 7,500 |
| Totals | $75,000 | 50.0 | $37,500 |
| | (5) | (6) | (7) |

Exhibit 5-1 illustrates sources of data, calculations, and output for this kind of management information process. In this illustration, column (2) represents the value assigned to each activity, the total of which, item (5), is the subcontract amount. Column (3) contains the percentage completed for each activity reported by field production. Column (4) is column (2) multiplied by the decimal equivalent of column (3) and is the value of product in place, the total of which, item (7), is the total value in place for the work package. This amount, by the way, is used for verifying subcontractor requisitions and vendor invoices.

these activities have been carefully budgeted and scheduled, the performance ratios may be quickly and easily calculated.

The same actual performance data has other important uses in converting raw data to useful management information. For example, the percentage of completion of any subcontractor work package can be calculated directly from actual performance data. This procedure is applicable when using either the completed activity (100%) for validating invoices from subcontractors, or percentages of completion for individual activities.

The percentage of completion for the entire subcontractor work package item 6 is equal to the total of the percentage values of all activities divided by the total value of the subcontract package. Many other similar applications may be developed for generating useful management information.

Similar calculations may be developed for both schedule control and production control and will be further developed.

## SUMMARY LEVELS AND AUDIT TRAILS

There are several basic issues which need to be dealt with in the design and development of management information systems. These involve management by exception procedures, summary levels, and audit trails. It is not correct to assume that the more information (or detail) available to management, the better. The efficient use of information presented to management requires that that information be presented at appropriate summary levels, that exceptions to planned or expected performance be highlighted, and that audit trails are available for investigation of exceptions by management.

Typically (perhaps unfortunately), good management must focus on bad performance since it is in these areas that corrective action is required and will be most effective. For example, in a typical job of 100 activities, 95% or more of the activities will usually perform in terms of cost, time, and production as they were planned to perform (within an acceptable tolerance). An effective management information system will confirm for management this satisfactory performance with no other management attention or action required. The remaining 5% or so of activities are those requiring management attention. Effective management information systems, then, will highlight these activities and will provide the audit trails through the informational processes so that management may identify the source of the problems, take corrective action, and consequently maintain effective control.

In addition to timely information, some structure for summarizing management information is especiallly important. There appears to be a structure for summarization which can provide a convenient way of presenting management information at an appropriate level of detail while retaining enough flexibility to permit further investigation.

## CORPORATE STRUCTURE

To help examine methods of summary, a typical simple corporate organization is presented in Exhibit 5-2.

This organizational structure assumes an overall corporate responsibility, managed through three regional offices, each of which has several satellite offices under its direction. Reporting to each satellite office are several project managers, each in turn responsible for several projects. Each project consists of a roster of activities which is included in the project's planning and scheduling.

Through this structure, it is clear that the information required by project management with respect to each job and the activities within each job is much more detailed than the management information required for the satellite office, primarily concerned with the overall work of each project manager. It is also clear that management information for the satellite office could, if desired, examine details of a given job under the direction of a specific project manager, should that be of interest. Similarly, the management information required at the regional level is not as detailed as that required at the satellite level. It is also true, however, that the regional office *could* examine detail

**Exhibit 5-2. Corporate Organization Related to Management Information from Many Levels**

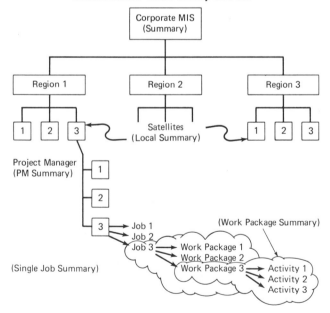

within any of its satellite offices or even individual jobs, should either level of examination be of interest. It follows that the amount of detail required by regional management relating to the satellite offices would not be the same as that required by corporate management pertaining to the performance of the regional offices.

This summary structure provides a key to effective management information systems which present summary detail but clearly retain an audit trail to whatever level of detail may be of interest to any specific level of management.

## MATRIX OF SUMMARY LEVELS

Based on the corporate organization just presented, a clear definition of summary levels emerges, beginning with five: project management, satellite office, regional, and corporate.

### Exhibit 5-3. Matrix of Summary Levels

| LEVEL | OFFICE | SUMMARY LEVEL | DETAIL INCLUDED | DETAIL IS PART OF | ENVIRONMENT | RESPONSIBILITY |
|---|---|---|---|---|---|---|
| 1 | Local | Lowest local | Individual activities | Work package | Individual job | Subcontractor/ vendor |
| 2 | Local | | Work package | Individual job | Local office | Project manager/ superintendent |
| 3 | Local | | Individual job | Individual project manager | Local office | Project manager |
| | | ascending | | | | |
| 4 | Local | | Multiple jobs | Individual project manager | Local office | Project manager |
| 5 | Local | | Multiple jobs | Individual subcontractor/ vendor | Local office | Production management |
| 6 | Local | Highest local | Multiple jobs | Local office jobs | Local office | Local management |
| 7 | Regional | Regional summary | All jobs in region | All local offices in region | Regional office | Regional management |
| 8 | Corporate | Corporate summary | All regional summaries | All regional offices in corporation | Corporate office | Corporate management |

**Exhibit 5-4. Summary Flow From Individual Activity to Corporate Management Information Summary**

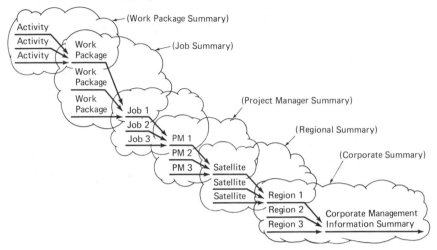

In addition to defining the summary levels, we need to focus on the level of detail required, where the information originates, the specific work environment, and the responsibility for management control in the office location. Exhibit 5-3 presents the matrix of summary level.

The matrix of summary levels provides a clear yet simple overview of the structure of summarization which should be a part of every management information system. While specific organizations may differ, the summary techniques remain essentially the same.

Presented in Exhibit 5-4 is a summary flow diagram for the accumulation of information.

Note that the point of performance and control begins with an individual activity within an individual job. The activity is summarized within the work package. The work package is summarized within a specific job by a project manager. This, in turn, may be absorbed into a multiple job summary by the project manager, and then combined with all jobs for all project managers to present a satellite office summary. All satellite offices within a region may be further summarized to present a regional performance summary. All regions then may be summarized to provide the corporate summary.

These eight levels of summarization present a clear picture concerning how management information may be accumulated. Timeliness of the accumulation of the information is, of course, a crucial factor. What is presented here is an expansion of a management information structure which is effective beginning at the level of an individual activity on a specific job and going all the way through the corporate summary.

## MANAGEMENT REVIEW OF INFORMATION

Control assumes the ability to influence outcome. Management at any summary level will need to attend to the exceptions being reported to determine what corrective actions may be necessary. The analysis flow presented in Exhibit 5-5 indicates the flow of analytical procedures to investigate problem areas for the purpose of implementing corrective action. The interest in investigation may begin at the corporate level. If the first subordinate summary level (going down) would be at the regional offices, management could identify a specific regional office requiring management attention. Within the regional office, a specific satellite office may require attention. Within the regional office, a specific project manager may be experiencing difficulty which in turn may be isolated on a specific job, or a work package within the job, or a certain activity within a work package.

Parallel to the eight levels of summary flow, there are available seven levels of analysis, to the individual activity on a specific job. This degree of flexibility in management information is important because it allows the greatest flexibility for management action.

## TRAIL OF INVESTIGATION

One of the more difficult areas of management action is determining the source of a problem, the nature of the problem, the magnitude of the problem, and what actions, if any, are appropriate to the resolution of the problem. The trail of investigation is an important management information resource which allows quick discovery or disclosure of *potential* problem areas on a

**Exhibit 5-5. Analysis Flow from Corporate Management Information Systems through an Individual Activity**

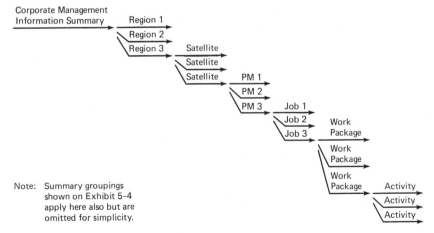

timely basis. The trail of investigation should be designed into the management information process.

## OTHER KINDS OF SUMMARY INFORMATION

While the information presented focuses on job related performance in a hierarchy, it should be apparent that other forms of summarization or other groupings of performance elements can be built conveniently into a management information system. For example, management at the project or satellite level may want to evaluate the performance of subcontractors and principal vendors. An additional summary level could be inserted (as done in Exhibit 5-2) into the matrix of summary levels to deal with multiple jobs which are part of the work of one subcontractor or vendor in the local or satellite environment under the control of production management. Obviously this kind of summary can reveal much about the performance of specific subcontractors and principal vendors. What makes these concepts of management information systems workable is anticipating the kind of management information which will be required, planning all of the production processes in anticipation of management information requirements, and capturing data for convenient summarization at any management level of interest.

## SUMMARY

Management information systems must provide information to management which is contemporary in nature, valid within an acceptable variance, presents only the kind of information required and in the form required. They also retain the capability of detailed audit examination at any level of interest to management. These kinds of management information systems become possible through the proper use of integrated cost and schedule control techniques which focus primarily on control of a project but contain sufficient detail level to allow structuring of management summaries.

The implementation of project strategies consistent with project planning and the data capture procedures are also crucial elements in management information systems. Where systems provide for a specific matching of planned performance data and actual performance data, management information may be presented in the form of performance ratios. These performance ratios may be generated in any area of interest. In construction, these areas of interest tend to be cost control, schedule control, and production control.

Management by exception procedures are especially useful when built into management information systems since they allow the systematic separation of acceptable performance and unacceptable performance. Since acceptable performance usually consists of 95% or more of all activities being performed, the management information system should provide confirmation of this performance. Exceptional performance, however, needs to be highlighted in some way for management focus. These features combined provide the structure for useful management information systems.

Management information systems utilized in the context of integrated cost and schedule control are contemporary in nature. They are distinctly separate from accounting information systems. They are designed to provide quick, accurate, and contemporary information on job performance and to take full advantage of the statement of expectations to measure performance. In manual information environments, the calculation of perfomance ratios clearly meets the requirement for simplicity. While computer environments are less sensitive in terms of the clerical effort associated with manipulation of data, the simple performance ratio, nonetheless provides direct and easily verifiable management information for all major parameters of performance. These are, of course, significant components of integrated cost and schedule control.

# 6

# COMPUTER APPLICATIONS

No treatment of integrated cost and schedule control would be complete without a discussion of computer systems. While this reference does not deal specifically with computer applications for integrated cost and schedule control, the techniques presented can be utilized manually or in a data processing environment.

Computers are used literally for everything. They represent the new high technology. Retail outlets throughout the country specialize in nothing but computer hardware, operating systems, applications programs. Even with senior management, the new status symbol is the personal computer.

Among the major capabilities of the computer is the facility for receiving and keeping track of large quantities of raw data, distributing that raw data through a wide range of work environments, manipulating the raw data, and converting it into useful management information. The computer has the capability of distributing both raw data and useful management information throughout a wide variety of work environments. The capabilities associated with single data capture have already been discussed.

Computer capability relating to integrating otherwise dissimilar processing requirements is of special interest. The full range of computer applications including data processing techniques, programming methodology, along with utilization of hardware, and its cost is so fluid that a detailed treatment of current hardware and software applications may prove unproductive. There are, nonetheless, certain basic issues of computer applications and data processing relating to integrated cost and schedule control which are generally unaffected by the technological development of hardware, programming and data processing.

For example, the basic elements of managing the construction enterprise remain essentially the same whether the information processing systems are manual or computer based. The financial management and accounting of a company are controlled by a mixture of management's needs for information, local and federal regulations, and tax requirements. Job cost accounting is standard in the industry. Payroll must be processed and the appropriate tax forms must be prepared and filed with the appropriate public agencies.

What the computer represents in all of these situations is a powerful tool having the capability of converting raw data into useful information quickly and expeditiously, leaving management and administration freer to deal with basic issues of conducting the enterprise.

## GENERAL USE OF COMPUTERS IN THE CONSTRUCTION INDUSTRY

When a construction organization seriously considers acquiring a computer resource, three computer applications are usually first considered. These are financial and accounting, payroll, and job cost. Computer capability for these areas is well established. Beyond these initial areas, specialized computer capability may be added for planning and scheduling, production control, and estimating. These three areas, however, tend to be more specialized and unique to each organization. The major concern which is only now being addressed is the full integration of all of the subsystems within a business environment. The special interest concerning systems integration relates to the computer resource which frequently has the capability of doing significantly more for the user than limited applications programs without integration would allow.

## COMPUTER SYSTEMS

For the uninitiated, the term "computer system" has a certain mystery associated with it. An examination of the basic elements of a computer system reveals that the computer mystique has developed more from lack of knowledge than from any inherent quality of the computer.

A computer system consists of three major elements: the computer hardware, the operating system and the applications programs. Each of these is essential to the system. Each is also affected in some measure by changes in technology. In older systems, for example, the operating system is wired in as an integral part of the computer hardware. Contemporary hardware design, however, separates the operating system, which is now largely a form of software, from the computer hardware itself. This advance is technology significantly improves the flexibility of computer hardware.

### The Computer Hardware

Computer hardware includes the input/output devices used by the operator to communicate with the computer and a central processor. Exhibit 6-1 shows a typical basic computer system.

**Exhibit 6-1. Basic Work Station Computer Hardware**

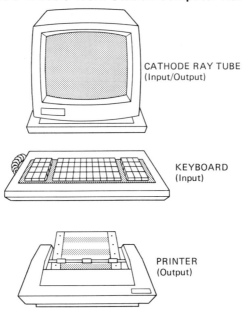

CATHODE RAY TUBE
(Input/Output)

KEYBOARD
(Input)

PRINTER
(Output)

A basic set of work station hardware consists of a cathode-ray tube (CRT or screen), a keyboard, and a printer. The CRT, or screen, is used to present a variety of information to the user. Information presented could include basic instructions for turning the system on, menu selections for different applications programs, or specific "screens" for data entry purposes. Any of these displays can be addressed through the keyboard. The keyboard and the screen are *input* devices, allowing the operator to communicate with the computer system. The printer is the primary *output* device, allowing the computer to provide certain information and instructions to the operator. The CRT, or screen, may also be used as an *output* device when certain information is presented to the operator by a screen display. We can say, then, that the keyboard and CRT are the primary input devices, and the CRT and printer are the primary output devices. In larger more complex systems, data entry may be done through other media such as punch cards.

Another major component of computer hardware is the *central processing unit* (*CPU*). The central processor addresses the application programs selected by the operator, accepts raw data into the program area for processing according to the programmed instructions, and delivers the processed information for storage or for output. In addition, most computer systems will have storage capacity which allows raw data to be received and stored for future access and which also allows processed information to be stored until it is required by the operator. Storage media may consist of storage built into the CPU itself, floppy disks, hard disks, and tapes. In summary, the hardware may consist of input and output and storage devices, and may include a CRT and keyboard, a printer, the central processing unit, and some forms of storage media.

## The Operating System

The operating system in a typical computer system consists of software contained on some storage media and accessible to the central processing unit. The operating system manages the computer resource, providing it with instructions on how to carry out the basic data processing functions required by the applications programs. The operating system controls the total resource, managing the resource's use. It also establishes priority of processing when demands for the computer resource may conflict. Operating systems are available in a wide variety of languages and formats, some of which are hardware discreet (work only with particular machines) and some which are more generic in nature.

## The Applications Programs

Applications programs make up that portion of the system which actually contains the instructions necessary to receive raw data, to manipulate raw data, to store both raw data and information, and to present the information to the output devices. Applications programs tend to relate to specific work environments or disciplines. For example, financial and accounting programs may include general ledger, payroll, accounts payable, and accounts receivable capabilities. Separate applications programs may be utilized for planning and scheduling, estimating, and budgeting. Job cost may be another separate applications program. Specialized programs for management and production control may also be developed. While some kinds of applications programs, such as financial and accounting, have a long history of development, others, such as estimating and subcontract control, tend to be user discreet.

Exhibit 6-2 provides a functional diagram of a computer system. The central processor, or CPU, consists of the control unit and the arithmetic logic unit. The main memory is associated with the CPU and provides space required for three specific systems functions. The first of these is the operating system, previously described, and the applications programs. These two combined are referred to as the *systems overhead,* or the space not available to the user for data and process. The third part of the main memory is the space available for data and processing.

Along with the central processing unit and the main memory is some

**Exhibit 6-2. Functional Diagram of a Computer System**

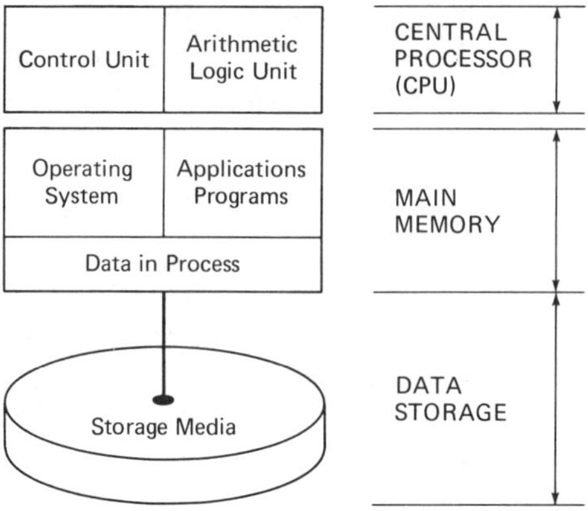

storage capacity. The storage capacity is used to retain raw data pending processing and processed information waiting delivery to the operator. Storage is also used to store applications programs not presently in use. Storage is also frequently used to maintain a copy of the operating system.

## INPUT/OUTPUT DEVICES AND METHODS

Input/output devices are used to maintain communication between the operator and the computer system. These range from punch cards to audio access and may also include plain keyboards without the CRT. The most common output device is a hard copy printer although many work environments find output through a screen display both appropriate and efficient for certain applications.

There are two general kinds of input/output worth noting. For bulk processing operations, such as large payrolls and certain accounting functions, *batch processing* is prefe. ᴄd. Batch processing accumulates relatively large quantities of raw data until the current time period for data processing expires. The data entry is then done in mass, the data processed a similar way, and the informational output delivered in the form required. With the current proliferation of computer systems both in the work environment and at home, *interactive processing (usually entering data through a CRT) is much more common; as the data is entered and accepted by the computer system, it is filed, stored, and/or processed immediately.*

The advantage of interactive processing in integrated computer systems, especially in integrated cost and schedule control, is that upon entry of data that data becomes immediately accessible to all parties of interest. In interactive data processing, raw data is entered through the keyboard and immediately displayed on the CRT in a preset format. Here, processing usually consists of a specificd data entry task in which the system acknowledges receipt of the raw data. Whether the raw data is processed immediately or simply stored for future processing depends on the program design and the need for subsequent processing of information.

Interactive processing has the additional advantage of immediate confirmation that the data has been received by the system. Input/output devices which allow interactive processing also allow for screen inquiry of certain information when hard copy is not required.

## HARDWARE SIZE, SPEED, AND CAPACITY

When a prospective user begins to consider a computer system seriously, he or she tends to focus on the computer hardware portion of the system and on the operating system to the exclusion of applications programs. The reason for this tendency is that most users consider data processing and hardware people to be experts in whole computer systems. From a user's point of view, however, the attributes of the *applications* programs will ultimately determine both the quality and the value of the computer system.

The technical niceties of computer hardware are less important than the ease with which the applications programs may be utilized. Such other issues as the speed of the computer are, in a large measure, controlled by the efficiency of the design and programming of the applications programs. The required size of the computer as measured by main memory storage capacity and other storage can be calculated once the general parameters of the applications programs to be used in the enterprise have been clearly established. The focus of a user seeking to acquire a computer system, therefore, should be the applications programs which will convert raw data to useful information and deliver that information when and where it is required. A crucial capability of the information delivery systems to be considered is the electronic transfer of data and information for use of interconnected terminals and systems. These interconnections are referred to as *networks* and should be considered as part of the master design of any computer system.

## MAJOR CONCERNS OF END USERS

Several specific areas should be examined when considering computer systems applications for the construction industry. A primary focus on the computer system as a tool to facilitate the handling of raw data and to prepare information for management purposes in the construction environment is essential. It is generally neither productive nor important to make data processing experts out of construction management and the clerical staff. Construction management and the clerical staff need to be trained in using the computer system as a tool to facilitate effective management.

It is unnecessary for a person to be a watchmaker to tell time nor is it necessary to be an expert mechanic to enjoy the comforts and convenience of the automobile. It is equally unecessary for computer systems users to become experts in data processing to use the computer effectively. In fact, diverting productive time to data processing areas when such diversion is not required can produce an unreasonable drain on an organizational resource.

Computer applications, then, should focus on the capture of raw data, the distribution of raw data to necessary work environments, the convenience of manipulating raw data, the conversion of raw data into useful information, and the distribution of that useful information to where it is needed. This "input/output" mentality can be one of the most effective tools available when the search is undertaken to find "the computer system." There are the three principal questions to ask when selecting a computer system:

1. What is the most efficient way to get raw data from present or existing source documents into the computer for data processing?
2. What is the most efficient way to process raw data?
3. What is the most convenient and efficient way to deliver the specific information required for management when that information is required?

It is necessary to examine complexities of programming for maximum computer efficiency. It is equally important to consider the demands placed upon the operator in utilizing the computer system. It is also necessary to understand the limitations placed on the computer resource by its processing and storage capacities. However, the technical niceties of computer hardware and programming should not be of primary concern to the user, nor should the selection of a programming language. What is important, is the ease with which raw data entry is accomplished from existing source documents, with the least amount of pre-data entry manipulation. Also important are the kinds of information needed and the form in which that information is required. Probably the major expectation of a prospective computer systems user is the timely deliver of information in keeping with management needs.

## USER FRIENDLY COMPUTER SYSTEMS

Significant advances in the way in which people interface with computer systems are making computer systems more accessible to the general public. A term used in the data processing environment describes computer systems which are comfortable and easy to use: *user friendly*. User friendly systems are easy to learn. The applications software in user friendly systems is designed to assist the operator in using the systems. The computer term "tutorial system" simply means that the applications program contain a subprogram to instruct the user in the use of the application. Interactive data processing through the CRT and keyboard has been a significant factor in making computers user friendly. Contemporary computer hardware, operating systems, and applications programs make it unnecessary to tolerate computer systems which require extensive training or are awkward and difficult to use. In the construction work environment, there is no reason for a user to accept a computer system or application programs which places unreasonable demands on the computer operator.

## HOW APPLICATIONS PROGRAMS ARE BUILT

Research and experimentation have developed some standard approaches to building applications programs. Since the applications programs focusing on input/output and timeliness of data processing are major concerns, these processes are worth examining. Exhibit 6-3 shows the functional personalities involved in system building processes.

The beginning point of any applications program design is to discover "the system" currently in use. Every organization has a "system" which is used in conducting its business. Whether or not this system is articulated in procedural manuals or the business of the company is conducted informally, the system, nonetheless, exists. Any sensible approach to applications programs design must begin with the determination of the system currently in use.

Procedures for developing some understanding of current systems have become well established over the past few years. The usual first step is to create a list of major informational areas which, in total, make up the system. The basic elements of the system within each of the major areas are then identified, followed by the development of a detailed flow diagram for the component parts of the system. Some documentation with regard to sources of raw data, ways data is collected, the ways data flows through the system, and the informational output of the system are all documented. A flow diagram is then generated to show the movement of information throughout the system, with a clear definition of the final output expected.

Documentation of the system currently used is prerequisite to any consideration of computer applications programs. It is an indispensable part of information where computer applications programs are to be built to the specific requirements of the user. It is equally important where there is to be an examination of existing software packages which may be applicable to the user's needs. Further, the process of discovering the system currently in use in most work environments creates significantly better understandings

**Exhibit 6-3. System Building Processes and Personalities**

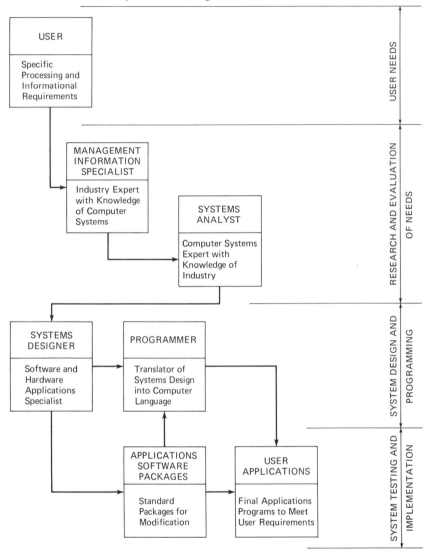

of how the routine functions of the business operation are carried out. Also uncovered are frequent differences between the perceptions of management and supervision concerning how tasks are carried out and the perceptions of those who physically perform the work. Any such differences must, of course, be resolved prior to addressing the issue of systems design.

## The Management Information Specialist

There are four speciality areas to be considered from a functional point of view in addressing the issue of systems design. Each of these has its own particular role to play in the development or selection of applications programs to meet specific needs.

The newest of these functional roles is that of a management information specialist. This role is relatively new to data processing and is a natural consequence of the communications difficulties between data processing experts and users. In industries as diverse as construction and data processing, achieving communications between these two groups have been difficult. That is, translating the processes and language of construction into data processing terminology and converting data processing processes and terminology into construction language has historically been a difficult task. The management information specialist is primarily an expert in a given industry—where his or her major skill lies—with sufficient knowledge of management information requirements and sufficient expertise in data processing to provide each side with understandings of the problems and opportunities of the other.

A management information specialist in construction might be a project manager or business manager who can initiate the requisite discovery process and confirm that the discovery task at conclusion is fairly reflected in the major programmatic flow diagrams and other essential documentation. Documenting existing manual and/or computer systems is also the responsibility of the management information specialist.

## The Systems Analyst

Since not all of the manual processes which may be used by a construction organization fit nicely into certain constraints or limitations of data processing, another kind of expertise is required. The second functional role is the systems analyst, whose responsibility is to take apart the documentation of the manual and existent computer systems prepared by the management information specialist and to examine these from a data processing point of view. (Contrary to the management information specialist, the systems analyst's primary skill is data processing with industry expertise secondary.) This examination will include such tasks as more clearly identifying data sources and the functional flow of raw data and information to determine the kind of input and output desired as compared to what is being provided. In essence, the systems analyst takes the user process apart and examine it in detail for completeness and validity.

## The Systems Designer

The next major undertaking in building an applications system is to reassemble the manual processes into a more convenient data processing form while at the same time maintaining the essential input/output elements of the original system. This step requires the capability of a third functional specialist— the systems designer. It is the responsibility of the systems designer to set the requirements identified by the management information specialist and the systems analyst into the data processing environment. It is *not* the function of the system designer to create anything new or novel as far as the end user is concerned, but is primarily to interpret systems requirements faithfully in the data processing environment. That is, the systems designer reassembles the dissected version of user processes created by the system analysis for appropriate induction into the data processing environment.

The systems designer gives due consideration to the programmatic capabilities, speed, and capacity of the equipment, size of the equipment with regard to a total systems requirement, and other essentials of data processing. At this point, the systems designer will consider such things as specific hardware to be utilized, programming languages to be used, speed of the equipment, and the operating systems.

## The Programmers

As the work of the systems designer progresses, information may be provided to the user concerning basic systems decisions. This information may include the work of the systems analyst in evaluating the cost effectiveness of systems, the cost of operations, the cost of installation, cost of maintenance, and organizational impact. A basic decision concerning the applications programs must now be made by the user based on systems designer's recommendation. This decision concerns a choice between assigning the work of a systems designer to programmers to write the applications programs, or to begin a search for existing applications programs which might be worth considering. How the applications programs will be acquired must be decided *before* it is appropriate to consider hardware. The management information specialist, the systems analyst, and the systems designer provide full expertise upon which the user may draw in making this basic decision.

Should the user elect to have applications programs written, the fourth function role, the data processing programmer, is required. It is the responsibility of the programmer to convert the processes specified by the systems designer to appropriate computer language as programs for computer applications. It is not the responsibility of the programmer to create new or novel programs but to write the exact programs which have been designed by the systems designer. In earlier methods of developing applications programs, it was not unusual for a user to acquire the services of a programmer for the purpose of performing *all* the major functional tasks described—management information specialist, systems analyst, systems designer, as well as programmer. This procedure did not work well in the past. Better understanding of user system requirements have led to better methods of system development.

## THE APPLICATIONS PROGRAMS

The applications programs from the point of view of the user are the highest priority in selecting any computer system. When applications are carefully matched to the needs and requirements of the user, the basis for a satisfactory computer system results. Hardware (and an operating system) adequate to run the chosen applications, but with an expansion capability, can then be installed if not already on hand. Conversely, the finest computer hardware with the most exotic operating system, but with poor, inappropriate, or difficult applications programs will result in an unsatisfactory computer system.

In specifying applications programs, all of the tasks which may be undertaken by the computer should be considered. An evaluation should be made of these computer applications to determine the appropriate level of computerization. Not all procedures which could be computerized should be made a part of the applications program. There will always be some procedures which function more efficiently in the original or manual environment. Once a decision has been made concerning programs to be placed in the computer environment, priorities should be set to establish a reasonable progression from the easiest to the more difficult programs.

Experience also indicates that each phase of an applications program successfully designed, installed, and brought into full operation will raise additional possibilities which if, in turn, successfully designed, installed, and brought into full operation will raise yet further possibilities for the later expansion of the computer applications program. These new opportunities can be rather exciting as computerization horizons are expanded.

A major outcome of the development of computer applications will be a much more detailed understanding of the processes and procedures of the business. Informal processes give way to structured procedures and a systematic and routine way will emerge for accomplishing similar tasks. This can only result in greater efficiency, which when related to the computer resource can make significant advances in efficiency and performance of the whole organization.

## PACKAGED SOFTWARE

The last ten years have been a proliferation of computer software packages for the construction industry. Virtually any application appropriate to the construction industry is available in some form. Packaged software should be a serious consideration for any user contemplating a computer system. The utmost care should be given to examining the capabilities of an available program and matching it to the corresponding program design prepared by the systems analyst; there might be a fit even if the assumption is that the programs will be coded from scratch. An exact match may prove to be difficult, however. For this reason, availability of source code and the convenience of revisions to tailor programs to specific application requirements are important considerations. The state of the art is such that certain applications programs are so far developed that it is unnecessary and unrealistic to consider writing these programmatic areas from scratch. Some such applications are likely to include financial and accounting, payroll, and, to a slightly lesser extent, accounts payable.

Great care must be taken in examining available software packages for construction so that unfortunate compromises will not occur in end user requirements in terms of input/output and management information.

## THE PROBLEMS OF SYSTEMS INTEGRATION

As noted, a major problem in computer applications for construction just now being addressed is that of systems integration. There are many good software applications programs available to the construction industry. Most tend to be very specialized in nature and very limited in application. For example, a freestanding program typically for estimating may be acquired. Another program may be acquired for planning and scheduling. A third program may be acquired for job costing and a package of programs may be acquired for financial, accounting, accounts payable, accounts receivable, and job cost. Subcontract and purchase control may be yet another available program.

A careful examination of how the industry conducts its business will reveal close interrelationships between the program areas mentioned and others as well. Clearly the issue of estimating is not separate from cost control, and planning and scheduling is not an issue separate from production control. Similarly, accounts receivable is a direct reflection of production management; job cost and accounts payable are inseparably related to subcontract control. And the list can go on. Since the computer system receives and maintains all job performance data and since it also has the capacity for handling all financial and accounting data, a major consideration, if not the most important, in looking at software applications is the ability of these separate free-standing applications for various work environments to communicate freely with each other.

Systems integration is without a doubt the data processing of the future. The ways in which data is managed and in which data is made available and shared among the disciplines of construction must, of necessity, be of the highest consideration in the construction industry. Traditional control of the computer resource by financial and accounting must give way to a sharing of that resource with all other work environments.

## SUMMARY

Computer systems for construction are essential to effective business management. The demands for contemporary and valid information make computer applications in the industry appealing to most companies. There is virtually no task within the construction industry which cannot be handled by computer applications. Computer systems, however, must be built with the end user in mind and with an input/output mentality which focuses on the use of the computer as a tool. End users should not abandon their responsibility for fully articulating the kinds and the sources of raw data to be processed nor can end users afford to be compromising in the demand for current valid information when it is needed and in the form required.

The user with the assistance of a management information specialist must clearly define system requirements. The primary role and function of data processing is to accurately and faithfully translate the user's requirements into the language of the computer. Teamwork in the building of a computer system can result in adding a significant resource in the form of computer applications to the construction organization.

# 7
# PROJECT STATEMENT OF EXPECTATIONS AND PERFORMANCE MODELS

Worthwhile projects are normally the result of a process which carefully moves from an idea, or a concept, through an executable design, through a specific plan for carrying out that design, the implementation of the plan, and its completion. There is usually a specific point of beginning, a specific objective or group of objectives to be realized, and specific activities or actions which must be carried out for the successful completion of the project.

Construction projects provide good examples of how projects of any type may be initiated, planned, and executed. All the components of good project management are demonstrated by the successful construction of a project. The successful completion of a project, however, requires preconstruction planning if the project is to be successfully completed.

## JOB PLANNING

The job planning process begins with the first involvement in the project. On a competitively bid project, the first involvement is a decision to bid the project. The planning processes progress through a series of well defined phases, each of which has its own special planning requirements. The phases of interest are:

1. Prebid and bid.
2. Award and preconstruction.
3. Weekly job cycle.
4. Monthly job cycle.
5. Substantial completion.
6. Final completion and job closeout.

The planning associated with the contract award and preconstruction phases of the project is, perhaps, the most important since it controls all the subsequent actions with respect to the project and establishes the expected boundaries of performance. The term "job planning" is general in nature and addresses three primary and accumulative questions or stages:

1. How will the job be constructed? (the plan)
2. When will it be constructed (the schedule)
3. What is it expected to cost? (the budget)

Detailed job planning is required in order to answer each question in a specific and objective way. The answer to one question leads to the answer for the next. An objective of the planning process, to provide performance models, is essential to subsequent measures of job performance.

While job planning covers these three areas, the "plan" for the job deals with the details of how the job will be constructed. Job planning in this sense is specific and limited to defining the activities required to complete the job. It determines physical relationships between certain activities; when there are no physical relationships, it makes certain choices as to the optimum sequence of the work. In this context, the term "job planning" comes from the planning and the scheduling procedures associated with critical path method and other logic network techniques. Its primary value and utility is that it permits the complete planning of the project in terms of (1) the *activities* required and (2) the relationships between those activities, without regard to the resources required and the actual scheduling of the project.

The project plan using the critical path method is first seen in the form of a logic network diagram. The logic network diagram provides a basis for resource planning which addresses the question of *who* will physically perform the work and what *resources* are available for the performance of each activity.

The planning of resources to be used for each activity and the rate at which those resources will be applied is the primary determining factor in *how long* (schedule) it will take to perform each activity. The duration of each activity involves availability of the resource to be used and must be established before the scheduling computations can commence.

The scheduling computations which result in establishing the time boundaries for each activity in the logic network diagram involves simple arithmetic calculations based on the duration of each activity as established by activity resource planning. The schedule review process involves an examination of the proposed schedule after the scheduling calculations have been completed to examine certain specific areas of interest.

The first of these areas of interest concerns multiple activities using the same resource at the same time. When such conflicting demands for a single resource become evident, these can be resolved by an additional planning process which looks at alternate resources, alternate relationships or priorities among activities. The result should be a plan for the project which, when calendar dated, provides the schedule for the project.

The first step in planning the project before it can be scheduled consists of identifying and listing those activities which make up (or which themselves must be completed in order to complete) the project. This is also conveniently the first step in creating an activity based budget. However, planning of project activities and the relationships between those activities may proceed without reference to budgets or cost. Budget or cost information is important in establishing or calculating the *duration* of activities and these must be determined before the scheduling calculations can realistically be completed.

The quantities, unit prices, and cost of the budgeting process are fixed by project requirements. The budget for each activity is determined by the quantity of that activity times its unit price. Quantities are fixed by specific job requirements. Unit prices are established by the fair value of the work. The budget, when completed, provides the expected cost for each activity and of the project as well.

The planning of the resource to be used (i.e. own labor, sub labor, etc.) and its rate of application (i.e. how many men) establishes the expected duration of each activity and enables completion of scheduling computations and, when a start date is known, calendarizing of the plan.

## THE PROJECT STATEMENT OF EXPECTATIONS

The project planning processes provide specific and detailed answers to three of the four central questions about any project, and justify the overall planning of the project. These (plan, schedule, and budget), however, are continually visible for control purposes. The fourth, cash impact, involves certain additional analytical routines discussed in Chapters 11, 12, and 13.

### The Schedule and Expectations

The plan concerns how the project will be constructed and how the activities making up the project relate to each other and are merged into the construction schedule. The resources to be utilized are also merged into the schedule since the *rate* of application of those resources is a controlling factor determining durations of activities and as a result the time required for the project. The schedule, therefore, not only presents the time statement of expectations for the project but in its best development shows how the project will be built and what resources will be utilized in its' construction. The schedule is derived from the plan for building the project and the resources to be utilized.

### The Budget and Expectations

The budget is the cost statement of expectations and provides the basis for the cost control of the project. The schedule as a time statement of expectations and the budget as a cost statement of expectations combine to form the basis for a series of other performance models for the project worth examining.

The project statement of expectations when properly prepared is a tool which can be effectively used to express exactly what project performance is to be expected. These combine to establish the expected (planned) performance in terms of cost, time, and production. Historically, each of these parameters of job performance has been considered more or less independently. Current management science, however, views the interrelationship between these separate performance parameters as essential to measuring overall performance and in predicting the outcome of the project. Traditionally, these have been passively measured and recorded as performance occurred but have provided no real basis for the integration of cost and schedule controls.

The statement of expectations for a project should be formally developed. It is essential, however, to consider how the statement of expectations will be used for cost and schedule control. It is also important to consider how the raw data on actual performance will be captured. Furthermore, consideration should be given to how management information can be generated in the most timely and convenient way.

Consideration of these factors will allow a rational statement of expectations to be developed. It will also allow performance models which anticipate integrated cost and schedule control to be generated from the statement of expectations. Both the statement of expectations and the performance models which derive from the statement of expectations should be designed to eliminate the need for constant recasting of raw data or informational output to provide useful management information.

## THE STRUCTURE AND RELATIONSHIPS OF THE STATEMENT OF EXPECTATIONS

A series of performance models make up the statement of expectations. The construction industry focus is on the budget and the schedule. The budget is the cost performance model and the schedule is the time performance model. Effective project management and project control can be carried out using only these two performance models. The budget and the schedule either completely establish or control the remaining performance models and are referred to as "primary performance models."

The remaining performance models are shown in Exhibit 7-1, which also shows the area of interest of each performance model, the data source, and the level of detail appropriate to that performance model.

Some further description of the budget and the schedule is appropriate

**Exhibit 7-1. The Statement of Expectations: Structure and Interrelationships of Performance Models**

PRIMARY PERFORMANCE MODELS

| | DESCRIPTION | PRESENTATION | CONTENT OR USE | DATA SOURCE | DETAIL |
|---|---|---|---|---|---|
| 1 | Budget | Tabular | Cost control | Refined estimate | Individual activity |
| 2 | Schedule | Logic network or time scale | Calendar dated schedule | Roster of activities and project plan | By activity distributed on time scale |

SECONDARY OR DERIVED PERFORMANCE MODELS

| | DESCRIPTION | PRESENTATION | CONTENT OR USE | DATA SOURCE | DETAIL |
|---|---|---|---|---|---|
| 1 | Cost curve | "S" curve | Cost/time | Budget and schedule | By activity distributed on time scale |
| 2 | Production curve | "S" curve | Production values | Cost curve plus budgeted overhead and fee | By activity distributed on time scale |
| 3 | Schedule of Values curve | Modified "S" curve | Values for requisitions | Weighted cost and/or production curve | By activity and work package distributed on time scale |
| 4 | Cash income curve | Histogram | Accounts receivable (A/R) | Schedule of values plus contract payment terms and conditions | By activity and work package distributed on time scale |
| 5 | Cash requirements curve | Modified histogram | Accounts payable (A/P) and payroll (P/R) | Subcontractor and vendor payment terms and payroll cycle | By activity and work package distributed on time scale |

in order to clarify the relationship of these two primary performance models to the derived performance models.

*Budget* data is generally presented in tabular form. Its concern is cost. The data source is the estimate and the detail is presented at the activity level. The *schedule* is usually presented as a time scale, either a Gantt chart or a network. The area of interest is time required for the project. The schedule is derived from the project plan using budget information to determine resource requirements and is generally presented by activity for production control purposes.

The derived performance models present additional information as a result of systems integration. In general, the derived performance models are either controlled by or established by other information or data. The preparation of the derived performance models normally involves some simple graphic techniques.

*Cost curve* data is presented in the form of a typical "S" curve. The cost curve integrates data from the budget with respect to cost and from the schedule with respect to time. The information is presented by activity and on a time scale with plotting intervals which may be daily, weekly, or monthly.

The *production curve* also takes the shape of a typical "S" curve and is concerned with the value of production. It is derived from the cost curve plus overhead, profit, and other indirect cost added to the cost curve and is normally presented by activity and on a time scale with selected plotting intervals.

Many construction contracts require a Schedule of Values to be prepared for payment purposes. The *Schedule of Values* is a listing of major elements of the project with dollar values assigned to each. These dollar values are used for billing purposes. The source of data for the Schedule of Values is the cost or production curve; certain weighting of the production curve data may be involved in preparing the Schedule of Values data.

The Schedule of Values data may be plotted as a modified "S" curve. The Schedule of Values, production and cost data, may also be presented in tabular form. The Schedule of Values data becomes the source of values for requisitioning processes. It represents the billable value of work in place but does not represent expected cash income. The detail presented is usually by trade groupings or work packages. The data, however, is maintained by activity.

The *cash income curve* presents significant data on the amount of cash which may be expected and when it may be expected. It represents the planned accounts receivable (A/R) for the project. Its presentation is in the form of a histogram although the data may also be presented in tabular form. The data source for the preparation of income curve is the schedule of values curve. In addition, owner/contractor contract terms and conditions of payment are required to determine when payments may be expected and the amount to be paid compared to the scheduled value of work in place. The difference between the scheduled value of work in place and the anticipated cash income is made up of retainages, holdbacks, or other deferred payments. The cash income data would normally be presented by activity and summarized by work package.

The *cash requirements curve* is normally presented as a modified histogram although the data may also be presented in tabular form. Its primary areas of interest are accounts payable and payroll. The sources of data for the cash requirements curve are the schedule, subcontracts, purchase orders, and payroll. It provides data to management on when cash will be required, the amount of the cash required, and is generally presented by activity and in the work package summary.

## REQUIREMENTS FOR INTEGRATION

The interrelationships of the primary controls of cost and schedule are essential to systems integration. While the requirements for systems integration through the project statement of expectations and performance models may appear complex, the means for anticipating systems integration has already been identified. It is clear that construction projects are managed in the field by assigning specific resources to specific activities, and that the critical path method for planning and scheduling conveniently relates to the way in which work is managed in the field environment. It is also clear that the data source for all primary and derived performance models utilizes the roster of activities as the basic unit of work.

The *roster of activities* has the unique characteristic necessary for systems integration expressed in the form of performance models. Each activity can be described so that it carries with it all the necessary data and the reference files for performance planning and for performance measurement. Additional data from the derived performance models as well as additional informational requirements for cost control, schedule and production control may be added. The significant key continues to be the common roster of activities.

The job planning processes which have been described result in a significant quantity of planned data for each activity. These are shown in tabular form in Exhibit 7-2.

The *planned performance data* includes the appropriate cost code references, descriptions, quantity unit price and budgeted cost, the resource identification, scheduled start, duration, and completion dates. Planned data establishes the plan for what is to be done, who is to do it, and when it is to be done. The question of *how* it is to be done is part of the job planning process and is not visible in the tabular data. From the planned data, specific perfor-

**Exhibit 7-2. Activity Data and Information**

## PLANNED OR EXPECTED ACTIVITY DATA AND INFORMATION

### PRIMARY PERFORMANCE MODELS

| ACTIVITY | | | | SECONDARY (DERIVED) PERFORMANCE MODELS |
|---|---|---|---|---|

**ACTIVITY**

| IDENTIFIERS | | PERFORMANCE MODELS | | | | | | | |
|---|---|---|---|---|---|---|---|---|---|
| COST CODE | DESCRIPTOR | 1 BUDGET | | RESOURCE | 2 SCHEDULE | | | | |
| | | QUAN. | UNIT PRICE | BUDGETED COST | COST TYPE | START DATE | DURATION | COMPLETION DATE | |
| | What | How Much | | | Who | | When | | |

**SECONDARY (DERIVED) PERFORMANCE MODELS**

| 1 | 2 | 3 | 4 | 5 |
|---|---|---|---|---|
| COST CURVE | PRODUCTION VALUES | SCHEDULED VALUES | CASH INCOME | CASH REQUIREMENTS |
| | | Cost/Values Distributed Over Time | | |

## ACTUAL ACTIVITY PERFORMANCE DATA AND INFORMATION

### ACTUAL PERFORMANCE DATA

**ACTIVITY**

| IDENTIFIERS | | PERFORMANCE ACTUAL MODELS | | | | | | | |
|---|---|---|---|---|---|---|---|---|---|
| COST CODE | DESCRIPTOR | 1 ACTUAL QUAN. UNIT PRICE, COST | | RESOURCE | 2 ACTUAL TIME DATA | | | | |
| | | QUAN. | UNIT PRICE | ACTUAL COST | SUBCONTR. VENDOR SUPPLIER | START DATE | DURATION | COMPLETION DATE | |

Note: For all planned or expected activity data and information, there is a matching set of data on actual performance. The comparison of actual data to planned data is basis upon which management information is developed.

**ACTUAL PERFORMANCE DATA**

| 1 | 2 | 3 | 4 | 5 |
|---|---|---|---|---|
| COST OVER TIME | PRODUCTION VALUES | BILLING VALUES | CASH INCOME | CASH EXPENDED |

## MANAGEMENT INFORMATION

| Cost Performance Ratio (CPR) | Time Performance Ratio (TPR) | Rate of Product Placement Ratio (RPP) | Production Performance Ratio (PPR) | Cash Income Ratio (CIR) | Cash Requirements Ratio (CRR) |
|---|---|---|---|---|---|

(Performance Ratio = Actual ÷ Planned)

89

mance models may be derived including performance expectations with regard to cost, production, schedule of values, cash income, and cash requirements. All of these include the information available on any activity, work package, trade grouping, or job.

When an activity is placed into construction, it should be apparent that actual performance data can be made available matching the planned or expected performance. Appropriate data capture procedures become essential so that the actual data matches exactly the planned data.

When *actual performance data* is compared to planned performance data, the result is a *performance ratio.* Management information systems requirements normally focuses on three specific performance ratios: the time performance ratio, the cost performance ratio, and the production performance ratio. These are also indicated in Figure 7-2.

One of the major questions to be resolved in integrated cost and schedule control procedures concerns the level of detail to be included in the planning, scheduling, and budgeting processes. Activities which are too large in scope and content will not be responsive to critical path analysis for planning and scheduling. Conversely, activities which are too finely described add additional detail to the integration process without proportionate benefit.

The key is to determine the level of control to be exerted in the field production of the project. The project is then divided into activities which are carefully matched to the level of control to be exerted. This procedure will result in workable performance models which are conveniently matched to data capture procedures and from which useful management information may be quickly and easily derived.

## SUMMARY REQUIREMENTS

A typical construction project may consist of hundreds of activities. Not all parties of interest or work environments are concerned with the level of detail required for field production purposes. As a result, there is the need for several convenient summary levels which provide data and information to several work environments and disciplines suited to their needs.

The appropriate summary capabilities allow the statement of expectations and performance models to be utilized conveniently in all work environments and by all disciplines. It further allows data capture procedures to be kept to the simplest form. Distribution of management information is also greatly facilitated.

## PRIMARY PERFORMANCE MODELS

The interrelationships of the primary and derived performance models are important to systems integration. It will now be useful to look at these performance models to examine how they are generated, the relationships that exist between them, and how they may be utilized in integrated cost and schedule control.

## The Budget

The budget is a primary performance model which may be generated directly or developed from the cost estimate. A variety of estimating budget techniques are used in the construction industry. Essential to integrated cost and schedule control is that the budget be expressed in terms of activities using the quantity unit price equals budgeted cost formula. Budgeting by activity represents a significant departure from more traditional estimating procedures in which estimates are prepared by trade groupings. While it is possible to accomplish some measure of cost and schedule control integration with trade groupings, trade groupings are much less appropriate than activity based budgets.

The key to systems integration is that the budget must be prepared utilizing the roster of activity which anticipates the schedule requirements. Conversely, the schedule should be derived using a list of activities required for budgeting purposes. In both cases, the primary concern is that both the budget and schedule contain the identical roster of activities for complete system integration.

## The Schedule

A wide variety of scheduling techniques are used in the construction industry. The quality of cost and schedule control is highly enhanced when one of the logic network techniques is utilized for planning and scheduling. When a common roster of activities is used for both budgeting and scheduling, logically derived schedules become very convenient. Since all resources are assigned in the field by activity and data capture procedures may be conveniently matched to these activities, the preferred method of planning and scheduling is the critical path method.

The critical path method has the capability of displaying the results of analysis in several ways; the most useful results for production control are the time scale network and the precedent list of activities. The time scale network may be utilized for monitoring production in the field. The precedent list of activities may be used to assign work to the appropriate resources in order of priority.

## DERIVED PERFORMANCE MODELS

The budget and the schedule provide data for the derived performance models. It is difficult to generate the derived performance models without a budget and schedule. The correlation of both time and cost of each activity is especially important.

## The Cost Curve

The cost curve is the most basic performance model and integrates both cost and schedule into a single display. It utilizes budget information generated for each activity, along with the calculated schedule for each activity, to distribute the expected cost of the project over its duration. A variety of summary levels are available for the cost curve. A typical well planned project will result in a cost curve with the classic "S" configuration.

## The Production Curve

The production curve evaluates product in place against the contract sum. It consists of the cost curve data plus the overhead and fee and any other indirect cost not included in the cost curve. The production value of product in place is unaffected by contract payment procedures, withholding of retainages, and similar items. The production value accumulates daily as product is placed, in much the same way as the cost accumulates.

## The Schedule of Values Curve

Most construction contracts require a Schedule of Values for requisitioning purposes. The Schedule of Values may use weighted values for certain items to suit the convenience of the owner and contractor. These weighted values become the basis for requisitioning and for determining value of work in place to enable payments. The scheduled value of work also accumulates on a daily basis similar to the production value and the cost of the work.

## The Cash Income Curve

Most construction contracts provide for various retainages, holdbacks, or deductions from the scheduled value of work in place for payment purposes. A typical payment retainage of 10% may be expected. Since in the contract payment for work is usually based on some specific time period, such as monthly, cash income may be planned on the basis of planned production and scheduled value of the work. The previous derived performance models take the shape of a standard or modified "S" curve. The cash income curve, however, is in the form of a histogram representing when payment may be expected according to the terms and conditions of the contract.

### The Cash Requirements Curve

The cash requirements curve is the most complex of the derived performance models since it represents a variety of cash requirements, each of which has its own peculiar set of prerequisites. For example, payroll is usually funded weekly. Suppliers may be paid by the tenth of the month. Principal vendors may be paid after receipt of payment from the owner without retainages. Subcontractors may be paid after payment by the owner but with a retainage applied. The cash requirements curve is derived from a number of sources and takes the shape of a modified histogram.

## COMPARATIVE DATA

Several of the performance models provide interesting data when compared to each other. One of the most critical of these comparisons relates to cash management; the information required becomes readily available when the cash income curve and the cash requirement curve are compared. This comparison may be made at any time during the project. The cash position of the project may be directly read from the cumulative curves for cash income and cash requirements. Where the cash requirements are below the cash income, the result is a positive cash delta. Where the cash requirements are above the cash income, as might be expected early in the job, there is a cash deficit, or negative delta.

These kinds of information are useful in determining capital requirements, requirements for credit accommodations, opportunities for investing surplus cash, and a variety of other cash management needs. The information from these comparisons allow for significant adjustments in production and scheduling a project to effectively manage cash flow. When the cash position is compared for all jobs in a business organization, very useful information for formulating cash management strategies results.

## ACTUAL PERFORMANCE DATA

The discussions thus far have all been directed toward planned performance models in the context of a planned statement of expectations. As the project is placed in production and proceeds toward completion, a similar set of data, now reflecting actual performance, is accumulated through appropriate data capture procedures. When the project plan is carefully matched to data capture procedures for actual performance, the basis for controlling the project with current valid information is established. Comparisons of actual performance to planned performance quickly result in a time performance ratio, a cost performance ratio, and a production performance ratio, as indicated in Figure 7-2. This kind of management information monitored throughout the course of the job provides early information on job trends. When adverse trends begin to develop, they become visible to management, enabling it to take corrective action wherever needed.

## OTHER USES OF THE STATEMENT OF EXPECTATIONS

The statement of expectations and the performance models have uses beyond control system integration. The performance models provide information to enhance the working relationship between the contractor and the owner. This sharing of information between contractor and owner may be especially useful. For example, the contractor's cash income curve is the owner's cash requirements curve.

## SUMMARY

The performance models which make up the project statement of expectations are an essential part of integrated cost and schedule control. The primary performance models of budget and schedule are standard operating procedures in most construction organizations. These differ only in detail and sophistication. The derived performance models are fixed when the budget and schedule are established. The techniques involved and the effort required to produce the derived performance models are less understood and less developed within the industry. The clerical effort required to derive the performance models and to maintain the actual performance data can be greatly reduced where an adequate computer resource is available for this purpose. The raw data to maintain actual performance information is already available. What is required are appropriate management directions and policies to enable and mandate the use of these techniques for integrated cost and schedule control.

# 8

# ESTIMATES AND BUDGETS AS PERFORMANCE MODELS

Estimating and the construction industry have always been closely related. With a few notable exceptions, those proposing to construct for themselves or for others have found it necessary to estimate the cost of the project to be undertaken. As far back as Biblical times, owners and builders saw the need to determine the cost of a project to determine whether there were sufficient resources for completion before committing to the project.

Estimating for construction has evolved over the centuries in several ways. The ancient master builder estimated materials to be required for the project (his quantity survey) and determined the cost to acquire those materials (his unit pricing) in order to establish the cost of the material. He then estimated the time required to place that material into the finished product using the workforce available to him. Lacking extensive sources of supply and price guidelines as now set by a competitive marketplace, these estimates, while well suited to their time and place, were crude by any contemporary measure. Since the work was done under the direct supervision of the master builder and a portion of the work may have actually been performed by the master builder, labor cost estimates were not considered critical. The primary focus was on earning a living rather than on our contemporary notion of profit. Low labor estimates resulted in lower pay (profit) to the master builder.

As projects have grown in size and complexity and as larger and more extensive resources have become necessary for the construction project, the need for better estimating procedures and better source data for both quantities and cost have developed.

Quantification methods and units of quantification have become relatively standard. Quantity surveying and pricing methods, while varying slightly among different trade groups, have also become relatively standard.

Even with these current facilities for quantification and estimating, cost estimating in preparation for the bidding process remains itself an expensive overhead cost item. The competitive bid process which makes up the method of awarding most of the construction projects in this country places a limitation on the amount of effort and resource which can be put into the estimating process. With a typical organization's success rate in competitive bidding ranging from a low of one in fifty to a high of one in five jobs, the problem of controlling the cost of estimating is an important item in overall business management. At the same time, there is the need for cost estimating which is at once sufficiently accurate to place the organization at an acceptably low risk in competitive bidding and sufficiently cost effective within the context of the organization's success rate.

Estimating relating to both competitive bidding and negotiated work is, for many organizations, the total extent of their sales and marketing capability. The conflicting demands for remaining active in the estimating and bidding areas while creating estimates which are valid continue to be major challenges to be addressed by each construction organization.

The limitations on time available for estimating purposes also places some limitation on the amount of detail which may be included in each estimate. As a result, the detail required for valid estimates for both competitively bid and negotiated work may not be sufficient for project planning and scheduling and may lack the fineness required for cost control purposes. In addition, bid strategy may require certain postbid adjustments to estimates when the organization is successful in the bidding process.

These industry requirements have resulted in a relatively common practice which separates estimates prepared for bidding purposes from the cost control mechanisms used by an organization when the bid is successful. By this process, the cost budget to be used for cost control having sufficient detail is prepared only for successfully bid projects.

The process of converting bid estimates to cost control budgets vary widely in the industry. Some construction organizations find it appropriate to use the bid estimate directly as a cost control budget. At the other extreme are those organizations which completely reevaluate projects from a cost control point of view, validating every aspect of each project in terms of both quantity and unit price. Taking into account these reevaluations is a reexamination of the resources to be utilized and the methods of production which might be considered most cost effective. In either case, what emerges for the successfully bid project is a budget for cost control purposes which is clearly a cost performance model.

The primary difference between the bid estimate and the cost control budget is in the level of detail and the fineness of the quantities and unit prices. The basic algorithm of quantity times unit price remains the same. For this reason, the remainder of this chapter will focus attention on the budget as a performance model, commenting on estimates only to the extent necessary for clarification.

## COST CONTROL AS A PRIMARY RESPONSIBILITY

The contracts for most construction projects in this country are awarded on a lump sum basis. Even where other forms of contracts may be used, such as a guaranteed maximum price or cost plus a fee, the control of the cost of the project is a primary function of the construction team. Assuming a lump sum contract and assuming the validity of the budget, the ability to control cost within the budget has direct and immediate consequences for the contractor. The contract amount fixes the amount to be paid by the owner for the work. Cost overruns must be absorbed within the fixed contract amount. Consequently, cost overruns directly reduce the overhead and fee dollar for dollar since these are the only variables available in a valid budget.

There are many factors which influence the ability of a contractor to control the cost of the work. The methods of acquiring the resources for the project is a significant factor in cost control. On construction projects where most of the work in the field is subcontracted and most of the materials are bought from vendors on a lump sum basis, the ability to positively influence cost, (that is, to reduce cost), once these awards have been made, is small indeed. Weaknesses, however, in the subcontractor and vendor negotiating processes may have a significant negative (that is, higher cost) effect on the cost of the work.

As a general rule, the ability to positively influence the cost of a project decreases with time. For this reason, early identification of adverse cost trends while corrections may be made is one of the most effective aids in cost control. The cost budget is an effective performance model since it provides for early visibility of cost trends. During purchasing of goods and services for a project, the budget provides a clear statement of expectations with respect to cost against which to measure subcontractor and vendor pricing. In the actual performance of the work by the contractor's own resources such as labor and equipment, it provides a basis for measuring performance in progress. It allows for the preparation of contemporary management information in the form of cost performance ratios, calculated by comparing actual cost performance to planned performance.

## DATA SOURCES FOR PREPARATION OF BUDGETS

A variety of data sources is available to assist in the preparation of the cost budget. Many industry references provide unit prices for every kind of work involved in construction and also provides rational methods for developing unit prices for special cases. Information is also available from subcontractor and vendor sources on current market pricing of goods and services.

However, there may be significant differences between market value data from subcontractors and vendors and the fair value of those products. Information and data must be available to show the differences between market value, a reflection of the law of supply and demand; and fair value, which represents the cost of the work plus a reasonable overhead and fee. The issue of fair value vs. market value is always a significant factor in preparing the cost budget; it involves the basic business philosophy of the construction organization and will be discussed further.

The third source of data for budgeting is the organization's own historic performance. This is, perhaps, the only valid source of data for items which will be performed by the organization's own resources. Labor and equipment costs should be available from historic job cost records. This assumes that a job cost data base exists within the organization to routinely and accurately provide such information. Even in the case of items bought from subcontractors and principal vendors, the organization's own historic records may prove to be the best source of data for budgeting purposes.

## JOB COST DATA BASE

Integrated cost and schedule control requires that job costing procedures not only record the cost incurred in producing product but also add to and maintain historic job cost performance data as part of the total management information system. This requirement includes all items which make up a construction project, whether performed by the organization's own resources or purchased or subcontracted from other sources.

Every construction organization has a particular performance profile which emerges as a result of the skill and expertise of the organization. Contractor A may be especially cost effective in performing concrete work. Contractor B may be especially cost effective in performing carpentry work. Accurate job cost records which convert historic performance to usable unit price data will allow each organization to take advantage of its particular expertise in enhancing its competitive position. It would be expected, then, that contractor A would be more competitive on a project with a proportunately large amount of concrete. Contractor B should expect to be more competitive on a project with a large amount of carpentry work. Historic job cost records provide the basis for each organization maximizing its competitive position based on its own special skills and expertise.

It follows that these variances in skill and expertise in different kinds of work have a very significant effect on the competitive bid positions of different organizations addressing the same project.

## MARKET VALUE VERSUS FAIR VALUE

Philosophically, each organization must resolve questions relating to market value and fair value as part of its budgeting processes. Decisions concerning market value and fair value are frequently made without a clear understanding of the processes which are involved. Some definition of fair value and market value are necessary in order to understand these questions.

Market value is established by the straightforward economic law of supply and demand. When there is an increased demand for a limited resource, its price goes up; when demand drops, the competition for available work increases and the price goes down. The nature of the construction industry is such that *market* value may, under a given set of circumstances, bear little or no resemblance to *fair* value. The kinds of product and services are a factor. The staying power of the organization providing those goods and services is an additional factor since pricing below fair value means operating at a reduced profit or even a loss. In construction, market value is best established by taking marketplace bids from subcontractors and vendors.

Some market fluctuations in construction are seasonally based. Exhibit 8-1 shows the seasonal influence on a weather sensitive construction product such as masonry.

**Exhibit 8-1. Seasonal Market Value of Masonry**

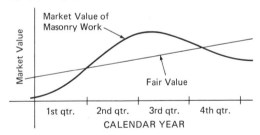

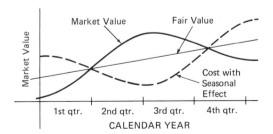

Exhibit 8-1 shows that a given masonry job will be priced lower during the winter months, when there is less demand for masonry work, and higher during the warmer months, when the demand for masonry construction is significantly higher. The construction organizations providing masonry goods and services choose to price those goods and services higher when there is a greater demand and lower when the competition for available work is more intense. This illustration becomes more interesting when the seasonal effect on the *cost* of the work, as a major component of fair value, is also considered.

The cost of the work for a given masonry job is also plotted in Exhibit 8-1 noting the *seasonal* effects (bottom diagram). A comparison between the market value of the work over a period of time and the cost of the work over the same period reveals a peculiarity of market value. The effects of winter weather on the products of masonry and the work force make the identical work package of masonry work more costly to construct in the winter than in the summer. When the cost and the market value of that work are compared over the course of a year, the *market value* of the work will be *lower* at a time the *cost* to produce that work is *higher.* Conversely, the market value will be higher at a time when the cost of the work is lower. Another significant issue to be considered and resolved in the course of preparing a construction project cost budget, then, is the time of the year for construction and its effect on market value.

Fair value on the other hand is more complex than market value but less volatile. First the definition and then some illustration for clarification. *Fair value* is defined as the cost of the work plus the effects of inflation, modified by the effects of technology plus a reasonable overhead and fee (profit). The *cost of the work* represents the cost of the work in labor, material, and indirect cost when performed by reasonably skilled and competent personnel. The cost of inflation represents the incremental increase in cost from the time the cost data was established through the current or desired period. The effect of technology represents a modifier to cost plus inflation; technology may have a positive or negative effect. For example, regulatory requirements such as environmental protection might add to the cost. Improvements in materials or methods of placement for products performing the same function might reduce the cost. An example of improved functional capability is increasing the fiber stress of structural steel from 18,000 pounds per square inch to 36,000 pounds per square inch. While the technological modifier may be either positive or negative, the normal expectation is that technology tends to reduce the cost for similar or equal functions. Exhibit 8-2 shows the effect of fair value over time (top) with the influence of technology separately indicated (bottom).

Further major influences on market value and fair value are overall regional

**Exhibit 8-2. Fair Value as a Function of Time**

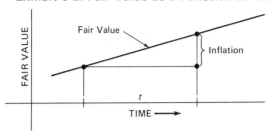

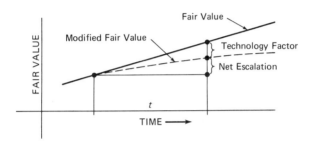

and national economic conditions, which act as modifiers to the normal effects of supply and demand at the local level.* These kinds of influences must be considered when examining market value for the purpose of setting up the project budget.

## PROJECT BUDGETING PHILOSOPHY

Before addressing the mechanics of creating the project budget as a performance model, the basic philosophy to be utilized in budget building needs to be established. There are two distinct budgeting philosophies from which each construction organization must choose. The first of these is market value budgeting; the second is fair value budgeting. This choice may be made whether or not knowingly.

*Market value budgeting* takes pricing from the subcontractor, vendor, and supplier marketplace at face value and uses these as the basis not only for estimating and bidding the project but also for budgeting the project when the bid and negotiations are successful. Since market value is a much more

---

* For example, a general slowdown in the national economy may result in lower marketplace pricing in a high demand local area when the goods and services may be produced and priced outside the local area. Also, demand for construction products may be stifled by excessive interest rates, which could actually reduce construction activity in a high demand situation.

volatile determinant of cost than fair value, superior knowledge of marketplace conditions is essential to successful market value cost budgeting. An additional component also becomes critical: market value cost budgeting must anticipate the *time frame* in which the project purchasing will take place and subcontracts and purchase orders will be let. This time consideration involves both seasonal or time of the year factors and factors relating to overall economic conditions. Exhibit 8-3 shows the effect of market value purchasing when the market value is in decline (top), and when it is on the rise (bottom).

When the market for the goods and services to be purchased is in decline, time is an ally in effective purchasing and in controlling cost as a function of purchasing. In a rising market, however, the reverse is true and time becomes an enemy of purchasing as a tool of cost control. In either case, intense knowledge of the marketplace is required in order to control the cost relating to the purchasing process. Exhibit 8-3 shows the delta, or increment, effect in each case. In a falling market, the delta effect is a cost reduction from delayed purchasing. In a rising market, the delta effect is an increase in cost from delays in purchasing.

The second major budgeting philosophy is embodied in fair value. Fair value as the basis for cost budgeting is much less volatile than market value.

**Exhibit 8-3. Effective Purchasing Related to Market Value over Time**

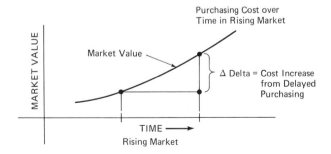

It takes into account the major factors which affect the cost of the work while placing less importance on market fluctuations caused by supply and demand, hence is less demanding of marketplace knowledge. Fair value takes into account the organization's cost experience in producing work of a specific kind at a specific point in time. The inflationary component takes into account changes in the cost of the work resulting from inflationary factors such as changes in wage rates, changes in the cost of materials, and changes in indirect costs such as payroll insurance, taxes, and similar items.

Fair value also takes into account the cost effects of technology, positive or negative. While technological improvements would tend to reduce cost, certain other technological factors, such as environmental protection regulations, may tend to increase costs. In any event, these are generally measurable components which may be applied to the historic cost of the work. A reasonable overhead and fee (profit) is then added to determine the fair value of the work. "Reasonable" overhead and fee is a function of a typical experience in a specific marketplace and is to be distinguished from the effect of the law of supply and demand on market value.

Certain assumptions are made when fair value is to be the basis for budgeting; these need to be considered. The first assumption is that the company, or organization, is well established in its marketplace. The second is that the company has established working relationships with subcontractors and principal vendors within that marketplace. The third, and perhaps the most critical, is that subcontract and purchase negotiations will be structured around fair value rather than market value. This purchasing philosophy means that when the market value is depressed fair value will be the basis for negotiations. It also means that when the market value is inflated, fair value will be the basis for negotiations.

Using fair value as the basis for cost budgeting and for negotiations should significantly modify market value fluctuations caused by the law of supply and demand. Fair value is a viable budgeting concept only to the extent that well established working relationships exist and that subcontractors and principal vendors are neither unduly penalized for depressions in market value nor are unreasonably enriched by rises in market value.

Some comparisons between fair value and market value are of interest. Recognizing that market value is established by a competitive and comparative pricing, a general contractor using a market value budgeting philosophy will simply add an appropriate markup to the marketplace pricing. A general contractor using a fair value budgeting philosophy, however, will independently establish the fair value of the work to determine, first, the differences which may exist between market value and fair value at that point in time and to separately assess the fair value of the work. Market value does not require a careful maintenance of historic costs, but carefully maintained and documented historic costs are essential to the use of fair value as the basis for cost budgeting.

As a practical matter, most well established and experienced construction firms will use a *mixture* of the two philosophies. The primary budgeting processes will be based on fair value concepts. The budget thus established then will be modified to some degree to reflect market conditions over which the company has no control. The result is a cost budget which is truely a reasonable statement of expectations with respect to the cost of a project.

## QUANTIFICATION AND CONSTRUCTION METHODS

The assumption is made that the construction organization has the technical skills, knowledge, and experience to properly determine the quantity of materials, goods, and services required for a construction project. To derive the historic informational base for fair value budgeting, it is important that the quantity surveys be related directly to the roster of activities which will be used for both budgeting and scheduling. Each organization will find that a pattern exists among jobs in the level of detail required for budgeting and scheduling and for the related quantification of job requirements. This pattern, or standardization, will serve as the framework for an effective job costing and historic job cost system which will provide the information necessary for fair-value-based budgets.

Also assumed is that the organization is appropriately familiar with and expert in using the most efficient production methods for the work required. Methods of production are significant factors in integrated cost and schedule control. Both the budget and the schedule must be prepared on the basis of specific production methods if they are to be valid.

## OTHER INFORMATIONAL REQUIREMENTS

The preparation of a job budget requires the accumulation of certain factual data. An analysis of the project to generate a list of activities required to complete the project is the first step in the budgeting process. The preparation of the roster of activities, of course, assumes some adequate technical description of the project, usually in the form of working drawings and specifications. The roster of activities for the job serves as the basis for developing both quantities and unit prices for budgeting. Note this process is the same for both estimates and budgets; the difference is that budgets tend to be more finely detailed and more thoroughly considered than estimates.

The quantities of materials needed ($Q_N$) are determined from the drawings by the standard quantity survey methods used by the organization. The quantities, however, are developed for each activity rather than by trade group or work summaries. Quantification by activities serves as the foundation for all other summary requirements and is an important part of the estimating or budgeting process. The unit prices for the work within each activity must be developed. Unit prices are generally a reflection of the quality ($Q_L$) of the work as defined in the specifications. The best source of unit price information is carefully maintained job cost records for activities of similar scope and quality. Where such unit prices are not available from historic data sources, they may be generated empirically to appropriately reflect the cost of each unit of work in terms of the quantities in each activity. The standard method of developing budgeted cost for each activity involves the formula

$$\text{Quantity } (Q_N) \times \text{Quality Unit Price } (Q_L) = \text{Cost}$$

The unit prices developed must reflect the current value of the work and must be based on the organization's budgeting philosophy. Where the budgeting philosophy is based on fair value, historic job costs are the best sources of information. Where the budgeting philosophy is based primarily on market value, current pricing must be obtained from the marketplace to develop the unit prices for the activity.

## SUMMARIZING ACTIVITIES BY WORK PACKAGES

Most construction organizations have, through experience, developed a reasonably consistent way of grouping the activities making up its typical project into work packages to be assigned to subcontractors, principal vendors, or suppliers. Developing the budget by activity allows summarization by these work packages when the work package may contain activities from a number of different trade groupings. Since the purpose of budgets is to provide cost control for the purchasing process, the grouping of activities by appropriate work packages will prove to be the most useful summary level. The trade summary levels of the standard sixteen divisions of work (similar to the Construction Specification Institute Master Format) are also available through the summarization process.*

---

\* The generally recognized Standard Sixteen Division Trade categories include:

| | | | |
|---|---|---|---|
| 01000 | General Requirements | 09000 | Finishes |
| 02000 | Site Work | 10000 | Specialties |
| 03000 | Concrete | 11000 | Equipment |
| 04000 | Masonry | 12000 | Furnishings |
| 05000 | Metals | 13000 | Special Construction |
| 06000 | Wood, Plastic, Carpentry | 14000 | Conveying Systems |
| 07000 | Thermal and Moisture Protection | 15000 | Mechanical |
| 08000 | Doors and Windows | 16000 | Electrical |

From *Construction Specification Institute Master Format,* 1983 Edition.

## ANTICIPATING ALL NEEDS FOR DATA AND INFORMATION

An important part of the preparatory work to begin the budgeting process is careful consideration of all the other needs for data and information to be generated as a result of budgeting. This consideration as a prerequisite to beginning the budget process will save some effort in recasting information subsequently generated. For example, certain portions of indirect cost are arithmetic functions of direct costs. Payroll insurances and taxes are determined as a percentage of direct labor; therefore, the ability to summarize total labor quickly and conveniently is important. Sales and use tax are determined as a percentage of material costs and equipment rental costs. The ability to summarize these costs is also important. These and certain other summary capabilities are related to the way in which job cost coding and resource identification may be handled.

Summaries required for estimating purposes should not only provide valid information for the bidding processes but also data for historic research and market analysis. There are a number of summary levels which are appropriate to the estimate and which are maintained in the budgeting process. The summary levels shown in Exhibit 8-4 allow summaries to be provided for a variety of purposes. The base cost level, summary 1, is particularly useful in doing comparison market analyses.

### Exhibit 8-4. Estimate and Budget Summary Levels

| SUMMARY LEVEL | DESCRIPTION | CONTENT |
|:---:|---|---|
| 1 | Base cost, | Divisions 2-16 |
|  | (plus) General requirements | Division 1 |
| 2 | Field cost | Total, Divisions 1-16 |
|  | (plus) Insurances, taxes, bonds |  |
|  | and other indirect costs | Indirect cost |
| 3 | Production costs | All direct and indirect costs |
|  | (plus) Contractor's general |  |
|  | overhead |  |
|  |  | Prorata overhead costs |
| 4 | General contractor's total | All direct and indirect costs plus |
|  | cost | overhead |
|  | (plus) General contractor's fee | Anticipated profit |
| 5 | Contract sum | All items included in the contract |
|  |  | sum |

## Exhibit 8-5. A Computer Generated Estimate

```
JUNE 11, 1984                          Frederick Wm. Mueller, CCE                        MONDAY 11:02 PM
                                      BUDGET WITH QUANTITY & UNIT PRICE
JC045J    JOB 0284  GETHSEMANE BAPTIST CHURCH   BUDGET    1 FLRS    5740 SQ FT  WOOD FRAME, BRK VENEER          PAGE 004
===================================================================================================================
```

| —JOB COST ACCOUNT— | | DESCRIPTION | COMMENT | QUANTITY | U/M @ | U/PRICE | U/M | = | BUDGET | DRAW |
|---|---|---|---|---|---|---|---|---|---|---|
| 10000 {9 | 1} | SPECIALTIES | BAPTISTRY | 1.00 | EA | 2075.000 | EA | | $2,075.00 | |
| 10000 {9 | 2} | SPECIALTIES | SPIRE | 1.00 | EA | 2563.000 | EA | | $2,563.00 | |
| 10300 {9 | 1} | MEDICINE CABINETS | TR ACC ALL | 2.00 | LS | 250.000 | LS | | $500.00 | |
| 10300 {9 | 2} | MEDICINE CABINETS | TOIL PARTS | 5.00 | EA | 250.000 | EA | | $1,250.00 | |
| 10300 {9 | 3} | MEDICINE CABINETS | U SCREENS | 1.00 | EA | 175.000 | EA | | $175.00 | |
| 10000 | | SPECIALTIES | | | | | | | $6,563.00 | |
| 11000 {9 | 1} | EQUIPMENT | FIRE EXT C | 3.00 | EA | 125.000 | EA | | $375.00 | |
| 11100 {9 | 1} | APPLIANCES | KIT EQUIP | 1.00 | NIC | | NIC | | $.00 | |
| 11000 | | EQUIPMENT | | | | | | | $375.00 | |
| 15400 {9 | 1} | PLUMBING SYSTEM | INT PLUMBG | 24.00 | FIX | 450.000 | FIX | | $10,800.00 | |
| 15400 {9 | 2} | PLUMBING SYSTEM | WATER PP | 1.00 | ALW | 750.000 | ALW | | $750.00 | |
| 15801 {9 | 1} | H.V.A.C. SYSTEMS | HEAT PUMPS | 17.50 | TNS | 1428.571 | TNS | | $25,000.00 | |
| 15000 | | MECHANICAL | | | | | | | $36,550.00 | |
| 16100 {9 | 1} | ELECTRICAL WIRING | ELECT RGH | 5,740.00 | SF | .750 | SF | | $4,305.00 | |
| 16100 {9 | 2} | ELECTRICAL WIRING | ELECT FIN | 5,740.00 | SF | 1.000 | SF | | $5,740.00 | |
| 16100 {9 | 3} | ELECTRICAL WIRING | TEMP ELECT | 1.00 | LS | 500.000 | LS | | $500.00 | |
| 16100 {9 | 4} | ELECTRICAL WIRING | 500A MDP | 500.00 | AMP | 4.500 | AMP | | $2,250.00 | |
| 16200 {9 | 1} | LIGHT FIXTURES | LIGHT FIXT | 100.00 | ALW | 35.000 | ALW | | $3,500.00 | |
| 16000 | | ELECTRICAL | | | | | | | $16,295.00 | |
| 80000 {8 | } | FEES AND MISC | CONST INT | 2.61 | LS | | LS | | $.00 | |
| 80000 | | FEES AND MISC | | | | | | | $.00 | |
| 82100 {4 | 1} | GENERAL CONTRACTOR'S OVERHEAD | GEN O HD | 291,232.62 | PCT | .045 | PCT | | $13,105.47 | |
| 82500 {4 | 1} | GENERAL CONTRACTOR'S FEE | G C FEE | 291,232.62 | PCT | .035 | PCT | | $10,193.14 | |
| 82600 {4 | } | CONTINGENCIES | ALLOWANCE | | PCT | | PCT | | $.00 | |
| 82000 | | CONTRACTOR'S OVERHEAD AND FEE | | | | | | | $23,298.61 | |
| | | TOTAL BUDGETED | | | | | | | $314,531.23 | |

## MODEL BUILDING METHODS

The data and information making up budgets and estimates are generally presented in tabular form. Older methods use multicolumn summaries with separate columns for labor, materials, subcontracts, and total. Contemporary methods where computer capability is available for summarization use the single column summary which contains the cost of each activity with the appropriate summary levels. The sample of a computer generated estimate confirms that the detail and content of information provided by both estimating and budgeting may be presented showing specific summary levels, all activities, or single line summaries for the entire project.

## PLANNED DATA AND PERFORMANCE EXPECTATIONS

Any performance model as essential as the cost budget or the estimate should be subjected to several levels of checking and validation. Checking consists primarily of arithmetically verifying the estimate for accuracy. Validations must be done by staff technically competent in dealing with such matters and knowledgable concerning quantifications and cost procedures relating to the project. Since several other parties of interest will need the data and information contained in the estimate and budget, an acknowledgement procedure can be an effective tool in administrative control of estimates and budgets. The functional groups which might be included in the acknowledgement (or sign-off) are project management, field supervision, administration, accounting, and management.

When the budget has been acknowledged by all parties, it is the cost performance model for the project. For use in calculating cost performance ratios, the budget becomes the planned data, which will be used for cost performance ratio calculations at any summary level from the individual activity to the entire project. When the performance model has been assembled in the way the project is expected to be constructed, a good match should be provided for comparing actual performance to planned performance.

## PERFORMANCE DATA CAPTURE

In the preparation of estimates and budgets, two major criteria require attention. The first is that the detail within the estimate or budget be assembled in the way the project is expected to be built. The second is that the data capture procedures should carefully match the way the project is expected to be built and the way in which the performance model is constructed. This careful matching of project cost expectations to the level of control to be exerted is the foundation of effective cost control techniques.

## EVALUATING PERFORMANCE

Evaluating actual performance against performance models in a contemporary time frame is an essential part of integrated cost and schedule control. Of particular importance is that measurements of cost performance and evaluation of performance be accomplished expeditiously. Comparing actual performance to planned performance as a ratio is one of the most effective and efficient ways of attending to this job control requirement. The cost performance ratio (CPR) provides immediate feedback to management and administration on job cost performance.

## IDENTIFYING PROBLEMS

The task of identifying problems becomes somewhat simpler with good management by exception procedures. Typically, better than 95% of the activities in a job will perform within planned criteria. Neither management attention nor corrective action is likely to be required on these activities. It is the remaining 5% which require action since these will be the source of cost control difficulties. Comparing actual performance to planned performance will provide the basis for designing corrective action while such is still possible. Cost control implies the ability to influence performance and outcome. Some strategy for identifying problems, for designing corrective action, for implementing corrective action and for monitoring the effects of correction is, therefore, an essential part of building performance models which relate to the budget.

## SUMMARY

The cost control budget is one of the two primary performance models. It deals with setting out in an organized and detailed way the costs which are expected to be incurred in the performance of the project. The cost control budget output is generally displayed in tabular form with appropriate summary levels. It represents the planned cost data for cost performance ratio calculations.

# 9
# THE SCHEDULE AS A PERFORMANCE MODEL

The schedule for a job is one of the most important control tools available. The proper scheduling of a job requires a detailed analysis of the project, establishing a plan for its construction and, based upon available resources, a determination of a realistic and valid schedule for the completion of the project.

The project schedule is one of the two primary performance models for a project upon which all other performance models are based. The validity of the schedule is directly related to the effort expended in planning how the job will be constructed. Historically, no clear distinction was made between job planning and job scheduling. In recent years, however, better understandings of the planning and scheduling processes have recognized planning and scheduling as two distinct and separate analytical processes. Certain planning and scheduling processes such as CPM are especially suited to their independent planning of a job prior to the creation of the schedule.

## THE CRITICAL PATH METHOD FOR PROJECT PLANNING AND SCHEDULING

The critical path method of project planning and scheduling is well suited for construction planning and scheduling. Jobs are built and resources are assigned by activities. The critical path method is activity oriented. The logic network, or arrow diagram, allows the complete planning of a project independently of and with little concern about scheduling.

In the critical path method, scheduling involves several basic arithmetic computations once the project plan is complete. The first of these computations addresses the resources available to perform each activity and the resulting duration of that activity. The second group of computations establishes the time boundaries for all events and activities throughout the project plan and allows the identification of the critical path. The scheduling computations provide the time boundaries for each activity which, when calendar dated, provide a calendar schedule for the job in its entirety and for each activity within the project. This kind of intense job analysis provides field construction and project management with a plethora of information concerning expected job performance.          *excess*

## CRITICAL PATH METHODOLOGY

There are a number of good references concerning the critical path method of planning and scheduling available to the interested reader. It will be useful, however, to review the basic techniques which are involved in CPM. Exhibit 9-1 shows the logic network diagram for a small project consisting of seven activities. It contains all of the basic elements in CPM diagram and logic. Within the context of the critical path method, a project is defined as any undertaking involving a group of related activities having a specific origin and a unique objective.

The logic network diagram for the project shown in Exhibit 9-1 contains all of the diagramming elements and relationships which are part of the critical path methodology. There are six regular activities: A, B, D, E, F, and G; an activity is a time and/or resource consuming task. Activities are shown in the logic network diagram as arrows. The flow of the work within an activity is from its tail to its head. For convenience, descriptions of the work within each activity are not provided. Each activity, however, is identified by a letter reference.

Note that activity C is shown as a dotted arrow and is referred to as a "dummy." The dummy activity is a logic transfer device within the arrow diagram. It transfers logic of limits from one event to another when no time or resource is involved.

### The Event

Each activity begins with an event and ends in an event. An *event* is a point in time representing the beginning or end of an activity or group of

**Exhibit 9-1. Logic Network Diagram**

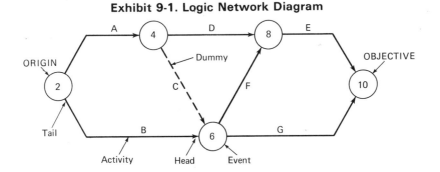

activities. Note that the origin of a project shown in Exhibit 9-1 is event number 2, and that the objective of the project is event number 10. Other events in the project network are numbered 4, 6, and 8, and along with 2 and 10 are shown as circles within the diagram.

## Relationships Between Activities

All relationships within a logic network diagram are described in terms of the relationships between activities. Events themselves are not considered within these definitions of relationships but serve only to mark the commencement of an activity and the end of an activity. There are four specific relationships essential to the logic network diagram. These relationships can be expressed in a number of different ways, but the following formal definitions are the most appropriate.

The first of these formal relationships is *precedence.* In Figure 9-1, A is precedent to D, D is precedent to E, B is precedent to F. B is also precedent to G, but C is precedent to G and F. Precedence simply means that one activity must be completed before subsequent activities may commence.

The second formal definition concerns *subsequence.* If A is precedent to D, then D is subsequent to A. E is subsequent to D, F and G are both subsequent to C, and so forth. The less formal way of expressing these same relationships might be: A must be completed before D can start, E cannot start before D is completed, F and G depend on B and C; all express the same form of precedence and subsequence.

The third formal relationship within the logic network diagram is *concurrence.* Activities are said to be concurrent when they share a common event at *either* the tail or head. For example, in Exhibit 9-1, A and B are concurrent, E and G are concurrent, D and F are concurrent, and so on. A further limitation on the definition of concurrence is that concurrent activities need not be performed in the same time frame.

A fourth formal relationships must be recognized in drawing the logic network diagram. If activities share a common event, they are *related.* If activities do not share a common event, they are unrelated. The point is that no relationship can be inferred between activities not explicit in the diagram. In philosophical logic, it may be appropriate to say: A is precedent to D, D is precedent to E; therefore A is precedent to E. But in logic network diagramming, if A is indeed precedent to E as well as to D, this must be clearly shown. We can infer no relationship between A and E since it is not explicit in the diagram. If some project modification would result in D being removed as a project activity, the inferred dependences on D to show a relationship between A and E also would be lost. If there is a necessary relationship to be shown between A and E, it must be explicit in the diagram. That is, they must share a common event.

## Absolute Logic of the Logic Network Diagram

The logic network diagram clearly depicts all relationships. As a consequence, there is a minimal chance of misinterpretation of what the logic network diagram in Exhibit 9-1 shows. A must be completed absolutely before D can start. There is no way of telling, however, from the information given whether the relationship between A and D is mandated by some physical characteristics of A and D or whether the planner has simply chosen to do A before D. In the one case, there is a physical constraint and, in the second case, there is a planner's choice which, within the context of the logic network diagram, operates the same as a physical constraint. This, perhaps subtle difference is essential to understanding both the power and limitations of the critical path method.

To illustrate this point, if activity A represents excavation for the footings and D represents the pouring of the footings, clearly the physical conditions involved require that the trenches be excavated before the footings are poured in those trenches. If, however, A represents the footings along the north wall and D represents the footings along the east wall, the planner has the option to pour *either* wall first. No physical constraint applies here.

In the planning process, where there is no physical constraint, many alternates are available. In fact, one of the utilities of CPM is in the examination of alternates where planner's choices are involved. However, within the context of the logic network diagram, physical constraints and planner's choices appear exactly the same and operate in exactly the same manner.

## Project Planning

Two important features of the logic network diagram for the project are shown in Figure 9-1. First, it represents the complete plan for the project; it clearly shows the relationships between all activities making up the project, and flows from the origin to the objective. The second feature is that it has been possible to complete the entire project plan or strategy with no reference of any kind to activity durations, scheduling computations, or a calendarized schedule. The complete separation of the project plan from the scheduling processes is a most useful characteristic of the critical path method.

## Scheduling Computations

Scheduling computations involve several arithmetic procedures and require certain data. Before scheduling computations can commence, some determination of the resource to be used in performing each activity must be made since this will have some effect on the duration of the activity.

The first determination which must be made is the duration of each activity within the logic network diagram. Of course, all dummy or dotted arrows will have a duration of zero. All others will need a duration assigned. While there are many exotic computational procedures, as a practical matter, activity duration is usually assigned by someone expert in the work and/or processes within each activity. Each planner must determine the ways in which activity duration will be established. Duration units have arbitrarily been assigned to the network shown in Exhibit 9-2. These are characteristically shown in the middle of the activity arrow immediately below the designation.

The next set of scheduling computations is referred to as the "forward pass." The forward pass consists of summing the durations of activities to each event and determining which activity controls the earliest time $(T_E)$ that event can occur. Any event cannot occur until all activities which end in that event are completed. The number shown in Exhibit 9-2, at the arrow-head, represents the early completion $(E_F)$ of each activity. Where several activities end in a common event, the *last* activity finished controls when the event can occur. For example, event 8 in Exhibit 9-2 cannot occur until both activities D and F are completed. The last of these two events to be completed will be F at duration unit 16. The earliest event 8 can occur, therefore, is 16. Sixteen is noted at the event as the time earliest $(T_E)$ for that event. Forward pass calculations continue from the origin to the objective. In most cases, calculations will begin with zero at the origin and end with the calculated longest duration.

## Exhibit 9-2. Scheduling Computations

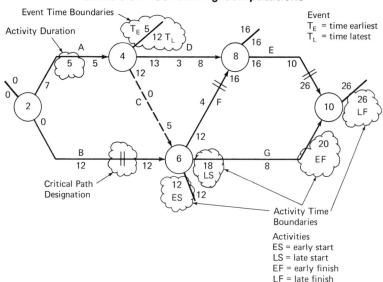

The next set of scheduling computations to be made is the "backward pass," which determines the latest each activity can start without affecting the schedule. It is similar to the forward pass except that the durations are subtracted from the event time to determine the late start ($L_S$) of each activity. For example, the late start for activity G in Exhibit 9-2 is determined by subtracting the activity duration of 8 from the time latest, 26, for event 10, resulting in a late start for activity G of 18. The latest event 6 can occur, however, is controlled by when each activity beginning in event 6 must start so that the schedule will not be delayed. The backward pass computations taking into account the duration of activity F indicates that F must start no later than duration unit 12. Therefore the latest duration unit at which event 6 must occur ($T_L$) is 12 rather than 18.

A partial confirmation that the backward pass and forward pass calculations had been made correctly is the confirmation that the time latest for event 2 at the origin is also zero (reached via activity B, the longer of A and B).

## Identifying the Critical Path

The scheduling computations are included in Exhibit 9-2, with the appropriate notational system for each. Those readers interested in a more detailed development of the critical path methodology are referred to other more complete treatments of the method. The purpose here is to show the simplicity of the method and the computational procedures.

The last of the scheduling computational procedures is to identify the *critical path*. The critical path is defined as the chain of activities which control the overall duration of the job. In nontechnical terms, it is the sequence of activities in which there is no room to move (float). This means that the events associated with the critical path have the same time earliest and time latest and that the activities between those events consume all of the time available between the events. In Exhibit 9-2, those events having the same time earliest and time latest are 2, 6, 8, and 10. It would be normal to assume that all activities between those events would be on the critical path. An examination of activity G, however, reveals this not to be the case. There is available between the time earliest for event 6 and the time latest for event 10 a total of 14 days ($26 - 12 = 14$). Since only 8 days, or duration units, are required for the performance of G, 6 days of *float* are available. It is clear, therefore, that G does not control the overall schedule. In the case of activity B, F, and E, however, each fully occupies the time available between events and are, therefore, critical activities. The sequence of activities, B, F, and E, are, therefore, identified as the critical path and are so noted by the double hash marks on the diagram.

These basic routines complete the critical path scheduling calculations and the identification of the critical path.

## THE TIME SCALE NETWORK

There are a number of advantages of using the logic network diagram for the basic planning and scheduling computational routines associated with the critical path method. As a production control tool, however, logic network diagrams for projects involving hundreds of activities tend to be confusing and complicated to the point of uselessness in the field production environment. Most field supervision is familiar with bar charts, Gantt charts, or other time scale presentations for project scheduling for production control. Fortunately, the conversion of the logic network diagram, once the scheduling computations have been completed, is a simple, straightforward task. Exhibit 9-3 shows the logic network diagram presented in Exhibits 9-1 and 9-2 converted to a time scale network.

Note that in the logic network diagrams in Exhibits 9-1 and 9-2 the direction of the arrows, and their lengths are unrelated to their durations. In Exhibit 9-3, however, each activity is drawn to a linear scale so that the length of each activity clearly represents its duration. For example, B is shown 12 units in length, F is shown 4 units in length, and E is shown 10 units in length. B, F, and E, represent the critical path and, combined, are 26 units in length—the duration of the project. Note there are no gaps, spaces or dotted lines in the critical path. A and D are similarly shown to scale but together do not require completion before the end of duration unit 16. The dotted line to duration unit 16 indicates the float available to the A and D chain of 8 days. Activity G may start on day 12 but does not require completion before day 26. The earliest completion is day 20 with 6 days of float. Activity C is a dummy. It has a duration of zero as indicated by the line showing 7 days of float.

As a practical matter, the conversion of a logic network diagram to a time scale network normally ignores the logic relationships and deals only with the durations of activity and the float of the activities. The typical method of plotting each activity is to begin with the early start ($E_S$) plus

### Exhibit 9-3. Time Scale Network

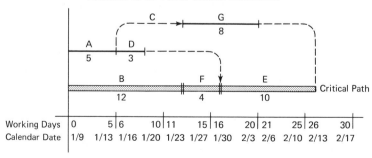

the duration. In this way, the float indicated in the time scale network is also the independent float available to that specific activity.

## CALENDAR DATING THE TIME SCALE NETWORK

The processes described thus far in this chapter have dealt with several specific details of creating a schedule for a job. The first of these is the independent planning of the project, consisting of structuring the relationships between the activities which make up a project. This is best accomplished through a logic network diagram which explicitly shows the planned relationship between activities which make up the project. The absolute nature of the logic of the logic network diagram makes this kind of diagram an especially powerful communications tool. The logic network diagram is, in fact, so explicit that it is difficult to misinterpret. The nature of expressing the logic, however, does not mean that the underlying planning processes are faultless.

Planning may be carried on without reference to resource requirements, duration, or time boundaries. A separate determination of duration allows the scheduling computations, including the forward pass, the backward pass, and the identification of the critical path, to be accomplished. These are all done in the framework of the logic network diagram. But logic network diagrams for projects of any complexity tend to become too cumbersome to be used directly as a production control tool. As a partial remedy, the conversion of the logic network diagram with scheduling computations to a time scale network places the expected performance information in a format where activity is time scaled, but the logic relationship among activities is less clearly evident.

Another major advantage of the critical path method of planning and scheduling now becomes apparent. All of the tasks thus far reviewed (including the scheduling computations which establish time boundaries for activities and events) may be carried out without knowing specifically when the project will be started or when it may be expected to be completed. Calendar dating of the project plan may be deferred until an actual start date has been established.

In Exhibit 9-3, assuming the duration unit to be one working day, the horizontal time scale shows the working days. Immediately below the working days, the calendar dates have been inserted. The insertion of any firm calendar date permits the calendarization of the entire project. In fact, in a purely technical sense, the project schedule does not exist until a calendar dated schedule has been established. However, once a time scale network is completed, the second performance model then exists; calendar dating may conveniently be added to the time scale network as soon as a fixed start date is established.

All other derived performance models are dependent on the time scale network and the budget. The many uses of the time scale network in the derivation of other performance models will be discussed in detail. The time scale network itself allows distribution of activities along a time scale. When each activity has been budgeted separately, all cost related job performance parameters such as budgeted costs, production values, schedule of values, cash income, and cash requirements may be correlated with the time scale network.

The time scale network serves as the foundation for integrated cost and schedule control; it becomes the project schedule when calendar dated. While the critical path methodology and logic network diagrams have been discussed to establish an adequate background for developing the time scale network, all further references to the project schedule as a performance model will refer specifically to the time scale network with or without calendar dating.

## THE ROSTER OF ACTIVITIES

As we have seen, the project schedule consisting of a time scale network may be logically derived through critical path methodology. This is true whether the time scale network attempts to maintain or show the logic or not. When project planning and scheduling matures to the point the project is ready for production, the time scale network becomes a much more useful production management tool. Planning and scheduling processes which have been discussed thus far are heavily dependent on the work, previously discussed, concerning the budgeting processes.

The basic unit of performance for measuring both cost performance and time performance is the activity. Proper budgeting procedures result in the creation of a roster of activities which reflect both the way in which the job will be built and the level of control to be exerted. This roster of activities, when properly set up for budgeting purposes, presents itself to the planning and scheduling environment for use in the critical path anaylsis of the project.

Historically, the most difficult work in the planning and scheduling processes using the critical path method has been the generation of the roster of activities. This task at this stage is almost eliminated when the roster of activities has already been generated for budgeting purposes. The planning and scheduling processes may result in some additions to the roster of activities for planning and scheduling purposes, but the bulk of the work on the roster has already been accomplished. When, to accommodate scheduling requirements, activities are added to the roster of activities already prepared for budgeting these added activities will also require budgeting. This budgeting process for new activity requires attention to the same quantity/unit price algorithms previously discussed (p. 105).

The importance of using a common roster of activities for both cost budgeting and for planning and scheduling must be emphasized continuously. Where the integrity of the job planning process is maintained, the process of integrated cost and schedule control is measurably simpler and easier to accomplish.

The criteria for detail in the roster of activities is also simple. The roster of activities should include the maximum detail required for planning, scheduling, or budgeting as a function of the complexity of the job. But, again, a point of caution is to avoid unnecessary detail in both the budget and the schedule, which adds complexity but not usefulness to project control.

## DATA SOURCES FOR PLANNING AND SCHEDULING

Several data sources are important in supplying information for the planning and scheduling processes. First, as we have seen, the complete roster of activities is essential to beginning the planning and scheduling processes. The roster of activities is initiated through the budgeting processes subject to modifications which may be required for planning and scheduling purposes. The roster of activities permits the commencement of the planning processes and allows the development of a logic network diagram appropriate to the project.

The second data source ties in with selecting the resources which will be utilized in performing each of the activities. There are four principal resources used in job construction. The first of these is the company's own resource consisting of its own labor, materials, and equipment. The second resource are the many suppliers who provide miscellaneous materials required for the completion of the project. The third resource is the subcontractors who perform significant portions of the work on site. The fourth resource is that of principal vendors who, while not installing work at the job site, provide significant portions of the work fabricated to job requirements and delivered to the job site.

Some preliminary determination of the resource to be used for each activity is essential to the planning and scheduling process. The principal concern from a planning and scheduling point of view is the duration of each activity.*

For field production activities, however, duration is a function of the resource available and the rate of application. The usual practical resource is the standard critical path test of resources to be applied to any particular

---

* For resource analysis leading to establishing duration, all activities are divided into two categories: those involving job site work (field production activities usually own or sub labor or equipment), and all others. The duration of all but field production activities usually involves setting delivery times for materials, supplies, or in some cases, data such as shop drawings; these serve as constraints or limitations on when field production may start.

activity. *Usual* refers to the workforce size or kind and capacity of equipment which would normally be used for the work being performed. *Practical* is the modification to that resource to meet the constrictions of time and space, and sometimes the need for a rate of placement beyond what might normally be considered appropriate.

While the traditional method of using knowledge of resource to determine duration of activity is empirical and is provided by a person knowledgeable in the work, there are more definitive ways of addressing or determining the duration of each activity. Duration is a function of *rate of placement* in the field environment. Exhibit 9-4, Duration Data Requirements and Calculations, shows the raw data required and methods of calculating duration based on specific budget data for the activity to be performed.

The budget for the rough-in includes 19 fixtures at $200 each or a total of $3,800, a workday of 8 hours, and a typical workforce of two plumbers and one helper which cost $38 per hour total, and 100 workforce hours which, when divided by 8 hours per day, gives 12.5 days required for the activity. Since duration is typically rounded to the next higher whole duration unit, the duration unit used in this illustration is 13 days. If, after inserting

## Exhibit 9-4. Duration Data Requirements and Calculations

| NO. | ITEM | | | DESCRIPTION |
|-----|------|------|------|-------------|
| 1. | Resource | | | A.B.C. Plumbing |
| 2. | Activity | | | Rough-in toilet rooms, 1st floor |
| 3. | Budget | | | |
| | COST TYPE | QUAN. | UNIT PRICE | COST |
| | Sub labor | 19 fixtures | $200.00 | $3,800.00 |
| 4. | Workday | 8 hours | | |
| 5. | Workforce: | | | |
| | CLASSIFICATION | NUMBER | RATE | COST/HOUR |
| | Plumbers | 2 | $15.00 | $30.00 |
| | Helpers | 1 | 8.00 | 8.00 |
| | | Total cost per workforce hour = | | $38.00 |

6.  Duration calculation:
    a. Duration = labor cost ÷ labor cost/day
    b.           = $3,800.00 ÷ $38.00/hr × $8.00 hrs/day
                 = 12.5 days
    c. Use duration = 13 days

this duration into the schedule and performing scheduling computations, it is determined that the duration of 13 days is longer than appropriate to the overall schedule, several options could be used. The workforce size could be increased, or the hours per day could be increased. Each of these would require separate examination to determine the best course of action.

Not all construction organizations have developed this expertise or would take the time for these rational calculations of activity duration. More typically, the direct supervision of the workforce is called upon to estimate (or approximate) the time required for each activity. While this empirical method is, perhaps, the most expeditious way to arrive at durations for each activity, each should be tested against budget constraints to confirm that the unit cost of the resource times the duration is at or less than the budget and that the estimated duration makes sense.

With the logic network completed and durations assigned to each activity, it is then possible to perform the scheduling computations. The scheduling computations establish a series of time boundaries for each activity and for each event throughout the logic network diagram. As we have seen, the event as a point in time has two boundaries. The first is time earliest ($T_E$), or the earliest the event *can* occur, and time latest ($T_L$), or the time the event *must* occur. While each activity has eight time boundaries determined as a result of scheduling computations, four are of particular interest. These are the earliest an activity can start ($E_S$), the latest ($L_S$) it must start, the earliest an activity can finish ($E_F$) and the latest it must finish ($L_F$). (Also see Exhibit 9-2).

The last piece of data required to complete the schedule as a performance model is the calendar dating of any significant event within the project plan. Typically, the calendar date of the origin of the project is expected to be set first though other dates may be used. It should be obvious, however, that the calendar dating of any event anywhere throughout the network determines the calendar date of all the remaining events and of all activities. Typically, the conversion of the logic network diagram to time scale occurs after the scheduling computations have been completed but before the calendar dates are inserted.

## OTHER INFORMATIONAL REQUIREMENTS

Preparing a job schedule as a primary performance model is a significant and noteworthy task. The data required for preparing the schedule is usually readily at hand. However, other informational requirements will impact on the planning process and the resulting schedule.

The terms and conditions of the contract concerning *contract time perfor-mance* are especially important. Specific start dates required by the contract must be recognized. The prerequisites to the start of construction must be carefully monitored from an administrative point of view. When the planning and scheduling processes clearly indicate that these may not be met for one reason or another, administrative action is required to document the reasons why the project cannot be started within contract requirements so that necessary schedule and legal adjustments may be made. The contract completion date is also of significance. Where penalties and liquidated damages pertain to the contract, planning decisions will need to be made to improve production or productivity to assure the completion of the contract within the contract schedule time.

Other major factors will have an impact on job scheduling. Seasonal considerations are significant in most parts of the country; the schedule may be more or less adversely affected depending on the kinds of construction and the materials to be used. Market conditions also have a significant effect on time schedules. In a rising market, limited resources within the subcontractor and vendor base significantly impact on what might otherwise be considered normal or reasonable performance of the work. Conversely, a falling market may present opportunities for accelerated construction with associated economies in both construction cost and the cost of money, enhancing profitability.

The schedule is an essential communication tool which presents to all parties of interest the time schedule for the expected performance of the project. In establishing terms and conditions of subcontracts, specific scheduling requirements should be a part of subcontract and vendor negotiations and certainly a primary consideration in the award of a subcontract. Properly prepared and presented schedules perform an important communications function and are indispensable to job control.

## ANTICIPATING ALL NEEDS FOR DATA AND INFORMATION

The continuing criteria for integrated cost and schedule control in building performance models is anticipating what raw data will be required to build a performance model and what raw data or information will be produced by the performance model. The schedule provides information to the owner on expected performance. It also provides information to both project management and field construction concerning the percentage of production capacity which will be consumed by the project. It allows management to examine the impact of the project on total corporate resources. It further allows the examination of cash flow, cash requirements, and working capital requirements. All data requirements both in and out of the scheduling processes and all information output from the scheduling processes should be anticipated in building the schedule as a primary performance model.

## MODEL BUILDING METHODS AND MODEL DISPLAY

Through the years, many different methods have been used to create and present the project construction schedule. Recently, a whole family of new methods referred to as logic network techniques has been developed to facilitate the planning and scheduling processes. This reference prefers the critical path method for two reasons. First, the critical path method is activity based. Since projects are built activity by activity, there is a convenient match between the critical path methodology and the way projects are built. Second, the critical path method is preferred for its analytical power. Properly used, the logic network diagram states unequivocally the plan or strategy for the construction of the project. Once the logic network has been developed and durations assigned, the scheduling computations are purely objective. Any person competent in the method will produce precisely the same result from the scheduling computations as all other competent persons.

There are, however, certain weaknesses in the critical path method. One is that the logic network diagram on a project of several hundred activities tends to be confusing. For this reason, and others as well, the logic network diagram is converted to a time scale network for production control purposes. The time scale network eliminates the confusing appearance of a logic network diagram and displays activities on a familiar time scale similar to a Gantt chart. At the same time, the time scale network opens the horizon for constructing a number of additional performance models which distribute cost and production value over time. The preferred display of the project schedule as a performance model is a calendar dated time scale network.

## PERFORMANCE EXPECTATIONS

The schedule is a project time statement of expectations. It presents specific information on the expected start and finish of every activity which makes up the project. It presents to all parties of interest the performance which is expected in their specific areas of responsibility. When activities are grouped into work packages and assigned to specific subcontractors and vendors, the information is available concerning detailed expectations of scheduled performance by subcontractor or vendor.

## PLANNED DATA

*Performance ratios* are the key to developing required management information quickly and conveniently. The schedule presents the planned data which will be used in calculating the time performance ratios (TPR). Each time performance ratio will provide management with ongoing information comparing planned performance to actual performance. Appropriate data capture procedures are required in order to convert comparisons between actual performance and planned performance into meaningful management information. When this information becomes available, however, the process of measuring and evaluating performance is a simple and direct ratio computation.

Integrated cost and schedule control, however, works on the assumption that performance measurements and the evaluation of performance will result in three specific actions:

1. Problems adversely affecting performance will be clearly identified in a time frame which will allow for adjustments.
2. Corrective actions will be designed based on knowledge of such problem.
3. Each corrective action will be implemented and monitored for satisfactory performance.

## SUMMARY

The schedule for a construction project is one of two primary performance models, the other being the budget. It presents the project statement of expectations with respect to time. The activity is the key to developing valid construction schedules which match the way in which the project is to be constructed and which are fully integrated with the budget for cost and schedule control. The critical path method is preferred for the planning and scheduling processes for its logic, analytical power, and for its precise capability in communicating the project plan. The critical path logic network diagram is less suited for production control purposes; the derivative time scale network prepared from the critical path analysis is much more effective as a production control tool, being more understandable in the field construction environment. The time scale network also opens the possibility for the development of a number of other performance models when it is combined with the project budget.

Neither the critical path logic network diagram nor the time scale network completely represents the schedule until each is calendar dated. The most effective production control tool is the calendar dated time scale network, which is the formal project production schedule. When the schedule is completed in this format, it also becomes the project statement of expectations with regard to time.

# 10

# THE COST CURVE AND THE PRODUCTION CURVE AS PERFORMANCE MODELS

The cost curve and the production curve are the first two derived performance models. They are considered within the context of a single chapter because of their similarities. Both require that (1) a cost budget for the project by activities has been prepared, (2) that a schedule for the project by activities has been prepared, and (3) that the roster of activities used for budgeting purposes and for scheduling purposes are the same.

The *cost curve* consists of the cost of the work distributed throughout all activities and along the time scale where those activities are placed by the scheduling processes. While the summary level used for the cost curve may vary from organization to organization, the recommended procedure is that the cost curve include all work through the field cost summary level. This summary level includes all work defined within the first sixteen standard divisions or trade groupings of work. This means that the cost would include all general requirements items but exclude such indirect costs as payroll insurances and taxes, other insurances, as well as performance and payment bonds. Table 10-1 presents the tabular budget data for the time scale network shown in Figure 9-3.

The *production curve* graphically represents the total of the contract sum. More specifically, it includes the cost of the work through the sixteen standard divisions *plus* the indirect cost such as insurances and taxes, *plus* the overhead and fee, the total of which equals the contract sum. The production curve is, then, nothing more than the cost curve plus the indirect cost plus the overhead and fee. In fact, the simplest way to prepare the production curve is to prepare the cost curve and to add to the cost curve at each plotting point the percentage increment which represents the indirect cost plus the overhead and fee (see Figure 10-3 for an example).

## DISTINCTIONS BETWEEN THE COST CURVE AND THE PRODUCTION CURVE

The cost curve shows the expected budgeted cost to be incurred continuously over the life of the project. It is common, however, to plot the cost curve

**Exhibit 10-1. Tabular Budget Data**
**(for Time Scale Network shown in Exhibit 9-3)**

| JOB COST CODE | RESOURCE | IDENTI- FIER | D | $Q_N$ | UP | COST | COST/D UNIT |
|---|---|---|---|---|---|---|---|
| 1000 | Labor | A | 5 | 10 | 100 | $ 1,000 | $ 200 |
| 2000 | Material | B* | 12 | 50 | 120 | 6,000 | 500 |
| 3000 | Dummy | C | 0 | 0 | 0 | 0 | — |
| 4000 | Labor | D | 3 | 300 | 5 | 1,500 | 500 |
| 5000 | Supplier | E* | 10 | 1000 | 5 | 5,000 | 500 |
| 6000 | Subcon- tract | F* | 4 | 40 | 100 | 4,000 | 1,000 |
| 7000 | Labor | G | 8 | 40 | 200 | 8,000 | 1,000 |

| | | |
|---|---|---|
| 100.0% Cost budget (production cost) | | $25,500 |
| 110.0% Overhead 10.0% | | 2,550 |
| 110.0% Total budget (contractor's cost) | | 28,050 |
| 5.5% Fee 5.0% | | 1,403 |
| Total budget (contract amount) | | $29,453 |

Percentage distribution based on production cost = 100.0%

* Critical activities.

at predetermined intervals such as weekly or monthly. It indicates the *direct cost of work expected to be placed* throughout the course of the job. From a management point of view, the cost curve finds its principal value in comparing planned cost with actual cost and serves as the basis for the cost performance ratios which may be calculated for any given time interval.

The production curve, on the other hand, depicts the *value of work* in contract dollars. It is the best measure available for rate of placement of product. Other measures of product placement have sensitivities built into them which do not affect production value. For example, cost does not take into account earned overhead and fee and reimbursement for indirect costs. Accounts receivable are modified by the effects of contract retainages and holdbacks. Cash requirements are similarly modified by retainages and holdbacks. Only production value provides a fair measure of the value of work placed in a specific time period.

Planned and actual production value comparisons or interpretations are also insensitive to factors of exact composition, unlike the ways that comparisons of planned and actual costs are affected. For example, a cost performance ratio calculated by comparing actual cost to planned cost is valid only to the extent that the composition of activities included in the actual cost are precisely matched to those included in the planned cost. The production performance ratio, however, is valid in the context of summing actual product in place in terms of its dollar value to planned product in place in terms of its dollar value.

Because of the similarities between the cost curve and the production curve, that is, the production curve being simply the cost curve plus the indirect cost and overhead and fee, the focus of attention will be on the processes and procedures relating to the cost curve as a performance model, with comparisons drawn to the production curve along the way when appropriate.

## DATA SOURCES

The cost curve and the production curve are the two easiest derived performance models to prepare. There are only two data sources required in order to create both curves: (1) the budget prepared by activities with a separate budgeted value assigned for each activity (see Table 10-1), and (2) the schedule prepared utilizing the same activities as included in the budget—the common roster of activities ever essential to integrated cost and schedule control.

It is also important that the schedule be displayed in a time scale format. The preferred method of plotting the time scale network from the logic network diagram is on the basis of early start of each activity plus duration. The time scale network prepared from the logic network diagram places each activity along the time scale based on its early start plus duration. When a separate budgeted cost is assigned to each of these activities, the basic data sources are available for preparing both the cost curve and the production curve.

While it is possible to plot the cost curve on a monthly basis to determine how level the cost is between time plotting intervals, it is generally less useful than the *cumulative* cost curve. The cumulative cost curve adds the current period cost to the cumulative cost through the previous period so that the total tabular data equals the total expected cost through the plotting period.

### Resource Notations and Cost Distribution

Exhibit 10-2 shows a notational system to be used to facilitate tabulating cost data and plotting of the cost curve.

Exhibit 10-2 takes the budget data for activity F and, using the notational system shown, provides both budget and resource data directly on the time scale network.

Exhibit 10-3 shows the same kind of data for the entire time scale network. These preparatory steps will greatly facilitate the tabulation of cost data and plotting the cost curve.

**Exhibit 10-2. Activity F from Time Scale Network (Exhibit 9-3)**

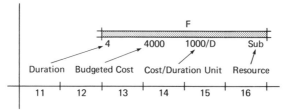

It is also helpful to anticipate the future need for separate information on different resource types. The subtotals by resource type will be carried separately for each plotting period for convenience in the plotting of the cost curve. It is obvious, however, that the cumulative tabular data could also be maintained by resource type. Recording and summarization of the tabular data and the actual plotting of the cost curve may be done directly on the time scale network by an overlay over the time scale network or as a completely separate plotting. Since the data tends to accumulate and proliferate, it is recommended that an overlay be used where manual plotting methods are chosen.

## Methods of Tabulating Data

The methods of tabulating the data required for plotting both the cost curve and the production curve are simple and direct when the time scale is used with the activities placed in the proper position along it. The last step of preparatory work is to select the plotting intervals to be used. Exhibit 10-3 assumes plotting intervals of five working days or one working week. The plotting intervals should be selected to provide the appropriate level of information required without the encumbrance of more detail than required to produce the cost and production curves.

Typically, one week plotting intervals work quite nicely. They are also appropriate to the plotting periods required for the cash requirements curve. Vertical lines representing the plotting period should be drawn through the time scale network extending from the top to the bottom of the sheet.

Provisions should be made for the separate period summaries of labor, materials, subcontractors and suppliers as these relate to the illustration in Exhibit 10-3. However, other resource designations may be more appropriate for a particular organization. Space should be provided in the tabular data for the subtotals and for the cumulative totals. It is also beneficial to anticipate the need for the production value by adding a separate line for cumulative production value figures.

These are all shown in Exhibit 10-4.

# Exhibit 10-3. Time Scale Network with Budget, Resource, and Cost Scales Added

| Value | Percent | | | | | | | | |
|---|---|---|---|---|---|---|---|---|---|
| $30,000 | 125% — (31,875) |
| 25,000 | 100% — (25,500) |
| 20,000 | 75% — (19,125) |
| 15,000 | 50% — (12,750) |
| 10,000 | 25% — (6,375) |
| 5,000 | |

COST

Critical Path

A  5 1000 200 Lab 3 1,500 (Lab)
Lab  Lab

C

D

G  8 8000 1000 (Lab)

B — 6000 — 500

F (Mat) — 4 · 4,000 — (Sub) · 10 — 5000 — 500 — (Sup)
1,000

E

| | TIME SCALES | | | | | | | | | | | |
|---|---|---|---|---|---|---|---|---|---|---|---|---|
| 1 | Duration units | 0 | 5 | 6 | 10 | 11 | 15 | 16 | 20 | 21 | 25 | 26 | 30 | 31 | 35 |
| | Calendar dates D unit = 1 day | 1/9 | 1/13 | 1/16 | 1/20 | 1/23 | 1/27 | 1/30 | 2/3 | 2/6 | 2/10 | 2/13 | 2/17 | |
| 2 | Production periods if D unit = 1 week | | 1 | 2 | | 3 | | 4 | | 5 | | 6 | | |
| | Calendar dates if D unit = 1 week | 1/1 | 1/29 | | 3/4 | | 4/8 | | 5/13 | | 6/17 | | 7/22 | |

Notes: 1. Time scale 1 is for duration unit equal to one day.
2. Time scale 2 is for duration unit equal to one 5 day week and is used to develop later performance models.

129

**Exhibit 10-4. Time Scale Network with Tabular Cost Data, Cost Curve, and Production Curve**

Note: Production curve values equal cost curve values multiplied by 1.155.

| Value | Percent | |
|---|---|---|
| $30,000 | 125% | (31,875) |
| 25,000 | 100% | (25,500) |
| 20,000 | 75% | (19,125) |
| 15,000 | 50% | (12,750) |
| 10,000 | 25% | (6,375) |
| 5,000 | | |

COST / TIME SCALES

Network labels: A — 5 1000 200 (Lab); B — 12 — 6000; D — 3 1500 (Lab) 500; G — 8 8000 1000 1000 (Lab); F — 500 (Mat)—4—4000—(Sub)—10—5000—500 (Sup); E — 1000

Planned End of Construction (Substantial Completion). Critical Path.

| | Duration units | 0 | 5 6 | 10 11 | 15 16 | 20 21 | 25 26 | 30 31 |
|---|---|---|---|---|---|---|---|---|
| 2 | Calendar dates | 1/1 | 1/29 | 3/4 | 4/8 | 5/13 | 6/17 | 7/22 |

| Tabular data | Production period | 1 | 2 | 3 | 4 | 5 | 6 |
|---|---|---|---|---|---|---|---|
| 1. | Labor (Lab) | (A) 1000 | (D) 1500 | (G) 3000 | (G) 5000 | — | — |
| 2. | Material (Mat) | (B) 2500 | (B) 2500 | (B) 1000 | — | — | — |
| 3. | Subcontractors (Sub) | — | — | (F) 3000 | (F) 1000 | — | — |
| 4. | Suppliers (Sup) | — | — | — | (E) 2000 | (E) 2500 | (E) 500 |
| 5. | Period cost totals | 3500 | 4000 | 7000 | 8000 | 2500 | 500 |
| 6. | Cumulative totals | 3500 | 7500 | 14500 | 22500 | 25000 | 25500 |
| 7. | Production value period Totals (line 5 X 1.155) | 4043 | 4620 | 8085 | 9240 | 2887 | 578 |
| 8. | Cumulative production values | 4043 | 8663 | 16748 | 25988 | 28875 | 29453 |

The methods for accumulating the necessary tabular data are elemental because all of the necessary preparatory work has been done. For plotting period 1, it should be apparent that all activities running through the space between the beginning and ending of period 1 represented by the two vertical lines are planned for construction within that time period. Activity A starts at the beginning of the first day, extends through the end of day 5, a total of 5 duration units. Its value per duration unit is $200 per day, which means that $1,000 in cost are planned to be incurred by Activity A during plotting period 1. The resource notation at the end of activity A indicates that A is a labor item; this means that $1,000 is to be inserted for plotting period 1 on the labor summary line. Activity reference A is shown for clarity in this illustration. Typically several labor activities may be performed in the same period and the total of these tabulated. For this reason a number of labor activities may make up the total to be inserted on the labor activity line.

Further reference to the time scale network indicates that activity B starts at the beginning of the plotting period and extends through the end of the plotting period, a total of 5 duration units. The cost per duration unit is $500; therefore, $2,500 is inserted for material. Note that activity reference B is shown like activity reference A.

Since A and B are the only activities performed with plotting period 1, no amount is to be inserted on the subcontract or supplier line. The period subtotal for plotting period 1 is $3,500 as indicated. Since period 1 has no predecessor, the cumulative total for period 1 is also $3,500. The same procedure continues for the remaining plotting periods. The first subcontract item occurring in plotting period 3 is activity F, with a period value of $3,000. Note that the subtotal for period 2 is $4,000 which, when added to the cumulative total for period 1, gives a cumulative total through period 2 of $7,500.

This process of tabulation is continued through the end of the time scale network. An arithmetic confirmation occurs when the tabular data is completed. The cumulative total for the last tabular period at the end of the job, or job cost, should equal the job cost budget. In the case of the illustration, the cumulative total is $25,500. This equals the budgeted tabular data for the cost budget at production cost. Assuming there are no offsetting errors, this would confirm that the distributions have been properly made.

The complete cumulative tabular data for the cost-curve is now available for calculating the production value of costs. The percentage distribution provided in Exhibit 10-1 indicates that the contract amount (production value) is equal to 115.5 percent of the cost budget, or a multiplier of 1.155. Each cumulative total for cost may be multiplied by this factor to provide the cumulative production value for the same production period throughout the job. Here a confirmation is provided for the contract amount, $29,453.

## Plotting the Cost and Production Curves

Cumulative cost and cumulative production curves are of the greatest interest as performance models. These are plotted directly from the tabular data and are shown in Exhibit 10-3. The relationship between the production curve and cost curve previously discussed is reflected in this illustration. In setting up the vertical scale for cost relating both to dollar amounts and percentages of completion, provision has been made to enable reading the percentage of completion of the project at any point along the time scale. The cost curve is used for this purpose because the vertical scale has been constructed on the basis of the budgeted cost being 100%.

It is worth noting that all of the raw data required for building the cost curve and the production curve as performance models is contained in the activity based budget and the activity based project schedule. The remaining processes and procedures to organize the data, to summarize it in a convenient form and to display the performance models, do not add *new* data.

## PLANNED DATA AND PERFORMANCE EXPECTATIONS

The production curve and the cost curve as performance models represent a portion of the project statement of expectations. It clearly indicates the way the costs are expected to accumulate during construction. It also represents the planned production during construction. These performance expectations can be used in overall production management and for serious sales and marketing strategies. Valid data is provided by the planned performance on what each project may be expected to contribute to overall production.

## PERFORMANCE DATA CAPTURE

Two important criteria apply to the data capture procedures which provide the necessary raw data on actual performance. The first of these is simplicity. In detail and quantity, the prepared data presented to field supervision concerning expected performance should be understandable and useful, and fairly represent how the job is to be built. It should also be summarized in ways that encourage the utilization of the planned information in managing production. The data capture procedures associated with performance measurements which result in performance ratios will be new and novel in most construction work environments, reinforcing the need for simplicity.

Of even greater importance than simplicity is that data captured on *actual* performance match exactly the way in which the data had been generated for *planned* performance. This matching of data is important for several reasons. The first of these, to repeat, concerns the use of performance ratios as a primary and early visibility set of management information. Where data capture on actual performance is not matched to planned performance, the calculation of performance ratios is measurably more difficult. Second, though both the production performance ratio and the cost performance ratio compare actual performance to planned performance, in the case of the *production* performance ratio, the comparison is the total dollar value of product placed during the reporting period vs. the planned value of product placement; there is no concern at all for composition or for what makes up value of product in place.

In the case of *cost* performance ratio, however, valid data may be generated only when the composition of the cost of work actually incurred is exactly the same as the composition of items, expressed as planned costs, to be placed during the period. To illustrate the problem with cost performance ratios, assume the cost curve at the end of the second reporting period shows cumulative actual cost to date to be $9,500 compared to the planned cost of the work through that period of $7,500. What conclusions could be drawn from this data with the cost performance ratio calculations ($9,500 divided by $7,500) indicating a cost performance ratio of 1.267?

If the question of *composition* is ignored, at least two conflicting conclusions may be drawn. The first is that we are incurring cost which are 26.7% over budget. But we might conclude instead that our *rate* of placement is 26% ahead of schedule. The first conclusion would represent disastrous performance; the second, exceptionally good performance. The data, however, is at best meaningless and at worst grossly misleading until confirmation is made that the activities making up the $7,500 planned cost are exactly the same as those making up the $9,500 actual cost. Only when these two are carefully matched does the cost performance ratio become meaningful, immediately validating one conclusion and rejecting the other.

## MEASURING AND EVALUATING PERFORMANCE

The cost performance ratio and the production performance ratio are significant management information statistics which of themselves provide early information on job performance. When these indicators show negative trends in the job and they predict these negative trends to continue, immediate action must be taken by management. First, the problems must be identified. With a clear understanding of the problems, corrective action can be designed and implemented. The correction action must then be monitored with new management information on continued performance in the form of production performance ratios and cost performance ratios to verify that the corrective measures are indeed effective.

## SUMMARY

The cost curve and the production curve are two performance models which are a significant part of the project statement of expectations. They are the easiest of the derived performance models to prepare since all the raw data necessary is contained within the project schedule and budget. Simple graphic and tabular techniques are required to create a display format for the two performance models. Both are early indicators of job performance. The cost curve is an indicator of cost control using the cost performance ratio. The production curve is an instrument of schedule control since it indicates whether the planned levels of production are being met on schedule.

In the *production* performance ratio, the composition of planned and actual data is unimportant; only the total dollar value of product in place is important. However, in the *cost* performance ratio, the composition of the planned data and actual data is important. Therefore, data capture procedures are especially important in creating accurate management information in the form of a cost performance ratio.

These two performance models will be utilized in the preparation of the remaining derived performance models. The production curve is arithmetically related to the cost curve since it represents directly the cost curve plus the indirect cost, overhead and fee. The shape of both the cost curve and the production curve is the classic "S" curve.

# 11

# THE PLANNED SCHEDULE OF VALUES AND CASH INCOME CURVES AS PERFORMANCE MODELS

One of the major responsibilities of management in construction is predicting cash income, cash requirements, and cash flow as a basis for effective cash management. Traditional accounting methods tend to focus on historic job cost, earned revenue, and similar items with little regard for informational systems required for cash management. The budget, as a primary performance model, concerns the expected cost of the work but does not deal with the issue of cash requirements; nor does the schedule as a primary performance model provide help concerning either income or cash requirements, both essential data in generating cash management information. For cash management, the production curve is similarly of little help since it addresses the contract value of work in place but does not take into account such things as contract retainages, holdbacks, or deposits made in advance of actual performance.

## INFORMATION REQUIRED FOR CASH MANAGEMENT

The information required for cash management is the amount of cash income expected, when it is expected, the amount of cash required, and when it is required. Management needs some routine process to obtain the kind of information required. When cash income can be forecast and when cash requirements can also be forecast, the data required to generate valid information for cash management is available. The methods for providing the required data and for converting this data into useful information need further discussion.

Other performance models are required which can forecast cash income and cash requirements. The performance models discussed provide some useful information in preparing these additional performance models which include the Schedule of Values curve, the cash income curve, and the cash requirements curve. This chapter presents the methods for generating the Schedule of Values curve and from it, the cash income curve. Chapter 12 considers the cash requirements curve.

## THE SCHEDULE OF VALUES CURVE

The production curve provides information on the contract value of work in place but, as we have seen, does not consider the amount of cash income or when that income will be received. It is obvious that more information is required in order to generate the cash income curve. The source of this additional information is the terms and conditions of payment in the contract. These stipulate the basis for payment, the amount to be paid and when payment is to be made. First, consider the basis for payment.

In most construction contracts, the basis for payments is a "Schedule of Values." The Schedule of Values is a summary of values assigned to significant portions of the work to be used as a basis for payment. Items included in the Schedule of Values typically include the standard 16 divisions of the work. A further breakdown may show the value of primary work packages where more than one work package is included in the standard division. Once agreed upon and approved by the parties of interest, the Schedule of Values becomes the exclusive basis for payment and controls the cash income of the project.

### Summary Groupings

The Schedule of Values summarizes various groups of activities from the budget for convenience in processing payments. Some items included in the budget, however, will not be shown as separate items but will be distributed by some process among the remaining items. For example, indirect cost, overhead, and fee may be distributed into other categories. Division 1, General Requirements, may also be distributed into the other categories. When this distribution is made on a straight line percentage basis, the Schedule of Values is a summary version of the budget. As a result, the Schedule of Values curve and the production curve are the same.

### Weighting of Scheduled Values

It is sometimes desirable to increase the cash income from a job in the early stages of the work to offset negative cash flow. This is usually accomplished during the course of preparing the Schedule of Values. There are two separate procedures which may be used. The first applies a disproportionate percentage of the items to be distributed, such as indirect cost, overhead, and fee, and, in some cases, general requirements, to those items scheduled for earlier completion in the project. The second procedure is to reassign some of the value from the items to be performed later in the project to those to be performed earlier in the project. The net effect of both procedures is to weight the values assigned to items occurring earlier in the project so that, when percentages of completion are assigned for requisitioning purposes, larger dollar amounts are included in the requisition.

## Spread Sheet Analysis

Preparation of the Schedule of Values curve begins with a spread sheet analysis of the budget which redistributes the costs of items shown or listed in the budget but not in the Schedule of Values (the transparent items) into the other budget items on an activity by activity basis. These redistributed values added to those budget items remaining become the Schedule of Values and, when graphically plotted on the time scale, the Schedule of Values curve. The tabulation is also extended to show the only summary levels to be included in the Schedule of Values since not all levels of detail in the budget are appropriate here. It is important to note, however, that the control processes and the data sources for requisitioning purposes are maintained at the individual activity level. The percentages of completion of each activity are the basis for calculating the percentage of completion for each item shown in the Schedule of Values.

The simple project used as an illustration in Chapters 9 and 10 will be used here to demonstrate how the Schedule of Values data is prepared and to show how the Schedule of Values curve is generated.

The spread sheet development of the Schedule of Values is shown in Exhibit 11-1 supported by the weighting calculations shown in Exhibit 11-2. The tabular data presented is used to prepare the Schedule of Values. Columns 1, 2, and 3 are taken directly from the budget (Exhibit 10-1), which is the primary performance model for the cost of the project. Column 4 represents the distributed values. For purposes of illustration, the overhead and fee are distributed through the remaining items in the budget. The percentage distribution of the budget from Exhibit 10-1 indicates that the budget total is 115.5% of the cost budget. To obtain the distributed values in Column 4, the budgeted costs in Column 3 are multiplied by 1.155 in each case. No distributed values are shown for the transparent items of overhead and fee. The distributed values check-total to the budget total of $29,453. Note that if the *distributed* values were used on the time scale network and plotted, the result would be the production curve.

If no further weighting of the scheduled values for billing purposes is desired, it is unnecessary to proceed further with the generation of the Schedule of Values curve as the basis for generating the cash income curve. The production curve, then, may be used in place of the Schedule of Values curve when—but only when—the distribution of nonvisible items in the Schedule of Values is made proportionately through all of the remaining items (as in column 4, Distributed Overhead and Fee, of Exhibit 11-1).

Exhibit 11-1. Spread Sheet Development of Schedule of Values

| 1 | | 2 | 3 | 4 | 5 | 6 | | |
|---|---|---|---|---|---|---|---|---|
| COST CODE | RESOURCE | ACTIVITY IDENTIFIER | BUDGET | DISTRIBUTED OVERHEAD AND FEE | WEIGHTED VALUES* | COST TYPES | SCHEDULED VALUES | % DIST. |
| 1000 | Lab. (1) | A | 1.000 | $ 1,155 | $ 1,386 | (1) Lab. (A + D + G) | $11,781 | 0.40 |
| 2000 | Mat. (2) | B | 6.000 | 6.930 | 8.316 | (2) Mat. (B) | 8,316 | 0.28 |
| 3000 | Dum. | C | — | — | — | (3) Sup. (E) | 5,198 | 0.18 |
| 4000 | Lab. (1) | D | 1.500 | 1.733 | 2.080 | (4) Sub. (F) | 4,158 | 0.14 |
| 5000 | Sup. (3) | E | 5.000 | 5.775 | 5.198 | | | |
| 6000 | Sub. (4) | F | 4.000 | 4.620 | 4.158 | | | |
| 7000 | Lab. (1) | G | 8.000 | 9.240 | 8.315 | | | |
| 100.0% | Subtotals | | $25,500 | — | — | | — | |
| 10.0% | Overhead | | 2.550 | — | — | | — | |
| 110.0% | Subtotal | | 28.050 | — | — | | — | |
| 5.5% | Fee | | 1.403 | — | — | | — | |
| 115.5% | Budget total | | $29,453 | $29,453 | $29,453 | | $29,453 | 1.00 |

* See Exhibit 11-2 for weighting factors and calculations.

## Exhibit 11-2. Weighting Calculations

| EARLY ACTIVITIES (FROM TIME SCALE, EXH. 10-3) | | | | LATE ACTIVITIES (ALL REMAINING ACTIVITIES) | | | |
|---|---|---|---|---|---|---|---|
| ACTIVITY | VALUE | WEIGHT FACTOR* | WEIGHTED VALUES | ACTIVITY | VALUE | WEIGHT FACTOR** | WEIGHTED VALUE |
| A | $1,155 | 1.200 | $ 1,386 | E | $ 5,775 | 0.900 | $ 5,198 |
| B | 6,930 | 1.200 | 8,316 | F | 4,620 | 0.900 | 4,158 |
| D | 1,733 | 1.200 | 2,080 | G | 9,240 | 0.900 | 8,315 |
| | $9,818 | 1.200 | $11,782 | | $19,635 | 0.900 | $17,671 |

\* Assumes a 20.0% weighting of early activities:
  Early activity weighting = weighted value less budgeted value
  $$= \$11,782 - \$9,818$$
  $$= \$1,964$$
\*\* See weighting calculations for late activities below:
  Late activity weighting factor = value less early activity weighting ÷ value
  $$= (\$19,635 - \$1,964)/\$19,635 = \$17,671/\$19,635$$
  $$= 0.900$$

## Distributing the Scheduled Values Over Time

On the assumption that some weighting of the early items in the job is also desired to minimize negative cash requirements, two steps are required. In the simplified sample project, a reference to the time scale network shown in Exhibit 9-3 indicates that three activities (A, B, D) essentially make up the first half of the project. For purposes of illustration, we weight these early items.

We then must decide the amount or percentage by which these are to be weighted. The decision concerning the percentage of weighting can go two ways. The first of these is an empirical decision based on past experience. The second can start in the same way but involves some testing: select a particular percentage, complete the preparation of the performance models, and complete a cash position analysis to determine whether that percentage accomplishes the objective of eliminating cash deficit early in the job. This additional analysis permits adjustments until the appropriate percentage is found.

## Weighting Calculations

To carry forward our illustration, a 20% weighting factor has been selected. Exhibit 11-2, Weighting Calculations, presents the processes by which the weighting of the early items is calculated and the calculations which show the effect and determine the impact of the weighting of the early items on the later items.

Each of the early items A, B, and D is multiplied by 1.2. The result is the weighted value of each item; this value is placed in column 5, Exhibit 11-1.

The effect of the weighting of the early items is calculated for the late items E, F, and G by first determining the dollar value of the weighting, which is the amount added in total to the early items (see footnote 1, Exhibit 11-2). This amount is deducted from the total value of the late items, the result of which is the weighted (reduced) value of the late items. When this amount is divided by the distributed value of the weighted items (footnote 2, Exhibit 11-2), the result is the weighted multiplier, 0.900, to be used in determining the weighted value of the late items. These calculations are completed for the late item and placed in column 5, Exhibit 11-1.

Column 5 represents the weighted value assigned to each activity in the project. For integrated cost and schedule control purposes, it is essential that these weighted values be calculated for each activity since they will be the basis for determining values of work in place for requisitioning purposes.

## Comparison of Scheduled Values and Production Values

Two features distinguish the Schedule of Values from the distributed production values. The first distinction is that the scheduled values may be weighted to favor certain items while reducing others. The second distinction is that the Schedule of Values, used in most construction contracts for requisitioning, is a summary form of the distributed values. The summary levels to be utilized in the Schedule of Values are the subject of negotiation and agreement between the owner and contractor, but are usually fixed at the work package and/ or division level. Of course, any other summary level may be used. Once the summary level is agreed upon, it then becomes fixed for the duration of the contract.

To illustrate how the Schedule of Values summaries are prepared using the tabular data in Exhibit 11-1, the summary level by resource type has been selected. There are four resource types used: type 1, labor; type 2, material; type 3, suppliers; and type 4, subcontractors. The Schedule of Values should show a separate line item for each resource type (summary level). Resource type 1, labor, consists of items A, D, and G, which, when totaled, equal $11,782. Resource type 2, material, contains only one item, B, whose weighted value is $8,316. Similarly, resource type 3, suppliers, contains one item only, E, with a weighted value of $5,198. There is also only one subcontractor item, F, with a weighted value of $4,158.

The Schedule of Values for the same project contains four line items with a scheduled value assigned to each item and a total equaling the contract sum (budget total). It is important to note that the source data down to the individual activity has been maintained since integrated cost and schedule control occurs at the level of the individual activity. The tabular data in Exhibit 11-1 represents planned expectations based on the primary performance models of budget and schedule, the decisions concerning the items to be distributed, the percent of weighting, and the summary level to be used in the Schedule of Values.

The next preparatory step for the Schedule of Values is to insert the weighted values into the time scale network for distribution along the time scale, as shown in Exhibit 11-3.

## THE SCHEDULE OF VALUES CURVE

Drawing the Schedule of Values curve is a process identical to that for the cost curve. On the time scale network, scheduled values are distributed along the time scale. In place of the budgeted values used in the cost curve, the weighted values (column 5 from Exhibit 11-1) are used for each activity. The time scale network with budget resources and cost scales added, and shown in Exhibit 10-2, is modified to reflect the weighted values provided in Exhibit 11-1. The only data changing is the value assigned and the result in cost per duration unit; these are shown in Exhibit 11-3.

All the information and data is now in hand to tabulate the Schedule of Values and to prepare the Schedule of Values curve. Exhibit 11-4 shows the time scale network with the tabular schedule values and the Schedule of Values curve. The accumulation of the tabular data and the plotting of

## Exhibit 11-3. Time Scale Network with Weighted Values from Exhibit 11-1

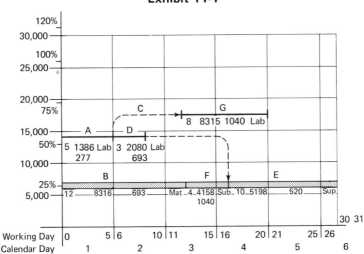

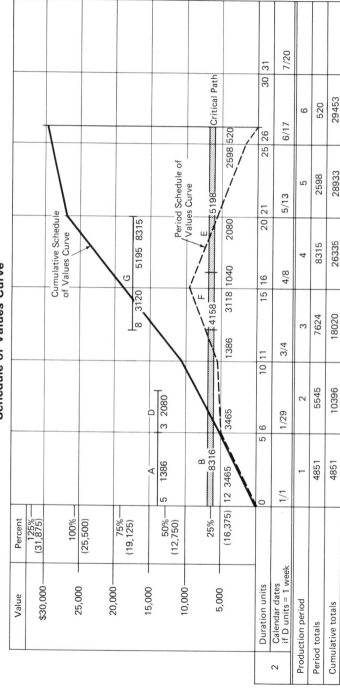

Exhibit 11-4. Time Scale Network with Tabular Scheduled Values and Schedule of Values Curve

the Schedule of Values curve proceeds exactly as for the cost and production curves. Note that the Schedule of Values curve has a modified "S" shape, similar to both the cost curve and the production curve but skewed to the left. When compared to the cost and production curves, the Schedule of Values curve will be above and to the left of both since time is plotted left to right and values are being moved to earlier placement on the time scale.

Exhibit 11-5 shows segments of the cost, production, and Schedule of Values curves on the same scale.

The relationship between the cost and production curve has already been discussed. The relationship between the Schedule of Values curve and the production curve is worth noting. The vertical distance, or delta, between the production curve and the Schedule of Values curve at any point in time represents exactly the extent to which the Schedule of Values curve has been weighted above the production value of the work in place. It represents what may be described as an artificially higher rate of production.

The Schedule of Values weighting has no effect at all on the contract sum. Its purpose is to reduce cash deficits resulting from significant time differences between work performance and payment for that work. It also is helpful in offsetting cash deficits resulting from significant differences between owner funding to the contractor and contractor funding for labor, material and principal vendor items. These disparities, which have serious cash implications, will be fully discussed in Chapter 13 on cash analysis.

Exhibit 11-6 shows the weighted values used in Exhibit 11-4. The cost/duration unit is also shown.

## Exhibit 11-5. Segments of Cost Curve, Production Curve, and Schedule of Values Curve for Comparison.

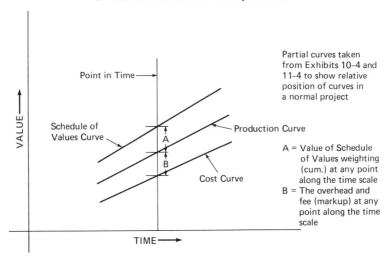

**Exhibit 11-6. Cost/Duration Unit of Weighted Values**

| IDENTIFIER | DURATION | WEIGHTED VALUE | COST/DURATION UNIT |
|---|---|---|---|
| A | 5 | $1,386 | $ 277 |
| B | 12 | 8,316 | 693 |
| C | 0 | — | — |
| D | 3 | 2,080 | 693 |
| E | 10 | 5,198 | 520 |
| F | 4 | 4,158 | 1,040 |
| G | 8 | 8,315 | 1,040 |

## THE CASH INCOME CURVE

Most of the data and information required to prepare the planned income curve has now been accumulated. To illustrate the contract factors which control cash income, the assumed conditions of the contract with regard to payment are, first, that the billing period will be five duration units; second, that a 10% retainage will be withheld; third, that payment may be expected at the end of the third duration unit after the billing period and that final payment will be made one full billing period after substantial completion.*

The most convenient way to prepare the cash income curve is to use the time scale network with the Schedule of Values curve. This procedure provides both the Schedule of Values curve and its underlying tabular data which also may be directly used in the preparation of the cash income curve. Exhibit 11-7 shows the Schedule of Values curve with tabular data on cash income, and also the cash income curve.

The first step in preparing the cash income curve is to determine the amount of the net cash income for each billing period based on the planned scheduled value of the work. Since the assumption has been made that the contract requires a 10% retainage, 10% is deducted from each cumulative scheduled value for each of the billing periods. When the retainage is deducted, the result is the net planned cash income for each billing or production period. With the net cash income determined for each billing and production period, the next step is to determine when the net cash will be received.

The assumption is that the billing and production period will be five duration units and that payment will be made at the end of the third duration unit following the billing period. This means that payment may be expected at the end of the 8th, 13th, 18th, 23rd and 28th duration unit and that final payment will be made (according to the terms of the contract) one

---

* In preparing the previous performance models, the duration unit was assumed to be one working day. To more effectively illustrate the primary features of the income curve, the production and billing periods are assumed to be five duration units in length (nominally representing one month). The same procedure will be used later for the cash requirements curve.

# Exhibit 11-7. Time Scale Network with Schedule of Values Curve, Tabular Cash Income, and Cash Income Curve

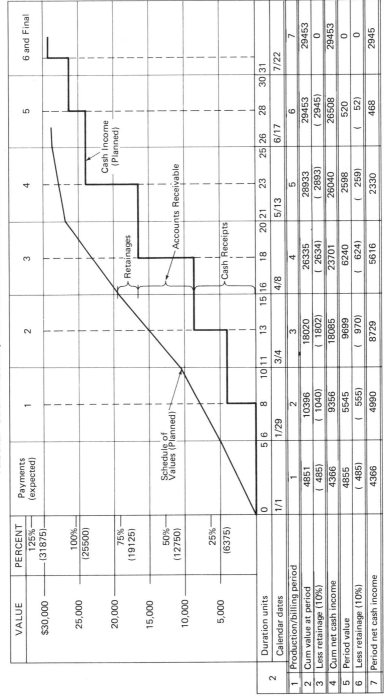

| VALUE | PERCENT |
|---|---|
| $30,000 | 125% (31875) |
| 25,000 | 100% (25500) |
| 20,000 | 75% (19125) |
| 15,000 | 50% (12750) |
| 10,000 | 25% (6375) |
| 5,000 | |

| | | | | | | | |
|---|---|---|---|---|---|---|---|
| **Duration units** | 0 | 5 6 | 8 | 10 11 | 13 | 15 16 | 18 |
| **Calendar dates** | 1/1 | 1/29 | 3/4 | 4/8 | 5/13 | 6/17 | 7/22 |

| | | 20 21 | 23 | 25 26 | 28 | 30 31 |
|---|---|---|---|---|---|---|

| | 1 | 2 | 3 | 4 | 5 | 6 and Final | 7 |
|---|---|---|---|---|---|---|---|
| 1 Production/billing period | 1 | 2 | 3 | 4 | 5 | 6 | 7 |
| 2 Cum value at period | 4851 | 10396 | 18020 | 26335 | 28933 | 29453 | 29453 |
| 3 Less retainage (10%) | ( 485) | ( 1040) | ( 1802) | ( 2634) | ( 2893) | ( 2945) | 0 |
| 4 Cum net cash income | 4366 | 9356 | 18085 | 23701 | 26040 | 26508 | 29453 |
| 5 Period value | 4855 | 5545 | 9699 | 6240 | 2598 | 520 | 0 |
| 6 Less retainage (10%) | ( 485) | ( 555) | ( 970) | ( 624) | ( 259) | ( 52) | 0 |
| 7 Period net cash income | 4366 | 4990 | 8729 | 5616 | 2330 | 468 | 2945 |

billing period, or five duration units, after substantial completion at the end of duration unit 31.

Exhibit 11-7 contains solid vertical lines representing the end of each billing period. Each time of expected payment is indicated as a dashed vertical line.

## PLOTTING THE CASH INCOME CURVE

As shown in Exhibit 11-7, the net cash income expected for billing period 1 is $4,366. This cash is expected at the end of the 8th duration unit. Since no income will be received until the end of the 8th unit, the income during that time period is represented by a horizontal line with a value of zero. At the end of the 8th duration unit, the planned income of $4,366 is plotted on the vertical scale. The value of work for the second billing period, cumulative, is $9,356 and a payment is expected at the end of the 13th duration unit. Since no income is expected between the end of the 8th duration unit and the end of the 13th duration unit, a horizontal line is drawn between the two. The cumulative figure $9,356 is plotted against the vertical scale. This process continues for each of the five billing periods, with the cumulative net income through the fifth billing period of $26,040 due at the end of the 28th duration unit. What remains open in terms of cash income is the work performed during the sixth billing period and all retainages held to date. Since the schedule for the project as indicated on the time scale network requires completion by the end of the 26th duration unit, and since the contract requires that final payment be made one billing period, or five duration units after substantial completion, the final payment, or the unpaid balance of the contract sum, may be expected at the end of the 31st duration unit.

## PLANNED DATA AND EXPECTED PERFORMANCE

The planned cash income curve for the project is a graphic representation of the planned accounts receivable (A/R) expected for the project. It is a result of the project statement of expectations including the schedule and the budget. It reflects the contract terms and conditions and the Schedule of Values.

Of the primary and derived performance models which have been discussed, the cash income curve is the first one that deals directly with the cash management of the project. When the cash income curve is used in conjunction with the cash requirements curve for project, the net cash position and, consequently, the capital requirements for the performance of the specific job may be identified at any time in the project. When multiple jobs are summarized, the result is an effective and valid projection of planned income which may be used in developing corporate cash management strategies. When compared to similar cash requirements summaries, the organization's future cash position within the planning horizon may be forecast accurately.

The planned cash income curve represents what is expected to happen based on the project statement of expectations. There is no appropriate equivalent to the cost performance ratio when dealing with cash management, although both the cash income ratio (CIR) and the cash requirements ratio (CRR) may be calculated. The primary interest is not in either cash income or cash requirements alone but in the *comparisons* between these two performance models, which provide the primary source of information on cash position.

## Cash Income and Accounts Receivable

Accounts receivable (A/R) is the vehicle by which cash income is generated. Cash income is the income expected based upon the accounts receivable invoices and requisitions. The planned cash income curve represents the planned data. The accounts receivable reflect the actual performance data capture for activities completed or in work. This source data, however, is provided through the integrated cost and schedule control processes in which the completion of individual activities are summarized in the form required for generating accounts receivable.

## MEASURING AND EVALUATING PERFORMANCE

One of the parameters for evaluating project performance is the comparison between planned cash income and actual cash income. The accounts receivable reflect the scheduled value for work in place and are the same as actual cash income when collected. When actual cash income, as reflected by the accounts receivable, is higher than planned cash income, higher than expected production is being maintained. When accounts receivable fall below planned income, it is a significant indicator that production is lower than expected. Lower than expected production represents a significant problem in project performance which must be identified and corrected.

## SUMMARY

The Schedule of Values curve represents the basis upon which payment will be made for the satisfactory performance of the work. It is related to, although not directly derived from, the cost curve and the production curve. It may be the same as the production curve if all concealed, or transparent, factors are proportionately distributed throughout the remaining activities. The Schedule of Values curve is prepared from the weighted values at the activity level.

The Schedule of Values used for accounts receivable purposes is usually summarized in the work package or trade summary groups for simplicity and convenience.

The planned cash income curve is developed from the Schedule of Values curve. The Schedule of Values curve provides information on the scheduled value of product in place at the end of each billing period. This value is modified (usually reduced) by retainages and other factors. Knowledge of the contract terms and conditions concerning payment determine when cash income may be expected. How much is determined by the scheduled value, less deductions. These two combined provide the information necessary to plot the planned cash income curve. The cash income curve provides two important pieces of information for cash management purposes: first, the planned amounts due as a result of the planned production of the project; second, when those cash receipts might be expected.

While worth noting, other factors which will affect planned cash income are not necessarily included in the planned cash income. For example, stored materials brought to the job in advance of the time needed may cause a temporary inflation in the planned cash income. This temporary increase is accounted for when the stored material is placed at the job site. Construction contracts may provide for a reduction of the percentage retained near the midpoint of the job if the work is progressing satisfactorily. This condition may be reflected in the planned cash income if it is realistically anticipated.

The planned cash income to be paid for any billing period is calculated by the following formulas:

1. Total scheduled value of work in place
   = sum of the scheduled value of each activity
   $\times$ percent completed of that activity
2. Expected cash income to date
   = total scheduled value of work in place $-$ retained percentage
3. Amount due for the current production period
   = expected cash income to date $-$ amounts previously requisitioned
4. Total amount currently due
   = amount due for current production period $+$ any past due balances

Planned cash income differs from planned production value in two significant ways. First, planned production value accumulates on a more or less continuous basis as product is placed in the field environment; planned cash income, however, occurs only at the points in time permitted by the contract based on billings or requisitions generated for specific periods of time. Second, the planned production value at any point in time will always be higher than the planned cash income, and, in any billing period, the planned cash income is equal to the scheduled value of product in place less retainages and holdbacks.

Retainages and holdbacks are clearly earned but they are not *billable*. These items, therefore, cannot be considered part of planned cash income until the payment terms and conditions of the contract allow these earned assets to become accounts receivable. The planned production between billing periods also represents the accumulation of earned assets which are not billable until the accounts receivable are generated. The vertical distance between the cash income curve and the planned production curve, at any point in time, provides the dollar value of the asset represented by retainages and accumulated but unbilled production.

# 12
# THE PLANNED CASH REQUIREMENTS CURVE AS A PERFORMANCE MODEL

Of great interest to management, in addition to an analysis of cash income, is an analysis of the *demands* for cash as a result of the ongoing business enterprise. Such cash requirements data is essential for effective cash management.

## THE NEED FOR VALID DATA

Several factors make valid data on cash requirements for payroll and accounts payable of interest. First, while job costs are incurred daily in any construction project, when the cash is required to pay for these costs is controlled by other factors including when the payroll must be funded and when bills must be paid. Issues such as the production/cash income lag and retainages under the general contract do not of themselves create a cash shortfall since cash required may also lag the overall accumulation of costs. It is the relationship between *cash income* and *cash required* which creates the basis for valid cash management information.

Still another factor (given current interest rates and short term investment opportunities) is the possibility of generating significant nonproduction income based upon valid information concerning cash requirements and cash income and knowing when demands for cash are likely to occur as a result of business operations. Closely related is the ability to earn discounts by timely payment for certain principal vendor and supplier's items on which cash discounts are allowed. Each of these in its own way contributes significantly to the success of the business enterprise and goes beyond the ability to construct projects economically.

## COMPLEXITY OF CASH REQUIREMENTS DATA

The cash requirements curve is the most complex to prepare of all of the derived performance models. It is, however, not difficult to generate assuming that the primary performance models are in place, that the different resources to be utilized in the performance of the work have been determined, and that the differing terms and conditions of payment to different kinds of resources have been established.

The validity of integrated cost and schedule control within each of the performance models (the primary and the derived as well) depends on the common roster of activities. The only exception concerns the Schedule of Values. Even in this case each line item is a summary level or a subset of activities from which the Schedule of Values is derived. The time scale network with the resources properly noted and the cost per duration unit calculated becomes one of the primary sources of information for generating the cash requirements curve.

## Cash Requirements and Time

The time scale network distributes the activities according to time of performance and simultaneously distributes the cost associated within each duration unit of each activity. In the case of the cost curve, simply tabulating the value of all activities extending through a particular duration unit summarizes the cost within that duration unit. In order to generate the cash requirements curve, however, the resources to be used must be examined to determine when payment will be required.

In examining the nine different resources or cost types (Exhibit 12-1),

**Exhibit 12-1. Payment Groups with Cost Type and Terms of Payment**

| PAYMENT GROUPS | | COST TYPE (RESOURCE) | | TYPICAL PAYMENT TERMS | |
|---|---|---|---|---|---|
| NO. | CLASSIFICATION | NO. | DESCRIPTION | PERIOD | TERMS |
| I | Labor (Lab.) | 1 | Own labor | Weekly | Net cash |
| II | Suppliers (Sup.) | 2 | Own material | Monthly | 2% discount |
| | | 3 | Own equipment | Monthly | Net 30 days |
| | | 4 | Own other | Monthly | |
| III | Subcontractors (Sub.) | 5 | Sub labor | Monthly | 5 Days after |
| | | 6 | Sub material | Monthly | Owner payment |
| | | 7 | Sub equipment | Monthly | to general contractor |
| | | 8 | Sub other | Monthly | less retainage* |
| | | 9 | Sub lump sum | Monthly | (usually 10%) |
| IV | Principal Vendors (P. Ven.) | 6 | Sub material, Principal vendors Only (F.O.B. job site) | Monthly | 5 Days after owner payment to general contractor with no retainage |

* Withholding of retainages as a guarantee of performance is a common construction industry practice generally applying to subcontractor payments which represent only a part of total cash requirements. Similarly a retainage is usually withheld from the total payment due to the general contractor.

most construction organizations have a reasonably standard policy with respect to the payment of each. The cost types may be divided into summary groups according to the ways in which payments are to be made and when payments, or funding, will be required.

### Payment Groups

In general, there appear to be four basic funding categories. Each of the nine cost types may be conveniently grouped within one of these. Cost type 1, own labor, for example, must be funded on a weekly basis. Cost type 2, material; cost type 3, own equipment; and cost type 4, own cost other, must be funded within a stipulated period in order to earn discounts. The third group is the subcontractor items. These are usually subject to the same retainage provisions as in the general contract; payment is keyed to the receipt of payment by the general contractor. Subcontractor items include the following: cost type 5, sub labor; cost type 6, sub material; cost type 7, sub equipment; cost type 8, sub cost other; cost type 9, sub contract lump sum. A fourth group is those principal vendor items, cost type 6, sub material (principal vendors only), which are not subject to retainages but where payment is keyed to the receipt of payment by the general contractor.

Exhibit 12-1 provides a summary of the typical terms and conditions of payment related to the four groupings which should be used by most general contractors.

To the extent that an individual general contractor has different requirements, these can be reflected in a similar grouping. The point is that the terms and conditions of payment establish the time schedule for payment and, consequently, control the cash requirements. It is necessary, therefore, to have the cost types in each of these four groups clearly identified as part of the resource information for each activity on the time scale network. In order to generate the cash requirements curve, it is now necessary to summarize the cost within each duration unit or payment period into those groups which have different terms of payment and, consequently, different cash requirements. In the case of the cost curve, we are not concerned with this breakdown since costs are incurred as product is put in place. In the cash requirements data, however, it is necessary to create the separation because of the different cash demands relating to different cost types.

## TIME RELATED CASH SUMMARIES

Two separate time related cash summaries may be made of cash requirements. The first of these is the current production period tabular cash summaries; these are shown for the sample project in Exhibit 12-2.

## Exhibit 12-2. Current Production Period Tabular Cash Summary

| Group | Classification | Paymt. due | Period 1 | Period 2 | Period 3 | Period 4 | Period 5 | Period 6 | Period 7 |
|---|---|---|---|---|---|---|---|---|---|
| I | Labor | Each duration unit | 1000 (200/wk) | 1500 (300/wk) | 3000 (600/wk) | 5000 (1000/wk) | | | |
| II | Suppliers | End of each period | 2500 | 2500 | | | | | |
| III | Subcontractors | See net subs | | | 3000 | 1000 | | | |
| | Less retainage | 10% | | | − 300 | − 100 | | | + 400 |
| | Net subs | | | | 2700 | 900 | 2500 | 500 | 400 |
| IV | Principal vendors | Fourth D-unit after period | | | — | 2000 | | | |
| | Total cash incurred (end of 4th D-unit payments) | Fourth D-unit after period | | | 2700 | 2900 | | | |
| | Total period cash incurred | | 3500 | 4000 | 6700 | 7900 | 2500 | 500 | 400 |
| | Cum. cash incurred | | 3500 | 7500 | 14200 | 22100 | 24600 | 25100 | 25500 |
| | Duration units (D) | | 1 2 3 4 5 | 5 6 … 10 | 10 11 … 15 | 15 16 … 20 | 20 21 … 25 | 25 26 … 30 | 30 31 … 35 |
| | Cumulative D-units | | 0 | 5 | 10 | 15 | 20 | 25 | 30 … 35 |

153

The current production period tabular cash summary data is useful in matching cash requirements to the production period. Comparing this data to the tabular cost data provides the value difference between cost incurred in a current production period and the cash required to pay for those cost. The current production period tabular cash summary, however, does not take into account *when* cash will be required. Since there are significant differences in when payment must be made and the amount of payment among the four payment groups, an additional analysis is required to provide tabular data as to when cash is required.

Exhibit 12-3 presents a tabular summary of payment period cash requirements.

Several additional items are required to understand the essential differences between Exhibit 12-2, Current Production Period Tabular Cash Summaries, and Exhibit 12-3, Tabular Payment Period Cash Requirements. First, note that the tabular data from Exhibit 12-3 is prepared primarily for the purpose of creating the cash requirements curve. It is important to note that each of the four payment groups has its own specific requirements which determine when cash is required for those payments. Group I is the labor items which must be funded weekly. Group II is the suppliers which are usually funded by the end of the month and have discounts associated with them. Group III is the subcontractors who are paid after owner payment to the general contractor and are also subject to a 10% retainage. Group IV is the principal vendors for whom the time of payment is the same as group III, Subcontractors, but who are not subject to the retainage requirements of subcontractors. An analysis of these payment groups indicate four different terms of payments but only three different payment or cash requirement dates.

### Tabular Data and the Time Scale

It must now be noted that the tabular data for both Exhibits 12-2 and 12-3 represent the same time scale as used for the time scale networks in the earlier performance models. This means that the typical cash payment points for each production period can be plotted on the time scale used in generating the payment period cash requirements and, subsequently, the cash requirements curve. This information is also shown in Exhibit 12-3. For clarity, the "Total 4th D Unit" line covering the total time period over which payments are made for the related production period 2 has been shaded in.

## PREPARATION OF TABULAR DATA FOR CASH REQUIREMENTS CURVE

Noting that the actual payment points for any production period extends generally over two production periods, some notational procedures will be

## Exhibit 12-3. Tabular Payment Period Cash Requirements

| Group | Classification | Paymt. due | 1 | 2 | 3 | 4 | 5 | 6 | 7 |
|---|---|---|---|---|---|---|---|---|---|
| I | Labor | Each duration unit | 800 / 200 | 1200 / 300 | 2400 / 600 | 4000 / 1000 | | | + 400 |
| II | Suppliers | End of each period | 2500 | 2500 | 1000 | | | | |
| III | Subcontractors | See net subs | | | 3000 | 1000 | | | |
| | Less retainage | 10% | | | − 300 | − 100 | | | + 400 |
| | Net subs | Fourth D unit after period | | | 2700 | 900 | | | 400 |
| IV | Principal vendors | Fourth D unit after period | | | − | 2000 | 2500 | 500 | |
| | Total cash incurred end of 4th D-unit payments | Fourth D unit after period | | | 2700 | 2900 | 2500 | 500 | 400 |
| | Total period cash required | | 3500 | 4000 | 4000 | 7700 | 2900 | 2500 | 900 |
| | Cum. cash required | | 3500 | 7500 | 11500 | 19200 | 22100 | 24600 | 25500 |
| | Duration units (D) | | 1 2 3 4 5 | 5 6 … | 10 11 … | 15 16 … | 20 21 … | 25 26 … | 30 31 … 35 |
| | Cumulative D-units | | 0 … | 5 … | 10 … | 15 … | 20 … | 25 … | 30 31 … 35 |

Notes:
1. Arrows indicate point in time cash is required.
2. The shaded area: Total cash incurred end of 4th D-unit payment, shifts amounts due to the end of the 4th D-unit in the subsequent period.

**155**

required to convert the current production period tabular cash summary, given in Exhibit 12-2, to the information contained in Exhibit 12-3.

Labor is indicated as a sloping line for clarity because of the frequency of funding and to distinguish labor funding requirements from all other cash requirements. The first production period of five duration units contains labor totaling $1,000. Since there is the possibility of other payments intervening at the end of the fourth duration unit, labor cash requirements are broken into two parts for production period 1: the first four duration units at $200 each or $800 in all, and the last duration unit at $200. These are indicated on the table in Exhibit 12-3 by the numerical values and an arrow pointing to the summary point at the end of the fourth duration unit, where there is a possibility of another cash requirement occurring.

Similarly, the material payments to suppliers, which are assumed to be due the first of the month, are noted. The amount for production period 1 is $2,500, which is the payment due at the end of the production period, indicated by the arrow.

Subcontractor payments are best illustrated by payment period 3 in Exhibit 12-3. The expected cost to be expended with subcontractors during that pay period is $3,000 (Exhibit 10-4). However, a retainage is to be held equal to 10% of the work performed, or $300. The net cash, which will be required to pay the subcontractor for the third production period work, is $2,700. Since there are no principal vendor costs planned for production period 3, the total cash requirement for subcontractors and principal vendors for production period 3 is $2,700. This amount, however, is not due and, consequently, cash is not required until the end of the fourth duration unit in production period 4. In order to minimize the confusion which might be possible due to the difference between when the cash requirements are incurred and when they must be paid, the line described as "Total Fourth Duration Unit" has been shaded. An arrow has been drawn indicating that the $2,700 cash requirement incurred in the third production unit is to be paid at the end of the fourth duration unit in the fourth production period.

In determining the amount of cash required for payment period 3, only the labor amount ($2,400 + $600) and the material amount ($1,000), totaling $4,000, is required during that time period. The cumulative cash payments required through the time period is $11,500. Note that both the total period payments and cumulative payments are carried in the tabular data.

## COMPARISON OF COST INCURRED, CASH REQUIREMENTS INCURRED, AND CASH REQUIRED

The data worth comparing is found in Exhibit 10-4, Exhibit 12-2, and Exhibit 12-3, and serve to fairly illustrate the differences between the accumulation of cost, when cash requirements are incurred, and when cash requirements are to be paid. Exhibit 10-4 shows the accumulated cost through production period 3 to be $14,500. Exhibit 12-2 shows the cash requirements incurred for the period to be $14,200, the difference being the $300 retainage applicable to subcontractor cost. Exhibit 12-3 reveals that the actual cash required, however, is only $11,500 during period 3 since the payment to subcontractors and principal vendors is not due until payment period 4.

## Summary Capability

It is apparent that a number of other summaries can be provided. As the number of summary categories expand, however, the amount of clerical work involved in maintaining these summaries manually increases significantly. It is apparent, however, that computer applications handle additional summary levels with considerable ease.

Having determined what summary groupings are appropriate, the tabular data required for the cash requirements curve may be generated from the data in Exhibit 12-2. Two separate curves may be generated resulting from the tabular data: total period cash requirements and cumulative cash requirements.

## OTHER PREPARATORY ITEMS

As a practial matter, the planned cash requirements curve should be the last of the derived performance models to be prepared. It is the most complex since it involves examining four different summaries of cost types for accuracy and validity. It is, of course, recommended that tabular data prepared for all of the derived performance models is the maximum level of detail required for any. This would indicate that the summary of the cost using the four groups of resource types would be the most appropriate.

Before the details of manually generating the cash requirements curve are dealt with, the specific demands for cash should be examined. The Group I type, own labor, is the most demanding since net payroll must be funded on a pay period (usually weekly) basis. Depending on the size of the payroll, employee withholdings, and employer contributions, the payroll taxes must be either funded directly on a weekly basis or held in escrow until time for payment. In either case, the cash demands for these additional payroll costs should be considered as immediate as those for the net payroll.

Group II cost types, including material and equipment, must normally be funded on a monthly basis. Payment is usually required in advance of the time cash income is expected. These costs, therefore, must be planned for from either a preceding monthly requisition or out of other cash available.

For suppliers and equipment rental vendors, discounts for timely payment may be available. Earned discounts, while not deducted from job costs, should be taken whenever possible. A proper understanding of cash flow and the demands for cash in relation to planned income provide this opportunity.

Neither labor or supplier items are funded in full by the owner. Both are subject to the same universal retainage deduction provided for in the general contract. It is obvious that both groups, therefore, represent, with each owner payment, a significant cash deficit equal to the retainage percentage.

Group III types of cost, generally lump sum contracts, are the easiest to administer from a cash point of view. The terms and conditions of payment and the time of payment are usually related to the general contract terms and conditions of payment. Payment to the subcontractor is usually made for a given requisition period after receipt of the cash funding for that period by the general contractor. As a general rule, most subcontracts provide for a retention in the same percentage as the general contract. Assuming no significant delays in the funding of the general contractor's requisition, the subcontractor's requisitions may be funded after receipt of cash by the general contractor. This represents an effective cash management strategy since it ties cash payment to receipt of cash.

Group IV cost types, which include principal vendors, usually allow for payment after receipt of payment by the general contractor. No discounts are available under most circumstances. The most significant factor in the principal vendor accounts payable is the retainage disparity. The general contractor may anticipate payment from the owner from which the retainage amount has been deducted. The requirement to fund principal vendors is on a gross basis without the privilege of deducting the retainage. This amount (usually 10% of the amount requisitioned) is to be funded by the general contractor either from the previous requisition funding, from working capital, or from other cash available. The principal vendor items therefore represent a potentially significant cash drain. A part of cash management strategy is to weigh the relative benefits and costs of buying vendor items as part of a lumpsum subcontract amount including both furnishing the materials and installing it, thereby eliminating the retainage disparity.

## PLOTTING THE CASH REQUIREMENTS CURVE

The tabular data, which is necessary to plot the cash requirements curve, is shown in Exhibit 12-3. It is derived from the tabular data shown in Exhibit 12-2 but adjusted to recognize when actual cash is required. The arithmetic verification of the tabular data appears on the cumulative payments line at the bottom of production, or pay, period 7. The $25,500 amount is the budgeted amount.

Exhibit 12-4 shows the cash requirements curve. It also shows the tabular data from Exhibit 12-3 as a matter of convenience.

When preparing to plot the planned cash requirements curve, it should be noted that there are *four* different groups of demands for cash payments. These are job related and have different time requirements.

The planned cash *income* curve should have already been prepared (see Exhibit 11-7). It indicates both the time the receipt of cash may be expected, and the cumulative amount of cash to be received. The planned cash income curve has been plotted over the time scale network. The cash income lines (vertical lines at payment points), indicating when each payment is expected throughout the job, have already been located and plotted along the same scale (see Exhibit 11-7). In much the same way, each component of the cash *requirements* curve is located and plotted. These funding lines should be indicated directly on the time scale network.

## Plotting Labor Cash Requirements

Since labor creates the most frequent demand for cash, it is best to plot the labor component for each cash funding cycle first. If payday occurs on Friday, the Friday funding line should be shown for each Friday through the total project payment cycle. For simplicity, only the methods of plotting the cumulative curve will be presented here since these represent the most useful level of information. The cumulative data is included in the tabular summaries in Exhibit 12-3 and is simply plotted on a vertical scale at the end of each pay period. Since payroll funding for both direct payroll and for the insurances and taxes component is required on a weekly basis, no adjustments need to be made in the tabular data.* In Exhibit 12-4 this results in the labor cost for the first period being shown as a sloping line rather than a vertical step at the end of the pay period.

There are two reasons for this technique. First, labor or payroll funding (usually weekly) may be considered (more or less) continuous compared to the less frequent (usually monthly) funding of suppliers, subcontractors, and principal vendors. The stepped (histogram) plotting of weekly payroll in much more tedious and, in the context of a basic monthly cash requirement cycle, adds no meaningful information. The second reason is that, since all other cash requirements are relatively large and occur on a monthly cycle, these

---

* The labor for the first pay period is graphically shown by connecting the plot points (zero at the beginning and $200 at the end of the pay period).

# Exhibit 12-4. Cash Requirements Curve

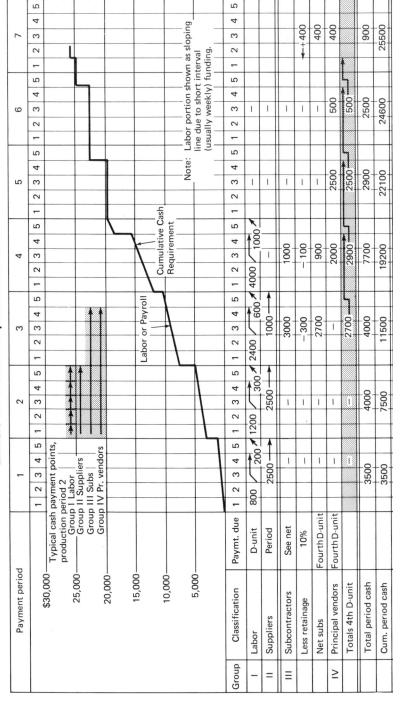

Typical cash payment points, production period 2
— Group I Labor
— Group II Suppliers
— Group III Subs
— Group IV Pr. vendors

Cumulative Cash Requirement

Labor or Payroll

Note: Labor portion shown as sloping line due to short interval (usually weekly) funding.

| Group | Classification | Paymt. due | ... |
|-------|----------------|------------|-----|
| I | Labor | D-unit | 800 · 200 · 1200 · 300 · 2400 · 600 · 4000 · 1000 |
| II | Suppliers | Period | 2500 · 2500 · 1000 |
| III | Subcontractors | See net | 3000 · 1000 |
| | Less retainage | 10% | −300 · −100 |
| | Net subs | | 2700 · 900 · 400 |
| IV | Principal vendors | Fourth D-unit | 2000 · 2500 · 500 · 400 |
| | Totals 4th D-unit | Fourth D-unit | 2700 · 2900 · 2500 · 500 |
| | Total period cash | | 3500 · 4000 · 4000 · 7700 · 2900 · 2500 · 900 |
| | Cum. period cash | | 3500 · 7500 · 11500 · 19200 · 22100 · 24600 · 25500 |

should appropriately be plotted as vertical steps (histogram) in the curve. This provides an instant graphic distinction between the more volatile payroll cash requirements and all others.

## Plotting Other Cash Requirements

The labor plotting procedure continues on a weekly basis until the next funding line (suppliers) is encountered and those payments are required. These would be payment of the group II activities, which are normally due on or about the first day of the month. Since this information has already been tabulated, it can be added as an increment above the cumulative labor cash requirements to that point. The cash required for suppliers ($2,500) is shown vertically at the end of production period 1 (not payroll period). Since retainages are not typically withheld and earned discounts are considered in a separate accounting category, no adjustments need to be made to the data already available. The payment of suppliers usually represents a significant step in the cash requirements curve.

Plotting continues with the next labor increment, or increments, until the next funding line is encountered and the next cash payments are due. These include the principal vendors (group IV), which are typically funded in full. No reduction of retainage is made for these items. They are, however, usually due at the same time as subcontractor payments (group III). Both the principal vendor and subcontract items are, therefore, plotted vertically, with the subcontract payment items from the tabular data reduced by the retainages. Note also that the $2,900 cash requirement incurred in production period 4 is not paid until production period 5.

This plotting method is continued until the cash requirements curve is complete. Retainage amounts, according to the provisions of the subcontracts, must be funded in the last payment period. Usually the retainage amounts are funded when payment is received for those items by the general contractor.

The plotting of the monthly payment cycle is repeated throughout the duration of the project by the procedures already described. The cash requirements cycle and the points in the cycle, when cash is required, may be summarized as follows:

1. Payroll: weekly.
2. Supplier accounts: according to the discount terms, normally on or about the first of the month.
3. Principal vendor accounts: gross amount, normally 5 days after cash income is expected.
4. Subcontractor accounts: normally 5 days after cash income is expected, with a 10% holdback, or retainage, deducted.

### Cash Required at Substantial Completion

The monthly cycle is repeated until the project reaches substantial completion. The terms and conditions of the contract determine when the collection of the retainages (to be included in accounts receivable) occurs. The retainages are usually released as cash income and included in the requisition for substantial completion. The cash income for retainages may be reduced by holdbacks for punch list or corrective work. Cash requirements funding for the current period work plus retainages to the subcontractors and principal vendors is subject to the same holdbacks for punch lists or corrective items to encourage timely completion.

### Final Completion

Final payment from the owner may normally be expected 30 days after the payment for substantial completion and the release of retainages, assuming that the punch list (see Exhibit 16-8) and corrective work, has been completed in the intervening 30 days. The cash requirements may be plotted anticipating the release of the final cash payments to the subcontractors and principal vendors involved.

## DISPARITIES BETWEEN CASH INCOME AND CASH REQUIREMENTS

There are several significant disparities in most construction contracts between the cash income (accounts receivable) to be paid by the owner to the contractor and the cash requirements (accounts payable) to be funded by the general contractor for payroll, supplier payments, subcontractor payments, and principal vendor payments. The retainage provisions of most general contracts are universal and apply to the total value of product in place. Labor, supplier invoices, and principal vendor invoices are normally not subject to retainages. This means that the general contractor must provide for the funding of cash deficits between cash income and cash requirements resulting from retainage provisions.

It is essential, therefore, that the contractor have valid information on the consequence of the resources planned for the work in light of the differing cash demands resulting from such choices. A significant cash management decision is whether to perform certain work with one's own work force or to include it as part of lumpsum subcontract transactions. It is important to note that these disparities do exist and do influence the net cash surplus or shortfall resulting from the project plan, schedule, and budget.

The only new information required to generate the cash requirements curve (beyond the original schedule and budget) are the terms and conditions of the subcontracts and purchase orders and the selection of the resources to be used to perform work.

## SIMPLICITY REQUIRED

For simplicity in generating the planned cash requirements curve, several minor details are ignored. For example, payroll costs should be plotted on a gross basis noting that there might be some minor time differences between the funding of net payroll and the funding of the employee and owner's contribution to payroll insurance and taxes. It is, of course, possible to plot the labor cost as the net payroll to be funded weekly, then to identify the payment periods for payroll insurances and taxes and to plot those as separate increments in the cash requirements curve.

Another minor element, as already noted, concerns payment discounts. These are technically not cash items though they reduce cash required. Since they are included in job cost and the discounts normally taken as earned income in other accounting areas, the earned discounts may be ignored for all practical purposes in generating the cash requirements curve.

It is, of course, possible to make the plotting of the cash requirements curve as technically precise as the user may desire. For example, the weekly labor cost may be plotted as a histogram rather than plotted as a sloped curve from week to week. This, in fact, may be a useful detail—a user's choice item—where the project under consideration contains a proportionately large amount of payroll. However implemented, the purpose of generating the cash requirements curve is to provide the maximum useful information in relationship to the amount of clerical effort required to provide that information.

Since the other elements of the plotting of the cash requirements curve are significant with respect to cash demands vs. anticipated cash income, it is best to stay with the four different kinds of demands for cash already discussed. There is, for example, a temptation to combine supplier payments with principal vendor payments since these are similar. However, the time phase difference between the two is sufficient of itself to warrant keeping the items separate. Supplier payments are tied to a calendar date, such as the first of the month, and normally occur *before* cash income may be expected. The principal vendor items, however, are usually tied to payment *after* cash income is received. The supplier items, therefore, are likely to cause a cash drain prior to receipt of income whereas the principal vendor items are specifically tied to funding after income is received.

## THE CUMULATIVE CASH REQUIREMENTS CURVE

The general shape of the planned cash requirements curve, as shown in Exhibit 12-4, is a series of similar and repetitive monthly cycles, consisting of a sloping week to week labor requirement, a vertical jump for the supplier payments about the first of the following month, a continuing slope for a period accounting for the continuing demands for cash funding of labor to another vertical jump occurring between the 15th and 20th of the month representing the combined principal vendor invoices and subcontractor requisitions net of retainage. These cycles repeat until substantial completion occurs, at which time the release of retainages creates another vertical jump in the curve with another minor jump occurring at final payment. To repeat, this is the *cumulative* cash requirements curve.

## MONTHLY CASH REQUIREMENTS CURVE

The *monthly* curve may also be plotted depending on the user's need for information and other requirements. Exhibit 12-5 shows the monthly cash requirements curve. It, along with and when compared to the monthly cash income curve, provides visual graphic information on net cash position.

### Similarities Between Cash Income and Cash Requirements

A monthly cash requirements curve (Exhibit 12-5) may be plotted from the tabular data or prepared by bringing the beginning of each pay period to the base line of zero.

The monthly tabular cash requirements (total payment cash required line, Exhibit 12-4) is especially valuable in making extended (six months or more) cash analyses. When combined with cash income, extended net cash projections may be made. The planned monthly cash income curve is also shown in Exhibit 12-5.

**Exhibit 12-5. Monthly Cash Requirements Curve and Monthly Cash Income Curve**

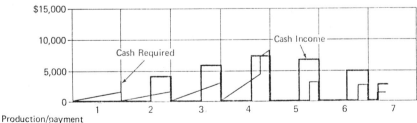

It is worth noting that when the monthly peaks of cash requirements curve are plotted, the result is the same general shape as the cost curve displaced below and to the right of the cost curve, extending beyond the end of the cost curve because of the time difference between the two. This simply means that the cash required to fund costs lags behind the occurrence of those costs by an average of about 30 days. The exact lag between costs and cash requirements is difficult to identify more precisely because the cost of work includes retainages normally held throughout the life of the job.

For example, the retainages from the first month of a subcontractor's work for a 12 month project may not be released until the thirteenth month of the project. The time phase relationship between incurring the cost of the work and the cash requirements to pay for that work is less important than the relationship between the cash requirements for the work and the cash *income*. This last relationship—between cash income and cash requirements— is the one of vital interest and is the foundation of valid cash management strategies both for individual projects and for all projects within a business enterprise.

## Uses for Planned Cash Requirements Information

The planned cash requirements curve has its important uses. It is, of course, instrumental in predicting the cash demands which will be made upon the cash resources of an organization resulting from field production. Valid information concerning cash requirements is essential to any sensible cash management strategy and is particularly useful where the cash requirements of all projects are summarized and calendarized. The time phase difference between the production curve and cash requirements is also interesting since the delta between a production curve and the cash requirements curve represents assets being generated as a result of ongoing production.

Perhaps the most interesting and exciting use of the cash requirements curve occurs in conjunction with income requirements. The planned schedule and budget for the project, along with the terms and conditions of payment of the contract and of the purchase orders and subcontracts, further determines the shape of these curves. When the planned cash income curve and the planned cash requirements curve are made on transparency overlays, and prepared on the same time scale network for ready comparison, the most interesting and vital cash management information becomes instantly available to management. The primary focus of interest is the delta, or increment, represented by the vertical displacement between the two curves, otherwise known as the *cash delta*. This is such an important aspect of the financial management of a project that it is dealt with in detail along with production forecasting and cash analysis in Chapter 13.

## SUMMARY

The cash requirements resulting from field production are of special interest to management. Preparation of an accurate cash requirements curve requires information from the project schedule and budget. It also requires specific information concerning the resources to be used and the terms and conditions from the subcontracts and purchase orders which control both the amounts of payments and when those payments will be made. Since the point at which payments must be made is related to the resources to be used in performing the work some predetermination of resources as reflected by cost types must also be made.

For convenience, both when payments must be made and how those amount of payments are determined may be grouped into four general categories. Group I concerns those items funded on a weekly basis. Group II covers those items funded on a monthly basis but unrelated to payment by the owner to the contractor. Group III are those items for which the time of payment is controlled in turn by payment from the owner to the general contractor less a retainage. Group IV is also controlled by payments from the owner to the general contractor but without consideration of retainages.

To distinguish items such as labor requiring weekly funding, they are plotted in the cash requirements curve as sloping lines. The remaining items are reflected in the cost curve as vertical lines representing the periodic occurrence of major cash requirements.

Plotting the cash requirements curve is greatly facilitated by the creation of tabular data, which is presented in two different ways. The first set of data relates cash requirements to when they are *incurred;* the second set relates cash requirements to when they must be *paid.* The latter data is used in plotting the cash requirements curve.

The cash requirements curve presents useful data to management when examined by itself. The information provided when the cash requirements curve is compared to the cash income curve is an essential cash management tool. The cash delta, or vertical distance between the cash income curve and the cash requirements curve, and which may be positive or negative, indicates the exact cash position planned for the job at any point in time.

Since there are significant differences in the cash demands placed upon a project according to the resources used to perform the work, a major task of cash management is the selecting of resources to be used.

Detailed cash analyses and six month cash projections are possible for individual jobs and for multiple jobs within a business environment.

Other items of managerial interest, but less vital to cash management, are the current assets represented by the assets delta between the cash requirements curve and the production curve. The asset delta represents the net value of work in place at any point in time in a project at which offsetting accounts receivable have not been generated.

In sum, the cash requirements curve can predict the cash required for the performance of the work. It is derived from the primary performance models of budget and schedule modified by (1) time related payment requirements associated with resource type (i.e. lab, sub, etc.) and (2) the subcontract and purchase order terms and conditions of payment. There is no other source available to management on the cash consequences of production.

# 13
# PRODUCTION FORECASTING AND ANALYSIS AND CASH ANALYSIS

Chapters 7 through 12 have dealt in detail with a number of performance models which are related to the construction of a project. These performance models, taken together, provide extremely important information concerning the overall planned performance of the project. The kind of information provided and the variety of ways in which it may be utilized is of particular interest in terms of project management. Before proceeding with the general topic of production forecasting and cash analysis using these performance models, it will be helpful to review each of the performance models to summarize the kind of information each provides.

Seven performance models have been discussed (eight if we consider the plan or strategy for the project separate from the schedule). The plan or strategy for the project is independent of schedule. The schedule, however, is a function of the plan and the rate of application of resource. For this reason the plan and schedule will be considered in a single context.

## PRESENTATION OF PERFORMANCE MODEL INFORMATION

The information which has been developed in the course of preparing the performance models is presented in two general forms. The first of these is tabular, or listing of information and data. The budget as a primary performance model is normally presented in tabular form since it deals primarily with activity rosters and the cost associated with each. The second form is graphic. The plan and schedule for the project are normally presented in graphic form since it is necessary both to establish a time scale and to relate the activities to it, done most effectively as a plotted image. The derived performance models which are prepared for integrated cost and schedule control purposes are generally two dimensional in nature, dealing with (1) activities distributed along the time scale and (2) the values associated with those activities represented along the vertical scale, as both dollar amounts and percentages of the job.

The chapters dealing with the preparation of the derived performance models present the data and information concerning each performance model in both tabular and graphic form. The uses of the information from the performance models can, in some instances, be presented more effectively as tabular data. Examples of the effectiveness of tabular data for comparisons abound in both production forecasting and cash analysis. An effective use of graphics is establishing the cash delta by comparing the cash income curve and the cash requirements curve.

The preparation of the graphic presentations for the derived performance model is greatly facilitated by the tabular data. As a matter of convenience, the tabular data is first prepared; from it, the graphic presentations are prepared. In this regard, it is worth noting that the use of standard time scales and standard value scales for all performance models allows the creation of overlays for quick and convenient comparisons of performance models.

## PRIMARY PERFORMANCE MODELS

There are two primary performance models from which all others are derived. They are:

1. The plan and schedule for the project.
2. The budget for the project.

### The Plan and Schedule

The plan and schedule for the project is, first of all, the strategy for the project. These present information on how the project is to be built, the sequence and interrelationship of activities, and the planner's choices where the relationships between activities are not physically constrained. Once the plan or strategy is in place, the duration of each activity is determined by identifying the resource to be utilized in the performance of the activity and the rate of application of the resource. The rate of application for a field production activity is controlled by the selection of crew size to be used in the performance of that activity. Once the crew size is determined, duration may be calculated by dividing the dollar value of the labor component by the periodic (daily, weekly, etc.) cost of the work force to be utilized in performing that labor. The result of this calculation is a duration for each activity.

Once the duration of each activity in the project plan has been set, the time boundaries for each activity may be arithmetically calculated using the critical path method of analysis. The goal is to convert the CPM logic network to a time scale network along which the activities making up the project are distributed. The time scale network is then the primary vehicle for the distribution of the budgeted cost (and all other values for each activity for the project) throughout the life of the project. In essence, development of the plan for the project, the determination of resources required, and the rate of application of those resources, allow the generation of a time scale distribution of those activities, essential to the preparation of all derived performance models.

## The Cost Budget

The second primary performance model is the cost budget. (In parallel, we could call the schedule the *time* budget.) When the cost budget, or simply budget, is applied to the time scale network, the costs to be incurred in the performance of the individual activities are distributed throughout the project, from the origin to the objective. In applying the notational systems previously discussed to each activity in the time scale network, several of the remaining derived performance models may be generated. The cost of each activity is initially developed in the project estimate and reassigned in the process of converting the estimate to a *budget*.

## Budget and Schedule Establish Other Models

When the primary performance models for any construction project have been set, the planned performance of the project is established. But the two primary performance models (budget and schedule) do not provide direct access to other performance information in several vital areas. It is necessary, therefore, to employ some special graphic techniques to create this additional information using the primary performance models and other data sources.

The primary performance models along with the cost curve and Schedule of Values curve distributes values along the time scale. The prime contract terms establish both how much of that value may be received (cash income) and when receipt may be expected. Terms of payment in the subcontracts and purchase orders establish how much of that same value must be paid out (cash requirements) and when payment is required.

All of the remaining derived performance models can be quickly and conveniently generated with this information in hand. When project planning precedes awarding of subcontracts, resource selection and usual terms and conditions allow the process to proceed nonetheless.

## THE DERIVED PERFORMANCE MODELS

The derived performance models examine specific details of the project resulting from the primary performance models. The derived models of interest are:

1. The cost curve ("S" curve).
2. The production curve ("S" curve).
3. The schedule of values curve (modified "S" curve).
4. The cash income curve (histogram).
5. The cash requirement's curve (modified histogram).

In each case, the information provided may be shown graphically or in tabular form. In either form, data may be summarized for specific time intervals such as daily, weekly, or monthly. Such data may also be summarized on a cumulative basis showing the effect of each time increment when added to the previous totals. The simplicity and convenience of such a presentation permits the selection of the graphic form and tabular summary best suited to the information required. These provide an effective basis for production forecasting and cash analysis.

## The Cost Curve

The cost curve is the simplest of the performance models to develop and shows in both graphic and tabular form the planned expenditures for the project. The cost curve is plotted directly from the time scale network and the resource information provided in the budget.

## The Production Curve

The production curve is closely related to the cost curve and differs only to the extent that the indirect costs and overhead and fee have been added to the cost curve. Its primary significance is that it represents the contract value of product in place and may be used for production forecasting.

## The Schedule of Values Curve

The Schedule of Values curve, though similar to the production curve, has several significant differences. First is that certain items such as indirect costs and overhead and fee may *not* be shown as separate items in the Schedule of Values. The second difference is that, for cash management, the Schedule of Values may be prepared by using an unequal distribution of the contract value for the activities making up the project. The third difference between the production curve and the Schedule of Values curve is found in the summary levels. The production curve contains all activities. The Schedule of Values also contains all items but these are usually grouped or summarized into work packages or into trade groupings such as 16 standard divisions of the work. The Schedule of Values for any contract, when approved, becomes the basis of payment and is, therefore, of particular interest in cash analysis.

### Other Data Required for Cash Analysis

The derived performance models discussed thus far require no information beyond that already provided in the primary performance models. But the remaining two, cash income and cash requirements, require additional information concerning contract terms and conditions of payment and subcontract and purchase order terms and conditions of payment. The cash income curve is based on when an owner will pay and how much will be paid. The cash requirements curve is determined by selection of resources (i.e. lab subs, etc), how much will be paid and when these resources must be paid.

### The Cash Income Curve

The cash income curve is closely related to the Schedule of Values except that it is modified by the terms and conditions of payment stipulated in the contract. These modifications include time of payment, certain retainages, holdbacks, and other adjustments which tend to reduce the actual cash income anticipated as a result of the ongoing production. Since most payments on construction contracts are made based upon monthly increments, the cash income curve takes on the appearance of a histogram reflecting these once a month cash income payments.

### The Cash Requirements Curve

The cash requirements curve provides information on when cash is required to fund payment for the cost incurred. The increments of cash demand tend to follow a weekly/monthly cycle throughout the life of the project, with special conditions occurring as the project nears completion related to the payment of retainages and holdbacks.

## PERFORMANCE MODEL COMPARISONS

Each of the seven performance models discussed thus far is of interest and value taken by itself. As useful as the individual performance models may be, however, maximum information concerning expected performance of the project is created when *comparisons* are made between the performance models. There are several ways of making these comparisons. The most effective is to prepare the individual performance models on transparencies or on some other visible media that allows each to be overlaid with the other for quick visual reference and comparison. These comparisons will be reviewed in detail in looking at production forecasting and cash analysis. Several observations at this point, however, are worthwhile.

A comparison of the cost curve to the production curve clearly indicates the growth of overhead and fee throughout the life of the project. The overhead and fee increment is directly related to the rate of production, which may vary only on a month to month basis. Comparing the production curve to the Schedule of Values curve will clearly indicate the weighting of the Schedule of Values via the differences between the two curves. When the Schedule of Values curve is above and to the left of the production curve, the Schedule of Values is weighted in favor of the early items. A Schedule of Values curve below and to the right of the production curve would indicate that the proposed billings for the project would fall short of the value of actual production in place. The cost curve may also be compared to the Schedule of Values curve, to indicate the relationship between the cost of the work and the value of that work to be used in determining the accounts receivable. These comparisons do not, however, relate to *cash* income or *cash* requirements.

## CASH POSITION

Of primary interest to management is hard data on the planned cash position of the project. The cash income curve takes into account the contract terms and conditions of payment. When it is compared to the production curve, the difference between the two curves represents the value of product in place which is not funded currently.

### Cash Income and Cash Requirements

From a cash management point of view, the most interesting comparison occurs between the cash income curve and the cash requirements curve. This comparison provides direct information at any point throughout the duration of the project on the cash position of the project. The cash delta, or increment, between the two may be positive or negative. For example, during the start-up phase of a project, the cash delta will almost certainly be negative since funding will be required for certain start-up costs and payroll during the initial weeks of the project. The cash funding for these items will not occur until the first payment is received, perhaps six weeks later. It is interesting to note that cash position of the project is established as (1) a result of the primary performance models and (2) the terms and conditions of payment for labor, subcontractors, and vendors. When the resulting cash delta is examined, alternate strategies for the performance of the work may be considered to minimize negative cash flow. In any event, comparisons between the derived performance models becomes essential management information which each manager may employ in a variety of ways.

## PRODUCTION FORECASTING AND ANALYSIS

Two important uses of the performance models require detailed consideration. The first of these is production forecasting and analysis; the second is cash analysis. Most construction business organizations have established an optimum level of production represented by a number of jobs or the dollar value of product to be placed weekly, monthly, quarterly, or annually. This information may take the form of elaborately prepared production goals and objectives, or an unwritten feeling or sense about the organization's capacity to do work.

In either case, management information which may provide some clear indication of the consequence of projects underway can be very useful. This kind of information is especially valuable as a sales and marketing tool. When production can be predicted, production overloads can be anticipated and managed. Similarly, forecasting can be used to plan sales and marketing strategies and to provide the horizon of visibility sufficient to allow specific action to be taken. Since the ability to obtain information and to examine the consequence of production is essential to sound business management, the techniques for production forecasting using the primary and derived performance models will be developed in some detail.

### Single Job Summary

Production forecasting and analysis involve a relatively simple utilization of the data provided by the production curve and the cost curve. The indicated monthly production for an individual job is the monthly total of the contract value of work in place between the billing cut-off periods. If the billing period is monthly, the contract value of production for an individual project is the value of the work placed during that period read directly from the production curve or tabular data. It is unnecessary to manipulate the data to arrive at this information.

Exhibit 13-1 shows the form to be utilized for both the single job summary and the multiple job summaries. The data used to prepare the production forecasting analysis for Job 1 is taken directly from the sample job and from the tabular data provided in Exhibit 10-3.

The production forecast and analysis for Job 1 presents in tabular form interesting information resulting from the plan and schedule and the budget for the sample job prepared earlier. Prior to a discussion of the specific production forecast and analysis, the general format of the tabular data should be reviewed. In Exhibit 13-1, each billing period is represented by a dual column, 1 through 7. In each case, the lefthand column is titled "Production"; the righthand column is titled "Cost." Production means the value of product

## Exhibit 13-1. Production Forecast, Single Job with Monthly and Cumulative Totals

| Production period / Month | 1 APRIL (PREV.) | | 2 MAY (CURRENT) | | 3 JUNE | | 4 JULY | | 5 AUGUST | | 6 SEPTEMBER | | 7 OCTOBER | |
|---|---|---|---|---|---|---|---|---|---|---|---|---|---|---|
| Job 1 | PROD. | COST | PROD. | COST | PROD. | COST | PROD. | COST | PROD. | COST | PROD. | COST | PROD. | COST |
| Monthly totals | $4,043 | $3,500 | $4,620 | $4,000 | $8,085 | $7,000 | $9,240 | $8,000 | $2,887 | $2,500 | $578 | $500 | — | — |
| Monthly net | 543 | | 620 | | 1,085 | | 1,240 | | 387 | | 78 | | | |
| Cumulative total | $4,043 | $3,500 | $8,663 | 7,500 | $16,748 | 14,500 | $25,988 | 22,500 | $28,875 | 25,000 | $29,453 | $25,500 | | |
| Cumulative net | 543 | | 1,163 | | 2,248 | | 3,488 | | 3,875 | | 3,953 | | | |

## Exhibit 13-2. Production Forecast, Multiple Jobs with Monthly and Cumulative Totals

| ($IN 1,000s) | 1 APRIL (PREV.) | | 2 MAY (CURRENT) | | 3 JUNE | | 4 JULY | | 5 AUGUST | | 6 SEPTEMBER | | 7 OCTOBER | |
|---|---|---|---|---|---|---|---|---|---|---|---|---|---|---|
| | PROD. | COST | PROD. | COST | PROD. | COST | PROD. | COST | PROD. | COST | PROD. | COST | PROD. | COST |
| Job 2 (1,000) @ 10% | $100 | $ 91 | $ 300 | $ 273 | 350 | $ 318 | 130 | $ 118 | 70 | $ 64 | 50 | $ 45 | — | $ — |
| Job 3 (5,000) @ 10% | 500 | 455 | 1,500 | 1,364 | 1,750 | 1,591 | 650 | 591 | 350 | 318 | 250 | 227 | 210 | 194 |
| Job 4 (3,000) @ 8% | | | | | 300 | 278 | 900 | 833 | 1,050 | 972 | 390 | 361 | 210 | 194 |
| Job 5 (2,500) @ 6% | | | | | | | 250 | 236 | 750 | 708 | 875 | 825 | 325 | 307 |
| Job 6 (8,000) @ 5% | | | | | | | | | 800 | 762 | 2,400 | 2,286 | 2,800 | 2,667 |
| Monthly totals | $600 | $546 | $1,800 | $1,637 | $2,400 | $2,187 | $1,930 | $1,778 | $3,020 | $2,824 | $3,965 | $3,744 | $3,335 | $3,168 |
| Monthly net before O.H. | 54 | | 163 | | 213 | | 152 | | 196 | | 221 | | 167 | |
| Monthly O.H. | | 100 | | 100 | | 100 | | 100 | | 100 | | 100 | | 100 |
| Monthly net income Δ | 46 | | 63 | | 113 | | 52 | | 96 | | 121 | | 67 | |
| Cumulative totals | $600 | $546 | $2,400 | $2,183 | $4,800 | $4,370 | $6,730 | $6,148 | $9,750 | $8,972 | $13,715 | $12,716 | $17,050 | $15,884 |
| Cum. net before O.H. | 54 | | 217 | | 430 | | 582 | | 778 | | 999 | | 1,166 | |
| Cum O.H. | | 100 | | 200 | | 300 | | 400 | | 500 | | 600 | | 700 |
| Cumulative net income Δ | 46 | | 17 | | 130 | | 182 | | 278 | | 399 | | 466 | |

planned to be placed in relationship to the contract amount. Cost means the cost planned to be incurred to produce that product within a specific billing period. There are several different descriptions which may be used for these two columns. Production may also be considered to be *planned income*. Cost could be considered to be *planned expenses*. The relationship between these two within any given billing period is especially important. The composition of the production value and the composition of the cost should match exactly. Keep in mind, however, that production forecasting and analysis is a significant management information tool rather than an accounting tool.

Exhibit 13-1 shows, on separate lines, monthly totals for both production and cost, the net difference between production and cost, and the cumulative net of these differences. There are, of course, many other totals and combinations which can be prepared once the basic information is provided. These will be added and the material presented in multijob and company summaries.

## Multijob Production Forecast Summaries

The individual job summary presented contains a wealth of information resulting from the planned production of a single job. From a management point of view, the multijob production forecast summaries become even more interesting. The data generated for each individual job production curve and cost curve is tabulated in a composite summary for all jobs in progress. An example of a multijob production forecast tabulation and analysis is shown in Exhibit 13-2.

To demonstrate the ways in which the information for production forecasting is accumulated, five hypothetical jobs titled Jobs 2 through 6 have been set up. Exhibit 13-3 provides the essential information required to create the tabular data.

In the absence of a production curve and cost curve for each of the jobs, certain minimum information is required to create the equivalent information.

### Exhibit 13-3. Job Related Data

| JOB | CONTRACT AMT. | CONTRACT TIME | OH & FEE (MARKUP) | START PERIOD | PERCENT ASSUMED FOR MONTHLY DIST. | | | | | | TOTAL |
|---|---|---|---|---|---|---|---|---|---|---|---|
| | | | | | 1 | 2 | 3 | 4 | 5 | 6 | |
| Job 2 | $1,000,000 | Six Months | 10.0% | 1 | 10 | 30 | 35 | 13 | 7 | 5 | 100% |
| Job 3 | 5,000,000 | Six Months | 10.0% | 1 | 10 | 30 | 35 | 13 | 7 | 5 | 100% |
| Job 4 | 3,000,000 | Six Months | 8.0% | 3 | 10 | 30 | 35 | 13 | 7 | 5 | 100% |
| Job 5 | 2,500,000 | Six Months | 6.0% | 4 | 10 | 30 | 35 | 13 | 7 | 5 | 100% |
| Job 6 | 8,000,000 | Six Months | 5.0% | 5 | 10 | 30 | 35 | 13 | 7 | 5 | 100% |

These include job identification, contract amount, contract time, overhead and fee or markup, and an assumed percentage monthly distribution of both production and cost. The assumed percentage monthly distribution over a six month period is the approximate distribution from a high production "S" curve.

Exhibit 13-2 presents these five jobs as tabular data with a start time indicated for each. Since an overhead and fee, or markup, has been assumed for each of the jobs, an arithmetic relationship is established between production and cost for tabular purposes. The cost in each case for each monthly period is calculated by the formula

$$\text{Cost} = \text{production value} \div (1 + \text{decimal markup})$$

For actual jobs, both the production curve and the cost curve would be available directly from the tabular data. After the production value and cost have been distributed for each job, several kinds of summaries may be created. The first of these is the monthly totals which indicate the planned production value for all jobs on a monthly or production basis. Production less cost is also calculated for each period. This figure is especially important since it represents the amount available for funding overhead and other indirect costs. As a matter of convention, the production column is considered positive and the cost column negative in each monthly summary. To distinguish between positive and negative amounts, net of production less cost is indicated in the *production* column, where the amount is positive; or in the *cost* column, where it is negative.

Each organization will have knowledge concerning its overhead. The next data inserted in the monthly totals is the monthly overhead expense. Since this represents a cost item, it is included in the cost column. These three lines permit a fourth line indicating monthly net income to be calculated.

For production period 1, production less cost is $54,000, overhead is $100,000, but net income (here, a loss) is $46,000. Note that this is indicated in the Cost column since it is a negative figure. The net monthly income is calculated for the remaining periods.

In addition to the monthly totals, cumulative totals may also be prepared for both monthly production and cost figures. These are also shown in Exhibit 13-2, on the Cumulative totals line. The nets of production less cost are also shown. Note that in each case the positive net is shown under production, and the negative net under cost. Production less cost are also cumulative figures as is the overhead. The last line of the tabular data in Exhibit 13-2 is the cumulative net income.

The production forecast and analysis may be prepared for any time interval. Beginning balances may also be added to the tabular data to present current position information. The effect of a new job on overall production capability and resources may be determined by adding its production value and cost figures to the summary for either monthly or cumulative totals. Production forecasting and analysis also take into account cumulative net income expected as a result of production within a specific production period. It is necessary to recognize, however, that cumulative net income or monthly net totals *do not* present information on either cash income or cash requirements. These are the subject of a separate analysis which will be developed in detail.

Analytical tools which allow reasonably valid production forecasting and analysis are essential to sound business management. The assumption made in the tabular data presented in Exhibits 13-1 and 13-2 is that the production curve and the cost curve have been prepared as a result of the planning and budgeting processes for the job. Production forecasting and analysis, however, frequently require production information and related cost information earlier in the job than this data is available. There are, fortunately, processes which allow the generation of production and cost information when the foregoing data, and the resultant curves, are not available. The required data is a contract amount with the overhead and fee shown as a separate item and an expected contract time or duration for the project. The procedures to be described are the same procedures used in the preparation of the tabular data for Jobs 2 through 6 presented in Exhibit 13-2.

## The "S" Curve Distribution of Production Values and Cost

An "S" curve distribution of the percentage of work completed over the time scale provides a simple method of generating production value and cost information for any job. Several sources of information can be used for the distribution. Each company or organization has, as a result of its ongoing operation, a typical pattern of work performance represented by a percentage of completion on the vertical scale against a percentage of time consumed on the horizontal scale. Where such historic records have been maintained or can be conveniently created for the organization, this represents the most valid information. This first method comes from each company's historic records.

The sixth column from Exhibit 13-3 contains assumed percentage monthly distributions, based on past performance patterns for similar projects, for six month jobs. The percentages indicated are for each monthly interval. Each six month distribution can be used, noting that it represents a slightly optimistic production curve for the work. This illustrates a second method which, while less accurate, can be very useful when no other data is available.

A third method of distributing values and cost over time involves the use of a normal (bell shaped) distribution. The area under the curve is unity (1.0). Beginning at −3.0 standard deviation units assumed to be zero, the area under the curve for each standard deviation is calculated from the Areas Under the Standard Normal Distribution Curve Table. These incremental values are summed on a cumulative basis and these data are used to plot the "S" curve. The horizontal scale is divided into six standard deviation units [3.0 − (−3.0) = 6.0] and the cumulative values plotted against a vertical scale representing 1.0 (area under the curve) = 100.0%. The best graphic display results when the horizontal axis and the vertical axis are the same length (i.e. 6 inches). The vertical scale is divided into 10 equal parts, each representing 10.0% of value or cost. The six divisions of the horizontal axis primarily accommodate the six standard deviation units, and this axis, in addition, is also divided into 10 equal parts, each part representing 10.0% of time. When the cumulative area under the curve is plotted for each standard deviation against its decimal value on the vertical scale, the standard "S" curve results.

This cost/production curve, derived from a normal distribution, is shown in Exhibit 13-4 as Cost/production curve 3, and is, perhaps, the most likely distribution in the absence of other information. Exhibit 13-4 also shows cost curve data plotted from two other sources. Cost/production curve 1 is plotted from Exhibit 13-2 and represents a more optimistic projection of

**Exhibit 13-4. Cost/Production Curve as Percentages of Cost and Time**

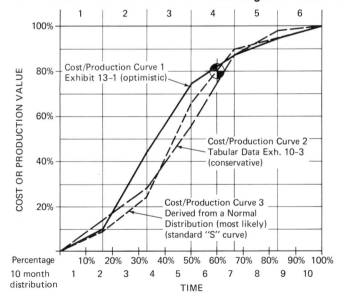

production. Cost/production curve 2 in Exhibit 13-4 is taken from the tabular data for the sample job used earlier and shown in Exhibit 10-3; it represents a more conservative projection of production.

## Using "S" Curves to Distribute Production Values and Costs

The three curves provided in Exhibit 13-4 as models will allow the distribution of both production values and costs for any job over its life on either a monthly or a cumulative basis when these distributions are needed but when the detailed plan schedule and budget have not been developed. The horizontal time scale is based on a six months' duration but with the total time equalling 100%. The six production months are indicated numerically across the top of Exhibit 13-4. The time across the bottom is shown as a percentage of the duration of the project.

To establish the appropriate time scale for some other job, the duration of the project is distributed along the percentage scale. For example, if a project has a planned completion time of ten months, each of the 10% intervals represents one production month. These intervals are used as the basis for obtaining the tabular data.

The cost or production value is plotted on the vertical scale and is similarly expressed as a percentage of either cost or production. The vertical scale may be converted to dollar values for a specific job. Both cost and production value are shown; however, the curve applies only to one or the other for a specific job. By definition, the cost curve plus the markup equals the production curve. In preparing tabular data, it is recommended that either production value or cost be graphically determined and the remaining value calculated based on the arithmetic relationship between the two. This arithmetic relationship is given by the formula

$$Cost + markup = production\ value$$

To illustrate the use of Exhibit 13-4 for determining production value and from it, cost of work, the following assumptions are made:

1. Production duration: 10 months.
2. Production value (contract amount): $1 million.
3. Markup over cost: 10%.

To determine the production value at the end of the six months, the following steps are used:

1. Convert the time scale to monthly production units. In this case, the project duration is 10 months, the end of the six months would be at 60% completion.

2. A selection is made concerning which of the three cost production curves are to be utilized, the one presenting the most optimistic production projection, the most conservative or the most likely. Assume the selection of the most likely.
3. The 10 month distribution has been indicated across the time scale. The end of the six months is identified which, for this illustration, coincides with the 60% time mark.
4. A vertical line is drawn from the end of the six months (at 60%) to intersect with cost/production curve 3, which represents the most likely distribution.
5. The intersection of cost/production curve 3 occurs at 80% of the production value. This means that the dollar value of production expected to be in place at the end of the six months is 80% or .8 times the contract amount of $1 million, or $800,000. Note that this is the cumulative production value through the end of the six months. Had the similar cumulative value been calculated at the end of the fifth month, the value of production for the sixth month is determined by deducting the value of work in place at the end of the fifth month from the value of work in place at the end of six months.
6. The expected cost equals the production value divided by the cost plus the markup. The cost plus the markup in this case is $1 + 10\%$, or 1.1.
7. Therefore, at the end of six months,

$$\text{Expected cost} = 800,000 \div /1.1 = \$727,272$$

When rounded to the nearest $1,000, the cost may be taken as $727,000 for the purpose of the production analysis. These processes are so simple and direct that there is no reason to be without an adequate production forecast and analysis even in the very earliest stages of a project, when only a gross parametric value may be available.

Production forecasting is directly related to product in place for a specific time interval. The production forecast for multiple projects in work may be extended over any selected period. The appropriate period is, however, related to the *average* duration of the projects being performed. A projection horizon equal to the average duration of a typical project may then be considered. In the absence of any other data, a six months' projection is recommended. Each project in work and each project under contract which is planned to start during the planning horizon, is listed and both production value and cost distributed throughout the planning horizon. These are totaled to indicate total production for each monthly interval. This portion of the analysis, however, is limited to projects under contract.

## Probability Analysis of Anticipated Construction

It is, on occasion, necessary or of interest to deal with projects which may not yet be under contract but which may be reasonably anticipated to start during the forecasting period. Some discretion, however, must be used in determining which proposed new projects not yet under contract should be added to the production forecast. While the implementation really depends on the needs of the individual organization, several suggestions may be made in forecasting production for jobs not yet under contract. The first of these is to determine a *probability level* to be used in adding potential new work to the production forecast. The probability of 0.8 would present an optimistic view of projected production; probability of 0.9 would present a conservative production forecast. The probability may be estimated, that is shrewdly guessed, of executing a contract for the work and of its startup within the forecasting time period. There is, however, a simple rational method which, adds some discipline to the process as compared with merely guessing at a probability factor. The rational method of estimating probability is shown in Exhibit 13-5.

The vertical component of the probability factor represents confidence in the owner/contractor relationship. For example, a newly introduced client may be rated at 0.1. A well-established client, for whom other work had been performed, may be rated at 0.8. The horizontal scale is the time scale from the first contact with respect to the project to its planned start date. For example, if the project is just in the beginning stages of planning, it may be viewed along the time scale at 0.1. If the project is nearing the completion of design, with pricing completed and financing in place, it may be rated at 0.9. The probability calculations are as follows:

$$\text{Owner factor} \times \text{time factor} = \text{probability}$$

For example, if the owner factor is 0.3 and the project is nearing the stages of final design completion at 0.8, the probability calculation produces the following result: $0.3 \times 0.8 = 0.24$. The contract award probability for this example, then, is 0.24.

As the probability threshold (such as 0.8 or 0.9) is selected for each projection, future projects not yet under contact may be added to the summary. It is recommended that actual projects in construction be summarized separately from anticipated new projects. The total projection may be shown as the total of projects under contract and projects anticipated. While the data generated by the production forecast may be shown in graphic form for certain special presentations, the tabular data is easy and simple to use.

Interesting management information may be derived by comparing the

**Exhibit 13-5. Contract Award Probability**

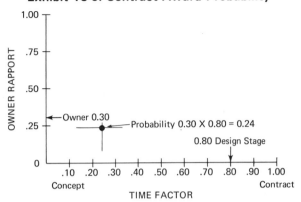

monthly totals of the production forecast to the monthly totals required to meet the production goals and objectives. The production level analysis (PLA) compares the forecasted production and the production goals and may be calculated for contracts in work as well as total forecast of production. The production level analysis is determined by the following simple calculation:

Production level analysis (PLA) = forecasted production/production goals

Assume a company's production goal is $2 million of product in place per month. Let's further assume that for a particular month the production forecast indicates production of $1.8 million. The production level analysis would be calculated as follows:

$$PLA = 1.8 \text{ million}/2 \text{ million} = 0.9$$

The result of this calculation (0.9) would clearly indicate that there is a production shortfall anticipated of 10%. Visibility to this problem several months in advance will allow adjustments to be made in the sales and marketing effort leading to the acquisition of new work to make up the shortfall. Information is immediately provided on when the shortfall may occur and the amount of the shortfall—extremely useful to management in maintaining effective business control of a construction organization.

## CASH MANAGEMENT INFORMATION

The long term health of the construction business enterprise is determined by production profitability and net income. The production forecast and analysis provides information on each. Short term consequences, however, are dictated by the issues of cash management, including cash income, cash requirements, net cash, and working capital requirements. It is essential that the construction business enterprise have good and valid information provided on a timely basis concerning the cash position of the company, opportunities to invest surplus cash profitably, and the availability of credit when cash requirements exceed cash available.

Construction is consistently among the highest industry business failures. Most of these can be attributed to under capitalization and/or poor cash management resulting from the absence of full, valid, and current information on cash position. Since cash position, cash flow, and cash management are of such vital concern and are so inseparably bound to the survival and prosperity of the construction enterprise, it is apparent that some effective cash analysis techniques must be developed to provide the kind of information required for firm cash management.

There is a tendency to think in terms of net cash as being equal to accounts receivable minus accounts payable. The factors, however, affecting net cash are far more complex than the accounts receivable and the accounts payable processes themselves would indicate. For example, the generation of accounts receivable related to production does *not* automatically result in providing the *cash* required to meet obligations for accounts payable and payroll. Accounts receivable must be *collected* to provide cash. Similarly, certain accounts payable may be managed in relationship to cash available. Some, however, such as payroll and insurances and payroll taxes, require cash for funding without regard to the status of accounts receivable. Also, unlike accounts receivable, which tend to concentrate during a relatively narrow time of the month, cash requirements tend to be distributed over the whole billing period and, in part, require funding prior to receipt of the related cash from accounts receivable. With the performance models already discussed in place, it is possible to generate the specific kinds of information which management requires to develop and pursue an effective cash management strategy.

## Cash Analysis

Detailed cash analysis may be made for any time period including one month, three months, six months, or a full year. Each company must determine the cash management planning horizon which will most effectively meet its requirements for management information. What is recommended here, however, is a six month cash analysis made up of one historic month, the current month, and a four months' analysis.

As discussed earlier, the cash income to be expected is not directly tied to the contract value of production but to the Schedule of Values established by most contracts for payment purposes. The cash income curve already discussed is, therefore, derived from the Schedule of Values since it provides for determining how much cash income may be expected. Similarly, the cash requirements are determined by the payment schedule which distributes the subcontract and vendor contract amounts through the activities which are to be performed by each resource. Since this detailed information may not be available at the time a cash analysis is to be made and since the purchased amounts will be similar to the budgeted amounts, the budgeted amounts may be used. The cash requirements, however, do not directly relate to when cost is incurred but rather when the cash will be required to fund those costs.

## Single Job Cash Analysis

The simpliest level of cash analysis may be made for the individual project. Assuming that the primary and derived performance models have already been prepared for a typical project, the cash analysis may be made by comparing the cash income to the cash requirements on a monthly basis. There are two methods of calculating the net cash requirements. The first is to deal with a calendar month and to take the data needed from the cash income curve and the cash requirements curve (or tabular data) at the end of each month. A second method is to deal with the cash requirements in terms of the production month and the receipts and disbursements associated with that production month. Either method will work. The former is normally more responsive to the actual dynamics in the construction environment. It maintains proper phasing and time relationships between cash income and cash requirements.

A tabular format is the most effective to generate cash information in completing the cash analysis. Once the monthly data had been selected and placed in the tabulation, it may be helpful to calculate the net cash position, also on a monthly basis. Exhibit 13-6 is a cash analysis tabulation for an individual job, showing monthly net cash.

The job is a sample job used earlier; the cash income data is taken from Exhibit 11-4, and the cash requirements data is from Exhibit 12-3. The format is essentially the same as for production forecasting using two columns for each month.

In cash analysis, the first month is noted as the previous month and should contain historic cash performance information. The second month should be the current month, for which the cash analysis should be most valid. Normally the projection would be carried for four additional months to extend for a total of six months. In the case of Job 1, however, the final payments and disbursements are scheduled for the seventh month. This month is therefore shown. As a rule, at least one additional month beyond the forecasting period is required to conclude the cash transactions associated with each job.

**Exhibit 13-6. Cash Analysis Showing Both**

| JOB 1 | 1 PREVIOUS APRIL | | 2 CURRENT MAY | | 3 JUNE | |
|---|---|---|---|---|---|---|
| CASH | RECEIPTS | DISBURS. | RECEIPTS | DISBURS. | RECEIPTS | DISBURS. |
| Monthly totals | — | $3,500 | $4,366 | $4,000 | $4,990 | $ 4,000 |
| Monthly net cash | | 3,500 | 366 | | 990 | |
| Cumulative totals | — | 3,500 | 4,366 | 7,500 | 9,356 | 11,500 |
| Cumulative net | | 3,500 | | 3,134 | | 2,144 |

Note that the terms "(cash) receipts and (cash) disbursements" are used in lieu of cash income and cash requirements. In calculating the monthly net cash for any month, the smaller number is deducted from the larger number. In the case of the cash shortfall for July, month 4, where the cash disbursements are planned to be $7,700 and the cash income $6,862, the net difference of $838 is shown under the disbursements column showing a cash demand for the project. Where the income is larger ($7,483 for month 5, August) and the cash requirements are smaller ($2,900), the net difference ($4,583) is then shown in the receipts column as positive cash generated from the project for that period. The analysis shown thus far, however, has not taken into account cash requirements other than those which are job related. Overhead and operating expense items are best held until a similar cash analysis is made for every project in construction.

## Multiple Job Cash Analysis

Most jobs to be included in the multiple job cash analysis will be contracts in work or contracts which have been executed and are planned to start. Under normal circumstances, all of these jobs will have been planned and scheduled, the formal budgets will have been prepared, and the cost curve and related production curve with tabular data will also have been prepared. Certain management decisions will have been made with respect to weighting of early items in the project. These decisions will be reflected in the Schedule of Values, which in turn will allow the preparation of the tabular data and Schedule of Values curve. With due regard to the contract terms and conditions which control the amount to be paid by the owner in relationship to the Schedule of Values of work in place, and when those payments may be expected, the cash income histogram will have been prepared along with the related tabular data. Finally, based upon the normal terms and conditions of the payment to subcontractors, material suppliers, and principal vendors, and the requirements for funding of payroll, the cash requirements curve along with the related tabular data will also have been prepared.

## Monthly and Cumulative Totals for a Single Job

| 4 | | 5 | | 6 | | 7 | |
|---|---|---|---|---|---|---|---|
| JULY | | AUGUST | | SEPTEMBER | | OCTOBER | |
| RECEIPTS | DISBURS. | RECEIPTS | DISBURS. | RECEIPTS | DISBURSE. | RECEIPTS | DISBURS. |
| $ 6,862 | $ 7,700 | $ 7,483 | $ 2,900 | $ 2,339 | $ 2,500 | $ 3,413 | $ 900 |
| | 838 | 4,583 | | | 161 | 2,513 | |
| 16,218 | 19,200 | 23,701 | 22,100 | 26,040 | 24,600 | 29,453 | 25,500 |
| | 2,982 | 1,601 | | 1,440 | | 3,953 | |

The normal preparation of a job for construction and the generation of the derived performance models create an abundance of data and information about the project. The data required for the cash analysis is immediately available and directly read from the cash income curve or tabular data, and the cash requirements curve or tabular data. For simplicity, the cash analysis data is taken in monthly intervals and shown as receipts and disbursements.

The tabular form used for cash analysis is essentially the same as the form used for the production forecasting and analysis. The added descriptive information, however, is helpful. The name of the calendar month may be inserted at the head of each column. A clear reference to the previous or historic month and the current month is also helpful. The use of the terms "receipts and disbursements" in lieu of cash income and cash requirements is more appropriate for the cash analysis format.

The cash income data comes directly from the cash income tabular data previously prepared. This data is inserted in the receipts column for the month in which payment is expected according to the terms and conditions of the contract. It is worth noting that the *receipt* of cash will usually be in the month following the actual *production* for which the income was earned. The careful tabulation of the cash income data, however, will routinely take the time lag in receipts into account.

The cash requirements data is also taken directly from the cash requirements tabular data and listed in the cash analysis form under disbursements. A note of caution concerns the relationship between expected receipt of cash and required disbursements. Where a cash requirement occurs earlier in the month than the cash receipts, that requirement should be included in the previous month since it must be funded out of cash which will be *received* in that previous month. For example, material suppliers are normally paid between the first and the tenth of the month so that payment discounts may be earned. Where payment is not anticipated from the owner until the 15th of the month, these items must be funded from cash receipts from the previous month and are, therefore, included in the previous month's totals.

## Rational Methods for Preparing Cash Income and Cash Requirements Data

The data required for cash analysis assumes that all of the preparatory work has been completed for each job during the job setup phase of the project. This may not always be the case, and on occasion it will be helpful to prepare cash analyses for tabular data when cash income and cash requirements are not available. In such cases, rational methods are available for creating cash data which is sufficiently accurate for cash analysis purposes. These techniques may also be used to complete cash analyses for projects not yet under contract to include them in production projections or in determining bid strategies.

There is certain information which must be provided, or which must be assumed, to prepare the raw data for cash analyses. This information includes the terms and conditions of payment in the general contract which establish both the amounts and times of payments resulting from field production. The information required also includes the normal terms and conditions of payment and funding for the four payment groups discussed earlier (pp. 151-158). Since the resource to be used controls the payment group for each item, some assumptions must be made concerning the distribution of production costs into the payment groups. Most organizations will have available, or can develop, a four payment group analysis of typical jobs which may be expressed as a percentage of production cost.

Job 2, included in Exhibit 13-4 and used as a part of a production forecasting analysis procedures, will be used to demonstrate methods of creating cash analysis data. These procedures may be somewhat tedious but are well worth the effort where valid cash information is required in the early stages of the development of a job.

Exhibit 13-7 provides initial information on Job 2. The contract amount is $1 million; the contract time is six months; and the markup over cost is 10%. The assumed percentage distribution for distribution of production

### Exhibit 13-7. Cost Distribution, Job 2

| PAYMENT GROUP | % P.C. | AMOUNT |
|---|---|---|
| I Labor | 15 | $ 136,350 |
| II Suppliers | 20 | 181,800 |
| III Subcontractors | 40 | 363,600 |
| IV Principal vendors | 25 | 227,250 |
| Product cost | 100 | $ 909,000 |
| Overhead and fee | 10 | 91,000 |
| Contract sum | 110 | $1,000,000 |

value and cost is shown in Exhibit 13-8. In the absence of other specific information, certain assumptions must be made to complete the analysis. These include the following distribution of production costs:

**Exhibit 13-8. Percentage Breakdown and Payment Terms, Job 2**

| ASSUMPTIONS FOR ANALYSIS | DISTRIB. |
| --- | --- |
| • Cash income or receipts by the 15th of the month | — |
| • Labor: 15% of production costs funded weekly | 15.0% |
| • Supplier: 20% of production costs funded by the 1st of the month | 20.0% |
| • Subcontractors: 40% of production costs funded by the 20th of the month with 10% retainage | 40.0% |
| • Principal vendors: 25% of the contract sum funded by the 20th of the month with no retainage | 25.0% |
| Total | 100.0% |

One additional calculation must be made before proceeding. The contract assumes a 10% markup. The cost of the work is calculated by dividing the contract sum by one plus the decimal equivalent of the markup:

Production cost = contract amount/1 + decimal markup
$$= \$1 \text{ million}/1.10 = \$909{,}090.90$$

For cash analysis purposes, all figures should be rounded initially to the nearest $1,000. The cost distribution for Job 2 is production costs, $909,000; overhead and fee, $91,000; total contract amount, $1 million.

The general form previously used in Exhibit 13-6 for a single job analysis is useful in preparing data required for cash analysis, allowing receipts and disbursements to be determined on a single form. The processes are, more specifically, a redistribution of production values and cost into cash income and cash requirements distributed on a monthly basis. Note especially that cash income is not based on production values but on scheduled values. It was pointed out earlier, however, that where no weighting or uneven distribution of values are made in the Schedule of Values, the schedule of values curve and the production curve are the same. Therefore production values may be used in place of scheduled values to determine cash income when no weighting factor is desired in the cash analysis. Using production values, however, gives a more conservative cash analysis.

Exhibit 13-9 shows the basic worksheet for the rational distribution of receipts and disbursements and cash analysis.

The first line entry for Job 2 is production/cost and is taken directly from

## Exhibit 13-9. Rational Distribution of Receipts and Disbursements and Cash Analysis

REDISTRIBUTION OF PRODUCTION AND COST INTO CASH INCOME AND CASH REQUIREMENTS ($1,000s)

| Job 2 (Exhibit 13-2) | 1 PROD | 1 COST | 2 PROD | 2 COST | 3 PROD | 3 COST | 4 PROD | 4 COST | 5 PROD | 5 COST | 6 PROD | 6 COST | 7 PROD | 7 COST |
|---|---|---|---|---|---|---|---|---|---|---|---|---|---|---|
| | 100.0¹ | 91.0² | 300 | 273 | 350 | 318 | 130 | 118 | 70 | 64 | 50 | 45 | — | — |
| | RECEIPTS | DISBURS. | RECEIPTS | DISBURS. | RECEIPTS | DISBURS. | RECEIPTS | DISBURS. | RECEIPTS | DISBURS. | RECEIPTS | DISBURS. | RECEIPTS | DISBURS. |
| Receipts—Prod. −10% | -0-¹ | | 90.0¹ | | 270.0 | | 315.0 | | 117.0 | | 63.0 | | 45.0 | |
| Ret. +10% | | | | | | | | | | | | | 100.0 | |
| Disbursements: | | | | | | | | | | | | | | |
| Group I Labor | | 13.7² | | 41.0 | | 47.7 | | 17.7⁴ | | 9.6 | | 6.8 | | |
| II Material | | 18.2² | | 54.6 | | 63.6 | | 23.6⁴ | | 12.8 | | 9.0 | | |
| III Subs | | | | 36.3² | | 109.1 | | 127.2 | | 47.2 | | 25.6 | | 18.0³ |
| Ret. -10% | | | | - 3.6³ | | - 10.9³ | | - 12.8³ | | - 4.7³ | | - 2.6³ | | - 1.8 |
| Net subs | | | | 32.7 | | 98.2 | | 114.4⁴ | | 42.5 | | 23.0 | | 16.2 |
| IV Vendor | | | | 22.8² | | 68.3 | | 79.5⁴ | | 29.5 | | 16.0 | | 11.2 |
| Pay sub retainage | | | | 91.0² | | 273.0 | | 318.0 | | 118.0 | | 64.0 | | 45.0 / 36.4³ |
| Monthly totals | -0- | 31.9 | 90.0 | 151.1 | 270.0 | 277.8 | 315.0 | 235.2⁴ | 117.0 | 94.4 | 63.0 | 54.8 | 145.0 | 63.8 |
| Monthly net cash | | 31.9 | | 61.1 | | 7.8 | 79.8 | | 22.6 | | 8.2 | | 81.2 | |
| Cumulative totals | -0- | 31.9 | 90.0 | 183.0 | 360.0 | 460.8 | 675.0 | 696.0 | 792.0 | 790.4 | 855.0 | 845.2 | 1000.0 | 909.0 |
| Cumulative net cash | | - 31.9 | | - 93.0 | | - 100.8 | | - 21.0 | 1.6 | | 9.8 | | 91.0 | |

Notes: 1. Receipts for period 1 equal period 1 production less 10% retainage (100% − 10% = 90%). The 90% is not received until period 2. The accumulating retainage while earned is not part of current cash income until paid at the end of the job.

2. Costs are distributed into the period when cash is required (31.9 in period 1 and 59.1 in period 2, or 91.0).

3. Sub retainage is a current cost but not a current cash requirement until it is paid at the end of the job.

4. Current period disbursements (cash requirements) are totaled for each period based on when cash will be required rather than when costs are incurred.

Exhibit 13-2. It indicates the production values and costs distributed over the life of the project.

## Distribution of Receipts

Cash income or receipts may be determined from production values by deducting the 10% retainage assumed for the contract. When payment may be expected is determined by the assumed point of payment of the 15th of the following month. The expected receipts for cash income for production period 1 are equal to $100,000 minus $10,000, or $90,000, which will not be received until the second production period. This means that no income may be expected in the first production month and the receipts are, therefore, zero. The receipts in the second production month will be limited to the cash income resulting from the first month's production. The $90,000 amount is, therefore, entered under receipts in the second production month. Similarly, the production value of work planned for the second month is $300,000 less 10%, or $270,000, which will be received in the third production month. Therefore, $270,000, is entered under receipts in the third production month. This process is continued one month beyond the planned production period since the payment for work in the last production period will be made the following month and since the retainages previously held will also then be released. These two last items, $45,000 and $100,000 (retainages), are therefore shown in the receipts column in the seventh production period. This process completes the distribution of receipts through the life of the project, the total of which is $1 million, equalling the contract sum.

## Distribution of Disbursements

Since the distribution relating to cash requirements or disbursements concerns both the amount to be paid and when it is to be paid, it is convenient to deal with the total production costs distributed into the four payment groups. The distribution will be that assumed earlier: 15% for labor, 20% for material suppliers, 40% for subcontractors, and 25% for principal vendors.

The planned costs for Job 2 in production period 1 is $91,000. Labor represents 15% of this amount, or $13,700. Since labor is funded weekly, it must be funded in full during the first production period. It is, therefore, recorded in the disbursements column as 13.7 in the first production period. The material suppliers represent 20% of the $91,000, or $18,200. Since these may be paid between the first and tenth of the following month but must be funded prior to the receipt of income, material suppliers are accounted for under disbursements in the first production period and are shown as 18.2 in the disbursements column.

Payment group III concerns the subcontractors. The assumption has been made that the subcontractors will be paid about the 20th of the month. The 20th was chosen since it is five days later than the expected payment from the owner. Subcontractors represent 40% of the $91,000, or $36,300,* which is to be paid for the first production period in the second production period, and are shown as 36.3 in the second period disbursements column. Subcontractors, however, are subject to a 10% retainage, which must be deducted to determine cash requirements. This retainage deduction of 3.6 results in a net due to the subs for the period, shown in the disbursements column as 32.7. The gross amount and the retainage amount for subs are shown in shaded areas in the tabulation to indicate that these are *not* added into the monthly totals.

Principal vendors, payment group IV, are the last to be considered. They also are to be paid on the 20th of the month following the production month. Payment, however, is the gross amount which is equal to 25% of the $91,000, or $22,800, shown under production period 2, disbursements column, as 22.8. The original $91,000 incurred during the production period may be verified by adding the figures which have been entered; this confirmation is shown in production period 2, disbursements column as check total 91.0.

In production periods 3 and 4, the cash requirements associated with the cost (disbursements) totaling $318,000 have been shaded in to clearly indicate how these figures are distributed for each monthly production period and cost over two payment periods.

The processes described are continued through the end of the production period and any subsequent payment periods. The next line entry is for the payment of subretainages, indicated in production period 7 in the disbursements column as $36,400.

The summary totals having the greatest interest for cash analysis are the monthly totals, the monthly net cash, the cumulative totals, and the cumulative net cash. The monthly totals are determined by adding the amounts in the receipts column and disbursements column for each of the payments periods. The net monthly cash is determined by deducting disbursements from receipts. When this result is negative, the amount is entered into the disbursements column. When receipts minus disbursements is a positive figure, as in the cash of payment period 4, the amount is entered in the receipts column. Cumulative totals are prepared by adding each subsequent month to the subtotal through the previous month. Cumulative net cash is calculated and recorded in the same way that monthly net cash is recorded.

---

* Actually $36,400 but included in period 2 case totals as 36.3 due to rounding of other items in the total of 91.0.

Several comments are of interest as part of the cash analysis for Job 2. This job will require cash input over the first three production periods accumulating to $100,800 or slightly more than 10% of the contract sum. Consequently, on a month by month basis the job will require cash input for more than 50% of its duration. Further, while the first positive cash on a monthly basis occurs in the fourth production period and is equal to $79,800, this is not sufficient to offset the accumulated cash deficits. As a result, a cumulative cash deficit is carried through the end of the fourth production period. The first positive cash on a cumulative basis occurs in the fifth production unit, or in the final third of the project.

## Major Factors Affecting Net Cash Position

There are two major factors which may be used to manage or manipulate net cash position on a specific job. The first of these concerns the selection of the resources and the resulting payment group into which these may fall. Only group III, subcontractor payments, relates directly to payments from the owner to the general contractor both in terms of the amount to be paid and when payments are to be made. These payment procedures for subcontractors maintain a balanced cash position. All other resources in one way or another will create a cash drain on the project. The most severe is labor because it must be funded on a weekly basis. In fact, with the assumptions made for Job 2, six weeks of labor must be funded before offsetting income is generated. The effect on net cash in jobs containing a higher percentage of labor is illustrated by the cash analysis for Job 1 shown in Exhibit 13-6. In Job 1, negative cash on a cumulative basis persists through the end of the fourth production period and, with the exception of the second and fifth months, continues on a monthly basis through the end of the sixth production period. Material suppliers and other items in payment group II require funding in advance of income especially if discounts are to be earned. Principal vendors included in payment group IV, while having payment tied to owner funding or cash income, are usually paid without deducting retainages. This means that payment group IV alone accounts for an accumulating 10% cash deficit relating to the amount of those payments. Liquidity and working capital, of course, need to be considered in developing cash management strategies, which in turn may, in fact, dictate the resources to be selected for the performance of the work.

The second major factor to be considered is the relative merit of weighting early project items as a cash management strategy. While this decision may involve certain basic business management philosophies, it is worth noting that in Job 1 the early items consisting of about 40% of the project are weighted by a hefty 20%. This weighting and the calculations associated with it are shown in Exhibits 11-1 and 11-2. This weighting, however, is still not sufficient to offset the cash deficits resulting from the relatively high percentage of labor (approximately 40%) and the normal cash deficit structure of typical construction contract payment procedures. The purpose in pointing out these variables is not to recommend a business management philosophy but to highlight the kinds of information which may be useful in dealing with the essential issues of cash management, net cash position and cash flow.

## The Rational Data for Multijob Cash Analysis

In order to use previous sample jobs numbered three through six to illustrate the multijob cash analysis, certain additional information needs to be calculated. Exhibit 13-10 shows the job cost breakdown between cost and overhead and fee previously discussed for Job 2.

The same method used to create the cash analysis information for Job 2 shown in Exhibit 13-8 could be used for each of the remaining jobs. However, since there are some commonalities, the information generated for Job 2 may be used to calculate the cash distributions for the remaining jobs. For Job 3, since the overhead and fee is the same as for Job 2, since the duration is the same, the simple ratio of job size may be used to calculate the tabular

### Exhibit 13-10. Job Cost Breakdown between Costs and Overhead and Fee

| | | OVERHEAD AND FEE | | | |
| JOB NO | CONTRACT AMOUNT | PERCENT | AMOUNT | COSTS | EXPECTED COMPLETION |
|--------|-----------------|---------|--------|-------|---------------------|
| 2 | $ 1,000,000 | 10.0 | $ 91,000 | $ 909,000 | 6th month |
| 3 | 5,000,000 | 10.0 | 455,000 | 4,565,000 | 6th month |
| 4 | 3,000,000 | 8.0 | 223,000 | 2,777,000 | 8th month |
| 5 | 2,500,000 | 6.0 | 142,000 | 2,358,000 | 9th month |
| 6 | 8,000,000 | 5.0 | 381,000 | 7,619,000 | 10th month |
| Totals | $19,500,000 | 7.1 | $1,292,000 | $18,208,000 | 10th month |

Other data, required for cash analysis:

1. Total duration of jobs, 10 months
2. Monthly overhead costs, $100,000/month
3. 10 months overhead = 10 × $100,000 = $1,000,000
4. Net cash after 10 months $1,292,000 − $1,000,000 = $292,000

cash receipts and disbursements data for Job 3 using a 5.000 multiplier with Job 2 data.

Job 4 is three times the size of Job 1. If the overhead and fee were the same, the multiplier which could be used would be 3.000. However, because of the difference in overhead and fee, a larger proportion of the contract sum must be distributed to the job cost portion of the contract. This modifier is equal to the ratio of one plus the markup for the two comparative jobs.

Cost multiplier = contract amount of Job B/contract amount of A,
  × cost factor of Contract A/cost factor of Contract B
      = $3 million/$1 million × 1.1/1.8 = 3.0555

Similar calculations of Job 5 result in a cost multiplier of 2.5943; and for Job 6, 8.3808. These multipliers are applied to cash receipts and cash disbursements for each production period for the remaining jobs. The data is tabulated as shown on Exhibit 13-11 with the appropriate shift reflecting the differences in scheduled start. To illustrate the cumulative effect of the multiple job cash analyses, these data are tabulated for 12 production periods. Normally, cash projections for routine management purposes would cover a six month or six production period interval. The first month would be the most recent historic month using historic data. The second month would be the current month with four additional months of projection.

Exhibit 13-11 provides an abundance of information concerning the expected performance of Jobs 2 through 6 including information on overall production, cash position, cash flow, and adjustments made for organizational overhead.

Exhibit 13-11 provides two sets of summary data, one for monthly totals, the second for cumulative totals. Both are of interest. Monthly totals provide direct information on both receipts and disbursements on a monthly basis. The difference between receipts and disbursements is monthly net cash before overhead. With an overhead factor inserted in the disbursements column for each month assumed at $100,000 per month, the net monthly cash can be calculated.

Cumulative receipts and disbursements totals progressing monthly are also shown; from these the cumulative net cash before overhead is calculated. From the monthly net cash before overhead, the cumulative overhead cost assumed at $100,000 per month is deducted, resulting in net cumulative cash for the period. The net cumulative cash for the period assumes a beginning balance of zero and provides clear exact data for the consequence of production and cash for the analysis period. Current cash can be calculated, however, by entering a positive or negative beginning balance and extending the cumulative figures throughout the 12 month production period.

# Exhibit 13-11. Multiple Job Cash Analysis Showing Both Monthly and Cumulative Totals and Net Cash

**Periods 1–6**

| PRODUCTION PERIOD (AMOUNT IN $1,000s) | 1 RECEIPTS | 1 DISBURS. | 2 RECEIPTS | 2 DISBURS. | 3 RECEIPTS | 3 DISBURS. | 4 RECEIPTS | 4 DISBURS. | 5 RECEIPTS | 5 DISBURS. | 6 RECEIPTS | 6 DISBURS. |
|---|---|---|---|---|---|---|---|---|---|---|---|---|
| Job #2 (1,000 10%) | — | 31.9 | 90.0 | 151.1 | 270.0 | 277.8 | 315.0 | 235.2 | 117.0 | 94.0 | 63.0 | 54.8 |
| 5.0000   3 (5,000 10%) | — | 159.5 | 450.0 | 755.5 | 1,350.0 | 1,389.0 | 1,575.0 | 1,176.0 | 585.0 | 470.0 | 315.0 | 274.0 |
| 2.5943   4 (3,000 8%) | — | — | — | — | — | 97.5 | 270.0 | 461.7 | 810.0 | 848.8 | 945.0 | 718.7 |
| 5 (2,500 6%) | — | — | — | — | — | — | — | 82.8 | 225.0 | 392.0 | 675.0 | 720.7 |
| 8.3808   6 (8,000 5%) | — | — | — | — | — | — | — | — | — | 267.3 | 720.0 | 1,266.3 |
| Monthly totals | — | 191.4 | 540.0 | 906.6 | 1,620.0 | 1,764.3 | 2,160.0 | 1,955.7 | 1,737.0 | 2,072.1 | 2,718.0 | 3,034.5 |
| Monthly cash before O.H. | — | 191.4 | — | 366.6 | — | 144.3 | 204.3 | — | — | 335.1 | — | 316.5 |
| less O.H. | — | 100.0 | — | 100.0 | — | 100.0 | — | 100.0 | — | 100.0 | — | 100.0 |
| Monthly net cash Δ | — | 291.4 | — | 466.6 | — | 244.3 | 104.3 | — | — | 435.1 | — | 416.5 |
| Cumulative totals | — | 191.4 | 540.0 | 1,098.0 | 2,160.0 | 2,862.3 | 4,320.0 | 4,818.0 | 6,057.0 | 6,890.1 | 8,775.0 | 9,924.6 |
| Cum. cash before O.H. | — | 191.4 | — | 558.0 | — | 702.3 | — | 498.0 | — | 833.1 | — | 1,149.6 |
| Cum O.H. | — | 100.0 | — | 200.0 | — | 300.0 | — | 400.0 | — | 500.0 | — | 600.0 (shaded) |
| Cum. net cash Δ | — | 291.4 | — | 758.0 | — | 1,002.3 | — | 898.0 | — | 1,333.1 | — | — |

**Periods 7–12**

| PRODUCTION PERIOD (AMOUNT IN $1,000s) | 7 RECEIPTS | 7 DISBURS. | 8 RECEIPTS | 8 DISBURS. | 9 RECEIPTS | 9 DISBURS. | 10 RECEIPTS | 10 DISBURS. | 11 RECEIPTS | 11 DISBURS. | 12 RECEIPTS | 12 DISBURS. |
|---|---|---|---|---|---|---|---|---|---|---|---|---|
| Job #2 (1,000 10%) | 145.0 | 63.8 | — | — | — | — | — | — | — | — | — | — |
| 5.0000   3 (5,000 10%) | 725.0 | 319.0 | 189.0 | 167.4 | — | 194.9 | — | — | — | — | — | — |
| 2.5943   4 (3,000 8%) | 351.0 | 287.2 | 292.5 | 243.9 | 435.0 | 142.2 | 362.5 | 165.5 | — | — | — | — |
| 5 (2,500 6%) | 787.5 | 610.2 | 2,520.0 | 1,971.2 | 157.5 | 787.8 | 504.0 | 459.3 | 1,160.0 | 542.8 | — | — |
| 8.3808   6 (8,000 5%) | 2,160.0 | 2,328.2 | — | — | 936.0 | — | — | — | — | — | — | — |
| Monthly totals | 4,168.5 | 3,608.4 | 3,001.5 | 2,382.5 | 1,528.5 | 1,124.9 | 866.5 | 624.8 | 1,160.0 | 542.8 | — | — |
| Monthly cash before O.H. | 560.1 | — | 619.0 | — | 403.6 | — | 241.7 | — | 625.0 | — | — | — |
| less O.H. | — | 100.0 | — | 100.0 | — | 100.0 | — | 100.0 | — | 100.0 | — | 100.0 |
| Monthly net cash Δ | 460.1 | — | 519.0 | — | 303.6 | — | 141.7 | — | 525.0 | — | — | — |
| Cumulative totals | (shaded) | 13,533.0 | 15,945.0 | 15,915.5 | 17,473.5 | 17,040.4 | 18,340.0 | 17,665.2 | 19,500.0 | 18,208.0 | — | — |
| Cum. cash before O.H. | — | 589.5 | 29.5 | — | 433.1 | — | 674.8 | — | 1,292.0 | — | 1,292.0 | — |
| Cum O.H. | — | 700.0 | — | 800.0 | — | 900.0 | — | 1,000.0 | — | 1,100.0 | — | 1,200.0 |
| Cum. net cash Δ | — | 1,289.6 | — | 770.5 | — | 466.9 | — | 325.2 | 192.0 | — | 92.0 | — |

## Observations and Comments from Multiple Job Cash Analysis

The best source of raw data for tabular cash analysis is the tabular data used for the cash income curve and the cash requirements curve for each job. When this data is not available, valid cash analysis data can be generated based on organizational historic performance and/or by making certain assumptions and by using distribution models (Exhibit 13-4) which will be close to expected performance.

The data generated for use in Exhibit 13-10 assumes no weighting of the production values to obtain the scheduled values which are the basis for cash income. Assuming, then, that the distribution of cost into the four payment groups is essentially valid for a particular contractor's operations (plus or minus 5%), it is possible to generate a standard cash income curve and a standard cash requirements curve for that contractor from the data by using the procedures employed in developing Exhibit 13-4, Cost/Production Curve as a Percentage of Cost and Time. Separate curves would be developed for cash income and for cash requirements, both plotted on the same graph using the same scales. The horizontal and vertical scales utilized in Exhibit 13-4 expressing values as percentages rather than time or cost units provide the standard reference for extrapolating data for any similar job. The further development of these standard cash income and cash requirements curves is left to each user since these are unique to the way in which each contractor subcontracts and buys work.

Since the data and analysis contain in Exhibit 13-10 has not been distorted by weighting factors and since the distribution into payment groups is reasonable, several significant conclusions may be drawn. The $19,500,000 total of five jobs extending over a period of ten months indicates an average monthly product placement of nearly $2 million per month. The rate of placement, however, varies from $540,000 in the second month to $4,168,500 in the seventh month. A nearly eightfold growth rate of production capacity in the six month interval would be difficult to manage. The effect on cash flow and profitability would also be difficult to manage. The overhead amount of $100,000 per month is appropriate for the *average* monthly production of $2 million but excessive for $540,000 and probably inadequate for the maximum production which is equivalent to an annualized rate of $50 million. Perhaps the statistic of greatest concern is that, with minor exception of the fourth month, all months through six operate at a monthly cash deficit.

From a *cumulative* point of view, the cash deficit continues unabated until the eleventh month, reaching a maximum of $1,749,600 (shaded cumulative total period 6 disbursements) in the sixth month. When compared to receipts resulting from the first six months of production of $12,943,500 (shaded cumulative receipts, period 7), net of the 10% earned retainage, the cash shortfall resulting from the business operation is a "whopping" 13.5%.

The cash analysis of an individual job or set of jobs is also interesting when compared to less formalized perceptions of cash generated by production and profitability. The five jobs included in Exhibit 13-9 have a market value of $19,500,000 of which $1,292,000 is markup for overhead and fee. Since overhead is estimated at $100,000 per month, the profit for the 10-month-period net cash appears to be $292,000. It would be easy to conclude that the net cash spread of $292,000 for 10 months of production, or approximately 1.5%, is reasonable considering the time distribution of the jobs. It is shocking, however, to realize that net cash on a cumulative basis does not become available until the 11th month, when it is reduced by an additional $100,000 in overhead to $192,000.

Several general conclusions from the cash analysis are appropriate:

1. The information presented in Exhibit 13-6 for a single job and Exhibit 13-11 for multiple jobs indicates that, under the standard terms and conditions of most contracts which control cash income and dictate cash requirements, a typical job in the start-up phase will run at a cash deficit on a monthly basis through about half of the job, and even longer on a cumulative basis.
2. Jobs in the closeout phase will tend to produce net cash on both a monthly and cumulative basis as earned retainages are released.
3. The lower the percentage of overhead and fee in the job, the longer the job will be in construction before net cash is realized.
4. Since jobs in the start-up phase tend to consume cash, establishing and maintaining a reasonably consistent level of production is necessary so that jobs in the closeout phase may support the cash requirements of jobs in the start-up phase.

Concerning the last item, it is interesting to note that the methods of cash analysis presented also provide significant information on the cash consequences of production and emphasize the necessity for maintaining production at a level appropriate to organizational capacity.

Without a doubt, one of the best uses for the information generated by cash analysis is in designing and implementing a cash management strategy. Cash analysis provides visibility to possible cash deficits six months in advance of the actual need for their funding. Similarly, cash surpluses may be anticipated, allowing planning for maximum investment opportunities. Both these data also underlie the larger sales and marketing issues which may address the need for new and profitable work within a specific time period not only as part of a production management strategy but a cash management strategy as well. Cash analysis not only provides cash information on a monthly basis for each job but it also allows the examination of cash position on individual jobs, permitting the use of the Schedule of Values and weighted distributions as part of the cash management strategy.

Two other components of information may be generated and used as part of the cash analysis. The probability of new projects has been previously discussed in the section dealing with production forecasting (pp. 182-183). It is possible to add items to the cash analysis for jobs above a certain probability threshold to create more complete management information rather than limiting the analysis only to actual construction in work, though it is not recommended that the anticipated projects be added until the complete cash position has been determined for contracts in work. The second component relates to examining cash position to provide additional information concerning the level of sales and marketing activity required to maintain production and, consequently, to maintain a positive cash flow.

For the best management information, the six month cash analysis should be prepared on a monthly basis. Monthly preparation of the six month cash analysis continually updates company cash position based on forecasted production. The historic month of current reports should be compared to the current month for the previous month's report. This comparison will determine the accuracy the cash analysis and cash position forecasted in the previous month. It is also useful to compare a projection for the current month to the previous month's first month of forecast to determine validity of projections. When desired, cash forecasting ratios may be calculated for both the historic month and the current month using the following formula:

Cash forecasting ratios (CFR)
= Historic month's net cash position ÷ current month net cash position from the previous report.

(These are the same months from two sequential reports.)

## OVERVIEW OF OTHER RELATIONSHIPS

The project statement of expectations and the related performance models discussed in detail from Chapter 7 through this chapter have all been predicated upon the project plan, schedule, and budget rather than actual performance. These performance models provide clear and concise management information on expected job performance and what the ultimate outcome of the projects will be if the project is carried out exactly as it has been planned. A serious management error occurs when it is assumed that extensive planning and scheduling of a project will automatically result in good project performance. This assumption is perhaps one of the most serious detriments to the critical path method for planning and scheduling of construction projects.

The early proponents of the critical path method held that complete and detailed project planning and scheduling would automatically result in the project being carried out according to plan. This, of course, simply is not correct. The most elaborately prepared project plan does not guarantee adequate performance of the project. The project plan must be fully implemented and carefully managed with adequate means to detect early adverse trends if project control is to be exerted. It is project control, through adequate project management, which assures the successful outcome of projects. Obviously *both* planning and control are required; that is, their complete and lasting marriage makes for the successful, flourishing construction enterprise.

Beyond its basic nuts and bolts, brick and mortar aspects, the successful project is the project that has been carefully managed in the field environment. Even this alone, however, is not sufficient. The successful project is also that project administered properly from inception through job closeout. Both project management and project administration are essential components of this total process. Chapters 14 through 19 will deal in considerable detail with project activity flow, especially those dealing with the administrative aspects of the project. Since a significant part of integrated cost and schedule control is the proper administration of projects, a full understanding of the flow of administrative activities throughout each project assures that those actions anticipated by the project plan, schedule, and budget will be carried out in their proper sequence and priority and with the level of attention which is essential.

## SUMMARY

Cash income and cash requirements are centers of management information. The information produced is vital to managing an ongoing and dynamic business enterprise. Clear and accurate knowledge of the greatly varying demands for cash based on an organization's capital resource is essential. Cash surpluses need to be invested to increase profitability. Cash shortfalls need to be provided for well in advance of the actual occurrence.

Production forecasting is also an essential part of good business management. With maintaining level production the key to effective management, the ability to forecast variations in production so that peaks and valleys may be leveled is especially important. Production forecasting and analysis are also essential in designing sales and marketing strategies for maintaining production.

The raw data for production forecasting and analysis and cash analysis may be taken directly from the cash and income curves prepared for each project. Techniques are available for production forecasting using a standard "S" curve and for cash analysis using prepared data sufficiently accurate for forecasting.

Each organization examining the processes and procedures described for production forecasting and analysis and for cash analysis will find that these techniques are, with practice, simple and easy to use and that the information provided is essential to sound business management.

Once the basic management information system is available, other creative uses will be discovered. These processes and procedures are also applicable to other analytical tasks and will result in a general improvement in the kind and quality of information available to management.

# II
# Administrative Project Activity Flow

The construction industry tends to focus on things like excavation, pouring concrete, erecting structural steel, and laying brick as its primary field of endeavor. What is frequently overlooked in the excitement of these physical construction processes is that the ability to carry these out efficiently in the context of integrated cost and schedule control is inseparably dependent on a whole series of supportive administrative activities and the efficiency and timeliness with which these administrative activities are completed. Part II, Administrative Activity Flow, covering Chapters 14 through 19, presents such administrative activities and their interrelationships through a series of activity flow diagrams similar in some respects to the logic network diagrams used for construction activities.

Each of the chapters in Part II presents and discusses in detail a particular subnetwork within the administrative flow diagrams and develops the primary concerns associated with each.

Chapter 14 deals with the prebid and bid activities. These lead up to the award of a contract. Since many of the cost control issues are determined in these very early stages of the development of the project, the crucial integrated cost and schedule control activities in the proposal phase of the project are developed and discussed in detail.

Chapter 15, on the assumption that a contract will be awarded, deals with the whole range of activities which are necessary to adequately and thoroughly prepare a project for construction. While these activities are not always given the attention and formality which they deserve, they are nonetheless major determinants in integrated cost and schedule control and in the ultimate outcome of the project.

The construction phase of the project and the associated flow diagrams are dealt with extensively in Chapter 16. It is in this environment that the project strategy, the statement of expectations, and the project game plan are implemented.

Chapter 17 deals with the activities associated with job closeout. The importance of bringing a job to a prompt and efficient conclusion and closing out all of the administrative tasks associated with the project is fully developed.

The primary thrust of Part II concerns overall project administration from inception through final completion. Chapter 18 is a detailed presentation of complete contract administrative procedures which may be utilized in addressing the plethora of administrative tasks associated with each construction project. The procedures set out in Chapter 18 are especially valuable in exerting appropriate administrative control of the project.

One of the major administrative responsibilities deals with the cash management of each project. Chapter 19 and its flow diagrams deal with cash management data and information and the necessary strategies associated with these. Chapter 19 also makes comparisons and distinctions to illustrate the differences between cash management and profitability.

# 14

# PREBID ACTIVITIES

In the contemporary American construction business enterprise, administrative project activities and their unimpeded flow are critical to the management process. Over the next decade, merely knowing how to build things will not result in the success of the organization. An adequate understanding of administrative processes, from the first notion that a project will be estimated or bid through its final closeout, is essential.

## PROJECT ACTIVITY FLOW

While the exact nature of the activity flow may vary from organization to organization, the foundations of effective cost and schedule control are very much the same. Significant organizational efficiency results from understanding the flow of administrative project activities in much the same way as efficiency is created by using a road map when traveling through unfamiliar territory. To the extent that these administrative activities are anticipated and addressed, organizational efficiency is enhanced and complete control of the project is maintained from its inception to its completion and closeout.

This chapter and the five which follow present the project flow diagram dividing these activities into the major administrative stages of the project. The first two, to be discussed in this chapter, are the prebid and bid activities. Chapter 15 will extend to the contract award and preconstruction activities. Chapter 16 then will survey the construction phase of the project and include a detailed examination of weekly and monthly project cycles and administrative actions required as the project reaches substantial completion. Chapter 17 addresses final completion and job closeout.

With these items in mind, Chapter 18 covers basic contract administrative procedures. Chapter 19 delineates and analyzes cash management strategies since their effective implementation continues to be a crucial element in successful business management.

With these understandings, it will be possible to explore the major elements of construction and production including the project game plan. Procedures for measuring performance, the necessary data capture processes and analytical routines which are related to these will also be developed. The question to be answered is: *When do the processes of cost and schedule control begin and how are they integrated?*

If these are to be effective and if they are to be accomplished with a minimum organizational resource, they must commence when a construction organization considers preparing and submitting a proposal for a job. The prebid and bid activities for every project should be carried out on the assumption that the organization may be required to construct that project. Every aspect of reviewing drawings and specifications, setting up of files, quantity surveying, pricing, and solicitation of subbids should undergird this basic assumption and anticipate future needs. Each of these activities should also anticipate the use of the information generated in the creation of the performance models, discussed in detail in Part I. The creation of the performance models from the bid and estimating processes should be accomplished with minimal recasting of information and data. In short, what is required is a clear understanding of what the subsequent or downstream processes will demand before the estimate is started so that future needs may be anticipated.

## THE LARGER CONTEXT

The prebid and bid activities and the actual process of submitting a bid should be considered in the larger context which goes beyond merely the pursuit of acquiring new work through the bid process.

### Sales and Marketing Inferences

There is, of course, a significant sales and marketing aspect of bidding work whether the bids are prepared for competitive bidding or negotiating purposes. Some other important aspects of the bid process include significant exposure to the subcontractor, principal vendor, and supplier base in the marketplace in which the bid is presented. The bidding process also provides exposure to the design and engineering community as well as the owner/client base for each particular kind of construction. These additional opportunities associated with the prebid and bid activities place special demands on the quality and validity of the bid process and the skills with which these are carried out. When a track record for success in both competitively bid and negotiated work is established, this successful record of itself becomes a worthwhile marketing tool in pursuing additional sales and marketing opportunities.

A flow diagram of the 20 or more prebid and bid activities, which will subsequently be discussed in detail, is shown in Exhibit 14-1.

**Exhibit 14-1. Prebid and Bid Activities Flow Diagram**

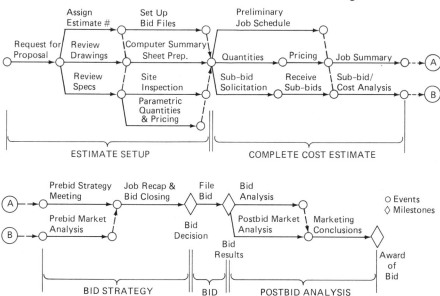

Selecting projects on which proposals are to be submitted, earning the privilege of bidding on selected projects and obtaining the opportunity to negotiate the contract for a job are very much a unique and individual function of each construction organization. Without regard to these uniqueness and internal variances, a series of more than 20 activities is common to the prebid and bid process. To the extent that these are carried out with technical competence, and in anticipation of future integrated cost and schedule control, the overall workload of the organization in acquiring new work is significantly reduced.

## SETTING UP THE ESTIMATE

The prebid and bid processes usually begin with some kind of invitation to bid or request for the organization to prepare a proposal for a project. The invitation to bid or request for proposal must be examined in detail and questions raised as quickly as possible concerning its completeness. When drawings and specifications are provided, these need to be examined as early as is practical. The review of the proposed contract documents will invariably produce questions which need to be answered. The earlier these technical and contractual questions are posed and the sooner answers are obtained, the less hectic will be the bid (or proposal) process.

## Estimate Numbers and the Master List

Most companies in the process of bidding work will have some reference system in the form of estimate numbers to maintain track of all estimates in progress. A master estimate number list may be maintained for this purpose. There is certain general information which may be maintained as a part of the master estimate number list and may be as simple as the estimate number, the name of the project, and when it is due. It may be expanded to include the name of the party within the organization assigned the responsibility for the project. The master estimate number list can be further expanded to include some data with respect to the results of the estimating or bidding process, including the amount of the organization's bid, an indication of where the organization placed in the bidding process, the dollar value at which the bid was awarded, whether or not the organization was low bidder, and the date and amount of the award of the contract. An example of a master estimate number list is provided in Exhibit 14-2.

There is some value in maintaining in a single summary all of the information about all estimates. This permits a quick review of the total bidding processes and provides information concerning the organization's success in acquiring work.

## An Estimate File System

For each project, a bid filing system should be set up appropriate to organizational requirements at the time the request for proposal is received and the decision is made to prepare or respond to the bid. It is suggested that a minimum of three separate files be set up for each estimate.

The first of these is the estimate file itself. The estimate file is to contain the estimate summary sheet along with all original pricing sheets and quantifications concerning the project. No other extraneous data should be contained in the estimate file since the original quantities, pricing, and estimate summaries have a longer useful life even on unsuccessful bids than the remaining files.

The sub-bid file should contain all requests for proposals to subcontractors, principal vendors, and suppliers arranged in some appropriate order such as the Construction Specification Institute's standard 16 divisions, or with a further breakdown into the logical work packages by which the project will be priced by the subcontractors and principal vendors. The sub-bid file provides for a complete, convenient repository of sub-bid data until it is required for the bid closeout procedures. A sub-bid tabulation sheet may be used for each work package to make a quick and convenient, competitive and comparative analysis of pricing. The more important aspect of the organi-

## Exhibit 14-2. Master Estimate Number List

MASTER ESTIMATE RECORD

| CODE | P.M. | Date Logged | E – # | Due Date | Job Name | Approx. Value | Quote Date | Contract Prepared/Mailed | Followup Date | Action Date | Action Proposal | % of Variance | Job # |
|------|------|-------------|-------|----------|----------|---------------|------------|--------------------------|---------------|-------------|-----------------|---------------|-------|
| B | JJ | 8/6/85 | 8591 | 8/27/85 | Office Building | $ 5,500,000 | 8/27/85 | Hand Del. | 8/28/85 | 8/28/85 | $ 5,650,000 $ | 2.3 % | — |
| N | FM | 8/22/86 | 8592 | 9/17/85 | Shopping Center | $ 3,750,000 | 9/7/85 | MTG | 9/7/85 | 9/18/85 | $ 3,600,000 $ | 0.0 % | 4412 |
| B | TP | 9/3/85 | 8593 | 9/25/85 | Spec. Warehouse | $ 2,250,000 | 9/25/85 | Hand Del. | 9/25/85 | / / | $ 2,327,000 $ | 0.0 % | 4413 |
| N | JJ | 1/7/86 | 8601 | 1/28/86 | Resid. Condos | $ 11,375,000 | / / | | / / | / / | $ $ | | |
| N | TP | 1/15/86 | 8602 | 2/11/86 | Bldg. Renov. | $ 4,300,000 | / / | | / / | / / | $ $ | | |
| | | | | | | | | | | | $ $ | | |

N – NEGOTIATED
B – BID

zation of the sub-bid file is that it should anticipate the need for organized subcontractor and principal vendor data for bid purposes and the subsequent purchasing process should the bid be successful. This is, of course, part of the integration of systems and anticipates future integration of cost and schedule control.

The third file is a memorandum and correspondence folder which may contain the original and (subsequently) the executed bid forms or proposals and all related correspondence and memoranda with respect to the project.

The three folders suggested allow for maintaining the original estimate work separate from the sub-bid data and from the memoranda and correspondence, all in a convenient and well organized package. A more elaborate file system may include separate files for the planning and scheduling work associated with the project and for cost engineering data generated with respect to the project. In any event, the file system should provide good organization for the estimate from inception through final disposition.

## Review of Drawings and Specifications

A thorough review of the drawings and specifications leads to the preparation of the estimate summary sheet. The preparation of the estimate summary sheet works in coordination with the organization of the sub-bid files. If the specification presents technical information on the project in the Construction Specification Institute's standard 16 divisions, the estimate summary sheet should contain the same major divisions. The industry practices in a given geographical area of the project will dictate a further breakdown of the standard 16 divisions into work package groupings.

For example, Division 15, Mechanical, may be broken down into plumbing, heating and ventilating, air conditioning, fire protection, and similar items.

The estimate summary sheet should be prepared as early in the bid process as practical for two reasons. First, it creates the organizational outline for the estimating process and for setting up of the sub-bid file. The early development of the estimate summary sheet also allows for continual editing and responsiveness which may affect job cost requirements. An item, of course, can be added to the estimate summary sheet at any time during the bid process as may become necessary.

## Site Inspections

An early inspection of the site of the work should be conducted and results of this site inspection appropriately recorded in the memo and correspondence file for future reference. Special conditions pertaining to the site, including such issues as access to the work, required temporary facilities and utilities, and other matters affecting either quantification or pricing or requiring special attention should be considered. The site inspection is frequently combined with a prebid meeting called at the request of the owner, architect, or engineers to review the details of the project and special project requirements with the prospective bidders. In any event, the site inspection should be conducted as early in the bid process as possible. Questions resulting from the site inspection should be addressed to the appropriate party as provided in the bid documents so that answers to these questions may be in hand well in advance of the bid due date.

## Parametric Quantities and Pricing

Parametric quantities and prices for the project should also be developed as early in the estimate process as practical. Parametric quantification and pricing may be as simple as a cost of the project per square foot of floor area with, perhaps, a separate pricing per acre of the required site development work. It could involve a much more elaborate parametric consideration of the project where 15 to 20 separate parameters describing the principal variables in the project may be considered. The parametric quantities and pricing provide a frame of reference for the value of the project as it proceeds through the bid process. The earlier these activities are completed, the more beneficial their value in the overall bidding process.

## THE FORMAL ESTIMATING PROCESSES

The activities described thus far comprise the basic setup of the project for bidding and pricing purposes. These should, of course, be accomplished early in the bid process so that they may be used as a foundation for the more formal quantification, pricing, and estimating routines. The next group of activities comprise the formal estimating processes and proceed along three concurrent flow tracks.

## The Preliminary Job Schedule

The first of these is the generation of a preliminary job schedule. The preparation of the preliminary job schedule is an important factor in the estimating and bidding processes though it is not always considered a serious or important part of the bid procedures. The generation of this schedule (even a simplistic bar chart) provides both a duration for the project and a calendarized reference for the performance of the job. With a calendar dated reference, factors such as weather conditions affecting the performance of the work and seasonal variations in the construction marketplace may be clearly identified and taken into account in the preparation and development of the estimate. The preliminary job schedule, however, is *not* intended to be a substitute for the more formal scheduling processes which have been discussed and which will likely use more elaborate analytical techniques, such as the critical path method. Although the preliminary job schedule does provide an early rational view of the time performance requirements of the project, it is not suited for production control.

## Quantity Surveying and Pricing

Along with the preparation of the preliminary job schedule, the necessary detailed trade quantifications should be prepared for the project. Prior to commencing the quantifications of the project, serious consideration should be given to the level of detail which will be required for subsequent integrated cost and schedule control. Careful identification of these requirements will significantly reduce the amount of effort required to convert the estimate to a budget which properly anticipates the planning and scheduling processes.

It is at this point that the first major difference between traditional methods of quantification and of quantification anticipating integrated cost and schedule control emerges. Traditional methods tend to take off quantities from drawings and specification for certain trades to be performed by the organization's own labor in greater detail than is necessary. Quantities for items of work to be performed by the subcontractor or principal vendors for the project, on the other hand, frequently lacks the detail appropriate for integrated cost and schedule control.

Several considerations that emerge with respect to quantification need to be firmly established. The first represents the second major shift from the traditional methods. Subsequent demands for integrated cost and schedule control require that estimating be done by *activities* rather than by trade groupings or work packages. This simply means that the roster of activities comprising a work package should be established before quantification begins. Both quantities and pricing may be easily and conveniently summarized by work packages for those activities to be included in each work package. In contrast, it is quite difficult to convert the trade-related estimate to an activity based budget with sufficient detail for integrated cost and schedule control.

Quantity surveying and pricing by activities is not at all difficult. It does require that the kinds of detail which will be necessary for integrated cost and schedule control be anticipated and accommodated. Working several projects through the estimating/scheduling processes will quickly develop an understanding of the right level of detail which is required. The level of detail is dictated by the requirements for either cost control or for planning and scheduling. The usual result is that on work packages being performed by one's own work force, less detail is required than for traditional methods. Where the work is to be performed by subcontractors and principal vendors, more detail than usual is required.

Another major principle anticipating integrated cost and schedule control emerges at this point. It involves the total quantification and total pricing of a project *independent* of pricing from the subcontractor and vendor marketplace. Time may not always permit such detailed pricing and quantification on all trades within a project. With experience, however, this estimating process is no more difficult and no more time consuming than the more traditional methods. It eliminates the initial dependency on external pricing, especially useful in negotiated work, which requires a high degree of confidentiality, and is, in part, a key to successful negotiations.

Complete independent pricing is very useful when the estimate and sub-bid files are used as part of the sales and marketing documentation in the negotiating process. As these procedures are developed in each organization, a data base quickly emerges, allowing these processes to proceed quickly and conveniently. The subsequent receipt and analysis of sub-bid pricing broken down by the activities for each work package provide continuing updating of the data base and confirmation of the validity of the complete independent project-pricing.

A note of caution at this point is worthwhile. Since most competitive projects require market place pricing from subcontractors and principal vendors, it is normally not prudent to depend exclusively on one's own pricing when dealing in the competitive bid marketplace. However, separately pricing each subcontract work package and confirming pricing validity through firm sub-bid pricing significantly enhances the ability to price work independently on future estimates.

Quantifications are completed on an activity by activity basis. These quantities are transferred to pricing sheets and the standard activity algorithm, quantity times unit price equals cost ($Q_N \times Q_L = C$), is applied. It should be possible, at this point, to proceed with an initial job summarization while awaiting the receipt of sub-bid and principal vendor proposals.

## Subcontractor and Vendor Pricing

Simultaneously with the quantification and pricing process, the necessary sub-bids should be solicited from the known subcontractor and vendor base in the marketplace. It is important to solicit sub-bids as early in the bid process as possible since time will be required for the subcontractors and principal vendors to prepare their proposals and to provide the data required

# Exhibit 14-3. Competitive, Comparative and Fair Value Sub-Bid Analysis

WORK PACKAGE: DIVISION 10,000. SPECIALTIES

| JOB COST | RESOURCE COST TYPE | RESOURCE CLASS | RESOURCE LEVEL OF IDENT. | DESCRIPTION OF ACTIVITY | BUDGET (FAIR VALUE)[1] QUAN. | BUDGET (FAIR VALUE)[1] QUAL. | BUDGET (FAIR VALUE)[1] COST | SUB A | SUB B | SUB C | BEST COMBINA-TION[3] |
|---|---|---|---|---|---|---|---|---|---|---|---|
| 10300 | 8 | — | 1 | Fireplaces | 1 | $3,400 | $3,400 | $3,200 | $3,600 | $3,400 | $3,200 |
| 10500 | 8 | — | 2 | Fire extinguishers | 6 | 100 | 600 | 600 | 700 | 550 | 550 |
| 10800 | 2 | — | 1 | Toilet room accessories | 10 | 506 | 5,060 | 5,200 | 5,060[2] | 4,900 | 4,900 |
| | | | | Division 10,000 work package totals | | | $9,060 | $9,000 | $9,360 | $8,850 | $8,650 |

Notes: 1. Budget is usually the same as fair value but must be independently validated before the purchase is made.
2. Value for items not included in that sub's pricing borrowed from budget (fair value) or from one of other sub's pricing in order to complete comparative totals and analysis for the total work package for subs A, B and C.
3. Best combination made up of lowest activity price from any sub as an indicator of market place variability.

MARKET VALUE CALCULATIONS:

1. Market value ratio (MVR) $= \dfrac{\text{lowest sub price (market value)}}{\text{budget (fair value)}}$

$= \dfrac{\$8,850}{\$9,060} = 0.977$ or 97.7%

2. Market value factor (MVF) $= \text{MVR} - 1.000 = 0.977 - 1.000 = -0.223$

or $-2.23\%$ (market value below fair value)

in their sub-bids and pricing. It is particularly useful to advise the subcontractor and vendor base that a breakdown of pricing will eventually be required by the list of the activities included in each work package, not by trade groupings.

## Job Summary

The contractor's job summary by the processes already described may be prepared independent of sub-bids and vendor pricing. As the sub-bids and vendor pricing are received, a competitive and comparative sub-bid pricing analysis should be made for each work package comprising the project, such as shown in Exhibit 14-3. This exhibit also shows a typical "best combination" analysis which gives some additional information on sub-pricing variability which compares to fair value and market value. Further, it shows one method of calculating the market value factor.

The competitive and comparative analysis of the work packages is especially important. *Competitive* pricing means that the pricing was taken in the same marketplace and at the same time. *Comparative* pricing means that each price is summarized at the work package level. *Fair value* is cost established from historic cost data independent of sub- or vendor marketplace pricing.

When a subcontractor may have excluded one of the activities making up a work package, the amount is borrowed from another sub-pricing breakdown and added to the subcontractor's pricing to complete the comparative total. Thus when the bidder has prepared pricing independently for each activity and work package, the comparative analysis may then be made. The method of calculating the market value factor (MVF) is also shown in Exhibit 14-3.

## BID STRATEGIES AND BID CLOSING

The activities which make up the actual bid preparation process include the preparation of preliminary schedules, quantities and pricing; solicitation and receipt of sub-bids; and analysis of those bids and results in the job cost summary.

Before discussing two separate issues (sales and marketing, and bid strategy), it is appropriate to establish the validity of the estimate at the base cost level. For bid analysis, prebid market analysis, and bid strategy, several summary levels are needed:

1. **Base cost:** the bricks and mortar cost of the project (Division 2–16) exclusive of general requirements, insurances

and taxes, bid and job surety, overhead, fee, and design and engineering.

Subtotal, base cost  _____

*General requirements:* the standard Construction Specifications Institute (C.S.I.) Division 1.

General requirements  _____

2. **Field cost:** actual cost estimated to produce the product at the field level.

Subtotal, field cost  _____

*Insurance and taxes:* payroll insurances and taxes, builder's risk insurance, bid surety, and project surety.

Insurance and taxes  _____

3. **Production cost:** the cost to produce the work exclusive of the general constructor's office overhead and fee.

Subtotal, production cost  _____

*Overhead:* general office operating expenses.

Overhead  _____

4. **General contractor's cost:** the general contractor's cost to produce the product, including a pro rata portion of general and administrative overhead.

Subtotal, general contractor's cost  _____

*Fee:* the anticipated markup of profit in the project above general overhead.

Fee  _____

5. **Construction contract sum:** self-explanatory.

*Design and engineering costs:* to be included where a specific design and engineering responsibility is assigned to the contractor. Note that this item may be included in advance of overhead or overhead and fee depending on the terms and conditions of the contract.

Construction contract total _____

6. **Total contract sum** _____

Prior to a discussion of the issues of marketplace, bid strategy and marketplace analysis, these summary levels need to be reviewed and understood.

## MARKETPLACE CONSIDERATIONS

Sensible and rational bidding requires two separate channels of information concerning the project in the context of its marketplace. These are shown graphically in Exhibit 14-4.

The first of these (fair value channel) is the cost estimate and its validity based upon the contractor's known ability to perform the work. This independent estimate establishes the contractor's "fair value" for the work. The

**Exhibit 14-4. Fair Value/Market Value Channels of Information**

| FAIR VALUE CHANNEL | MARKET VALUE CHANNEL |
|---|---|
| Quanties ($Q_N$) <br> Unit Prices ($Q_L$) <br><br> Base Cost (Fair Value) <br><br> Indirect Costs <br> Markup (Overhead and Fee) <br><br> Fair Value Estimate | Subcontractor Prices <br> Vendor Prices <br> Competitive and Comparative Analysis <br><br> Indirect Cost <br> Historic Market Data <br> Calculated Markup <br> Market Value Estimate |

| Fair Value Analysis <br><br> Completed Job Costs | Market Value Analysis <br><br> Competitive Bid Results |
|---|---|
| BIDDING STRATEGIES ||
| BIDDING DECISION ||

second of these channels (market value) is the value of the project in the marketplace based upon historic data on similar jobs bid in the same marketplace, with adjustments to account for changing costs of materials, services, and labor; or bid in a different contemporary marketplace with adjustments to reflect local values.

We must assume that any organization involved in the competitive or negotiated bid processes has sufficient technical competence to prepare valid cost estimates of the work to be bid. The importance of a valid cost estimate is emphasized by the fact that one's own cost estimate at the base cost level may be used as a statistical base for analyzing the marketplace.

## Base Cost

In examining the summary levels, it should be apparent that the least variability among bidders at any summary level occurs at the base cost level. Greater cost variables such as on-site organization, insurances and taxes, overhead and fee and similar items are not included in the base cost.

For postbid analyses, it would be beneficial to know the base cost the other bidders have used for the project, especially when the organization was not the successful bidder. This kind of information is, unfortunately, usually not available. It is possible nonetheless to use one's own base cost as a statistical measure of marketplace, assuming the validity of the base cost.

## Prebid Market Analysis

Having established some basic parameters for the project, it is possible to do a prebid market analysis. This analysis involves examining several similar jobs bid in the same marketplace and at the same time as the project under consideration. One's own base cost for the previously bid projects is used as a statistical base. The bid price reflected as a percentage of one's own base cost (using base cost = 100%) may be calculated. An example of a prebid analysis is shown in Exhibit 14-5.

The purpose of this analysis is to determine where similar projects have been bid in the marketplace as a percentage of one's own base cost for these projects. When four or five separate projects previously bid by ABC Construction are analyzed, two important pieces of data emerge. The first of these is the *average value* of these projects in the marketplace as a percentage over (or under) ABC's base cost. Note that ABC Construction's own base cost (not that of another low bidder if ABC was not the low bidder) is always known to ABC. The second is where the organization placed in the bidding projects as indicated by the percentage of base cost. In those bids

## Exhibit 14-5. Prebid Project Analysis

ABC CONSTRUCTION COMPANY, INC.

Prebid Project Analysis

Date: _10/15/85_

Job Name: _Metropolitan Business Center_ Due Date: _10/22/85_ Time: _2ºº PM_

Location: _Balt./Wash. Corridor_ Est. No.: _8528_ Market Area: _Baltimore_

Parameters: Bldg: _150,000_ S.F. Unit Cost: _48ºº_ Approx. Value: _7,200,000_

Site: _12.0_ Acres: Unit Cost: _75,000ºº_ Approx. Value: _900,000_

Cost/S.F. _54ºº_ Total Value _8,100,000_

### Comparatives

| | E.No. | Job | Bid Date | Low Bid Cost/SF | Low Bid Bid Amt. | %ABC's | ABC Bid Cost SF | ABC Bid Bid Amt. | % of ABC's |
|---|---|---|---|---|---|---|---|---|---|
| 1. | 8505 | Office Park | 4/16/85 | 56ºº | 6,500,000 | 106.25 | 57.90 | 6,720,000 | 109.63 |
| 2. | 8510 | Jones Tower | 5/21/85 | 62ºº | 12,200,000 | 105.78 | 62.00 | 12,200,000 | 105.78* |
| 3. | 8512 | Industrial | 6/25/85 | 46ºº | 3,650,000 | 108.30 | 47.89 | 3,800,000 | 112.41 |
| 4. | 8516 | Whse/Showroom | 7/30/85 | 44ºº | 4,320,000 | 107.90 | 44.31 | 4,350,000 | 108.59 |
| 5. | 8521 | Office Bldg. | 8/6/85 | 52ºº | 7,200,000 | 107.30 | 54.89 | 7,600,000 | 112.86 |
| * ABC low bidder | | TOTALS: | — | — | | 537.53 | — | — | 549.27 |
| | | MEANS: | — | — | | 107.11 | — | — | 109.85 |

Comparative: Low Bids 1. Low _105.78 %_
                    2. High _108.30 %_
                    3. Med/Avg _107.11 %_ (1)

Distribution
(Calculated)

ABC Base Cost: _100.00_ %
General Reqmts.: _3.48_ %
Ins. and Taxes: _1.12_ %
Production Cost: _104.60_ %
Markup: (Calculated) _2.51_ % (2)
Low Bid: _107.11_ %

### % Markup Over Production Cost

Low _107.11 %_
ABC Bid _109.85%_

Market Factors
Expected Low Bid % Over Base Cost _107.5%_

#### Bid Markup Computations

| | Post bid | Prebid (Base Cost) |
|---|---|---|
| ABC | | |
| Base Cost | | 100.00% |
| Gen'l Reqmts. | | 3.76 % |
| Ins. & Taxes | | 1.46 % |
| Est. Prod. Costs | | 105.22% |
| Bid Strategy | | |
| Overhead & Fee | | 2.28 % |
| Bid Amount | | 107.50% |

Markup over
  Estimated Prod. Cost _2.28 %_
Buy Range _3.33 %_ _6.17 %_ _4.72 %_
Budget for Overhead & Fee _7.00 %_

(1) Use Average or Median, whichever most representative.
(2) Calculated markup based on ABC's Base Cost, General
    Requirements, and Insurance and Taxes.

on which ABC was low bidder, the low bid calculation and the ABC data are the same since ABC's bid set the market value.

This analysis provides sufficient information for a rational judgment concerning where the current project is likely to be bid in the marketplace. With some experience in the use of this analytical procedure, it is nonetheless quite surprising how predictable the construction competitive bid marketplace becomes. An important assumption is that there are no significant errors or omissions on the part of any of the bidders; it follows that the "assumed validity" of one's own base cost estimate is crucial.

## The Prebid Strategy

The estimate for the project may be completed with an assumed or desired markup. This initial bid estimate is then compared to the marketplace value for the job, expressed as a function of markup over predicted base cost. With this information in hand, a bid strategy may now be set for the project and this should be done several days in advance of the actual bid. The management question is: *Where do we as a company want to bid the project with knowledge of and compared to its predicted marketplace value?*

An additional factor may be taken into account in determining the actual bid strategy. With the contract amount (owner's cost) fixed by the bid price, it is relatively common for the contractor's cost of the work to be reduced by the successful bidder in the postbid environment as a result of converting the bid estimate into a cost budget. These reductions in the estimated cost to arrive at budgeted cost may occur for a number of reasons. The bid estimate may be cleaned up by eliminating unnecessary risk factors and duplication of costs. Also, a better understanding of the time/duration requirements for the project frequently results in the budgeted cost reduction.

Another area of cost reduction occurs in what is commonly referred to as "the buy." The buy is the reduction of estimated cost or bid prices by subcontractors and vendors as the successful bidder becomes known to them, their understanding of the project requirements become more complete and the actual scheduled start and completion of the project become more certain. The buy may also reflect differences between market value and fair value especially in an inflated or rising market. The buy reduction in estimated cost should not be confused with cost reduction anticipated as a result of the practice of shopping sub-bids.

With experience in the marketplace, maintenance of good records concerning the "buy" in a project, and other cost reductions resulting in the postbid conversion of the estimate into a budget, it is possible to consider these cost reductions as part of the bid strategy. There are several rational methods which may be used to generate better information on the amount of the buy cost reduction which may be anticipated on a particular job; one is illustrated in Exhibit 14-6.

The buy analysis requires at least three competent and independent estimates of the buy. Each estimate considers the most optimistic, the least optimistic, and the most likely percentage of reduction of base cost. These are each averaged by groupings and the results are then averaged. This process produces a reasonably accurate estimate of the reduction of base cost which may be anticipated in the project under consideration as a result of historic performance. The percentage factor may be either applied directly to the reduction of the base cost or applied as an added contribution to overhead and fee. Either procedure provides the basis for a reduction of the proposed bid price while maintaining an acceptable level of markup in the project.

The purpose of these analyses is to make a decision prior to the bid closing concerning the markup strategy. The markup strategy may be expressed as a percentage of the job over base cost plus general requirements, insurances and taxes, or it may be expressed as a percentage of markup over the estimated production costs.

At this point, it is worth noting that a significant variable occurs in the general requirements cost; this variable is a function of the time required to produce the project. The time component of general requirements may be calculated by using the average time for the historic job and adjusting the general requirements for the project under consideration by an appropriate percentage to account for the time differential. But this additional routine of calculation can be avoided if the base cost is used as a statistical base

## Exhibit 14-6. Buying Market and Estimated Buy Percentage

| Buying Market | POOR 1 | 2 | 3 | 4 | 5 | 6 | 7 | 8 | GOOD 9 | 10 | Totals |
|---|---|---|---|---|---|---|---|---|---|---|---|
| 1. Sub-bid Coverage | | | | | | | | X | | | 8 |
| 2. Vendor Coverage | | | | | | X | | | | | 6 |
| 3. Known Subs & Vendors | | | | | | | X | | | | 7 |
| 4. Known Marketplace | | | | | | | | | X | | 9 |
| 5. Named Subs & Vendors | | | | | | | X | | | | 7 |

*Good range — total percentage over 50%*

| | | 1 | 2 | 3 | 4 | 5 | 6 | 7 | 8 | 9 | 10 | Totals |
|---|---|---|---|---|---|---|---|---|---|---|---|---|
| | TOTALS | | | | | | | | | | | |
| | RAW SCORE | | | | | | 6 | 14 | 8 | 9 | | 37 |
| | % | | | | | | | | | | | 74% |

| Estimated Buy Percentage | 1 | 2 | 3 | Totals | Means | Recap |
|---|---|---|---|---|---|---|
| 1. Most Optimistic | 5.0 | 6.0 | 7.5 | 18.5 | 6.17 | |
| 2. Least Optimistic | 3.0 | 3.0 | 4.0 | 10.0 | 3.33 | 14.17 |
| 3. Most Likely | 4.0 | 5.0 | 5.0 | 14.0 | 4.67 | 4.72 |

for analysis and for determining planned markup. The markup is actually applied to the production costs, taking into account estimated changes over time including the estimated general requirements costs.

## The Bid Decision

The prebid market analysis and the setting of the prebid strategy should result in establishing a budgeted overhead and fee expressed as a percentage of job cost in the event the bid process is successful. This avoids postbid negotiations on the percentage of overhead and fee to be anticipated in the project and simplifies the conversion of the estimate to a budget for integrated cost and schedule control.

Exhibit 14-6 also contains a job quality analysis which allows a bidder to examine the quality of a bid job on five essential parameters.

## Job Recap and Bid Closing

At this point, the job should be ready for recap and bid closing. The appropriate markup for the job has been agreed upon. The remaining tasks are to complete the competitive and comparative analysis for sub-bid and contractor pricing, to see that these are appropriately inserted in the bid close summary sheet, and that the project is totaled with the appropriate inclusions and markups. Each organization will use its own process with respect to bid closings, the preparation of bid documents, and other items.

## Bid Analysis

When the bid is filed and the successful bidder is known, it is useful to conduct several analyses of the bid. These include recording the number of bids filed, the grouping of the bids, calculating percentage spreads within the groupings, and the percentage spread in total.

## Marketing Conclusions

The postbid market analysis is especially important in preparation for future bidding. The post bid analysis includes distributing or calculating the percentage above base cost for each of the summary categories presented earlier. This information becomes quite valuable when doing the market analysis for subsequent projects. In addition, the low bid of the project should be calculated in terms of the contractor's own base cost. This should be compared to the prebid analysis and to the bid strategy. The number of bids offered for the project and other marketing conclusions should be reduced to writing for future reference.

## Award of Contract

Assuming that the bid is successful, the project should now be prepared to move into the award and preconstruction phase of the project with a minimum of recasting or restructuring of the job data. In the event that the bid is *not* successful, the bid analysis, the postbid market analysis, and the marketing conclusions become even more valuable in providing specific information with regard to market conditions. As part of the postbid closeout, all sub-bid files, including comparative and competitive analyses of sub-bids, should be handled with dispatch so that the record of each job will be final and complete.

## SUMMARY

Integrated cost and schedule control places additional clerical burdens on the organization in order to accumulate and distribute data, and to convert it into useful management information. The organizational impact of these additional requirements are directly related to the extent to which the need for subsequent data and information is fully anticipated and prepared for from the inception of a job through its final disposition.

The proper time to anticipate future informational and data requirements for any job is in the prebid and bid phases of the project. Some anticipation and preparation for subsequent requirements will pay substantial dividends for each job successfully bid. Even when a bid may not be successful, an abundance of essential information concerning the marketplace in which the job has been bid, the subcontractor and principal vendor base bidding that kind of job, the value of the job in the marketplace, and how competitive the organization is in the marketplace all become available as a result of the bidding process.

This process begins with a proper setting up of the estimate including dealing with the appropriate issues of estimate organization which result in credibility as to the validity of the estimate.

Successful estimating and bidding similarly require a clear understanding of the two separate informational channels which provide information independently on fair value and market value of each project. Included in the estimating and bidding processes is the proper grouping of activities in the project into work packages customary in the marketplace. Quantity surveying and estimating by activities are unique features of integrated cost and schedule control which require organizational experience for proper development.

Sub-bids and bids from principal vendors need to be properly analyzed on a competitive and comparative basis within the context of each work package. These analytical processes are an integral part of setting up the job for subsequent cost and schedule control and assuring the completeness and validity of the estimate.

Bid strategies are built upon a thorough prebid market analysis and an understanding of the organization's capability as well as limitations in producing the kind of work required by the job. The result of the bid analysis and the comparison of these results to the fair value estimate provides the basis for establishing an adequate bid strategy. The bid strategy can and should be established several days in advance of the final job recap and bid closing so that the bid strategy (i.e., the percentage for overhead and fee) is not left to the last moments of the bidding period.

A thorough analysis of the bid immediately after bid results are known is an especially important way of gaining significant insights into the nature and composition of the competitive bid marketplace, of market value and, as a result of comparing the two, marketplace trends. An adequate and thorough understanding of the composition of the marketplace, the present condition of the marketplace, and the trends within the marketplace are all essential to the successful addressing of the marketplace in either the competitive bid or negotiated work arena.

# 15
# PRECONSTRUCTION ACTIVITIES

The preconstruction activities in the overall project flow diagram (Exhibit 15-1) are those which make the transition from the prebid and bid activities to the physical construction of the job in the field environment. The notice of award of the contract sets in motion several groups of actions to prepare the project for construction. These actions are predominately administrative and analytical but are especially important since they will set the framework of cost and schedule control for the project. Each will be discussed in some detail to establish its relative importance to the overall project and to develop more specifically its impact on integrated cost and schedule control.

Reviewing the actions required from the time a notice of award is received until construction can start in the field reveals four subsets of activities within the larger area of preconstruction actions. These four groups are:

1. Job setup.
2. Purchasing.
3. Performance models.
4. Job start-up.

It is within these four major areas that the essential opportunities for cost and schedule control will be realized or lost. Essential to management information processes is each control vehicle which allows project management early visibility to project trends (particularly adverse), and the design and implementation of corrective actions where required.

*Control* is the ability to influence *outcome.* Valid information on job trends in a contemporary time frame is essential to exerting project control. The preconstruction activities do not consist of "winding up the project" and allowing it to run its own natural course. Rather they involve a careful examination of the performance of the project which considers both positive and negative trends: allowing or encouraging positive trends to continue, while reacting to negative trends with positive corrective actions. These processes are commonly referred to as "management by exception," an unfortu-

**Exhibit 15-1. Contract Award and Preconstruction Activities**

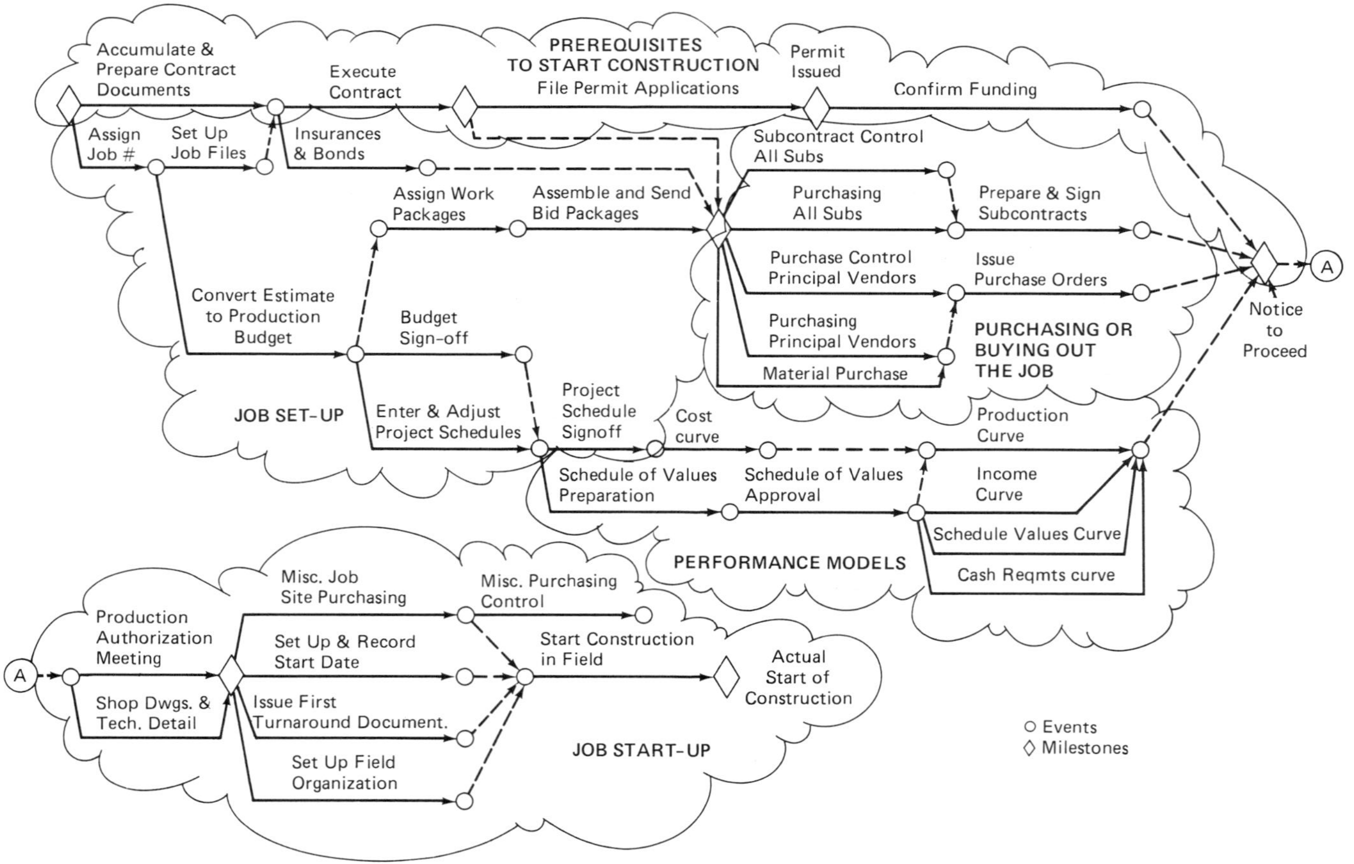

226

nate characteristic of which is a tendency to focus on the negative aspects of project performance.

For cost and schedule control purposes, however, the plain fact is that between 90% and 95% of all activities making up a typical job *will* perform as planned or better. These certainly do not constitute an area of concern nor do they require corrective action. All that is required is maintaining sufficient current information to confirm that these indeed are performing as expected. The difficulty lies, of course, in those 5% to 10% of activities which will not perform as expected and which require management help in the form of corrective action.

## CONTRACTS AND JOB SETUP

Each of the four major action areas will be considered in the context of setting up the job in specific ways to allow its careful monitoring when placed in construction.

### Contract Documents

The accumulation and preparation of the contract documents and the identification of those documents should commence with the beginning of the estimating process. All related technical data used in the preparation of the estimate, including such things as soils reports, addenda prepared during the course of the bidding processes, and so forth, should be included in the contract document package. While the bid documents may assign a responsibility to one of the parties of interest for the accumulation of the contract documents, all parties to the contract must participate in the identification of the contract documents and formally acknowledge those documents when the contract is executed.

The contract itself should be carefully reviewed to make certain that all essential areas affecting the completion of the work and full performance of the contract are fully understood. There are four major considerations dealing with the rights and responsibilities of the parties to the contract, listed below. Typically, in a two party contract, the rights of one party make up the responsibilities of the second party. The rights of the second party constitute the responsibilities of the first party. An appropriate review of the contract itself includes confirming that all rights to be expected by either party are defined in terms of the responsibilities of the other party. A stipulated right which does not assign the responsibility relating to it is not only meaningless but is almost certain to cause a contract dispute.

Assuming that other contract documents provide an adequate technical description of the project and of the product to be delivered, the four major issues which need to be carefully reviewed and acknowledged are:

1. The contract sum.
2. The time of performance.
3. The terms and conditions of the contract.
4. The physical conditions under which the work is to be performed.

A general tendency in the preparation and execution of the contracts is to focus on the contract sum and the contract time. These are, of course, important and are the most visible. The terms and conditions of the contract, however, are equally important and include such essential items as provisions for payment, conditions upon which payment will be made, prerequisites to payment, release of retainages and holdbacks, and the processes by which the project milestones are documented including contract commencement time, and substantial and final completion.

Among the important terms and conditions to be addressed are adequate provisions to resolve disputes which typically develop after the execution of construction contracts. These provisions should provide an orderly, efficient, and equitable way to settle disputes so that the performance of the work will not be disrupted.

Of equal importance are clear provisions in the contract concerning the conditions under which the work is to be performed. Changed conditions at the site of the work may represent a significant factor in the ability of the parties to the contract to perform what is required by the contract. Recognition of the effect of changed conditions at the site on the work should, then, be provided for in the contract.

Once the contract has been executed and all related contract documents acknowledged, it is recommended that one set of the original contract documents be packaged, sealed, and stored for safekeeping. This will preserve the integrity of the original documents, especially important in projects where the drawings and specifications are frequently revised after the contract has been executed.

The process of setting the job up for construction involves a series of more or less routine actions including assigning a job number, setting up the complete job filing system, ordering and obtaining necessary insurances and performance and payment bonds, filing of permit applications, obtaining the necessary permits, and confirming that funding for the project is in place and that payment will be made according to the terms and conditions of the contract. These actions should be attended to as quickly as possible after the notice of award.

## Prerequisites to Start of Construction

In most construction contracts, there are five prerequisites which constrain the actual start of construction and establish the start date of construction for contract purposes. These include:

1. Execution of the contract.
2. Issuance of all necessary permits.
3. Completion of the drawings, specifications, and other contract documents.
4. Funding of the project.
5. The notice to proceed.

The contract should provide that the contract time will commence when these five prerequisites are completed. The last to be completed is the notice to proceed, which should be in confirmation that the contract has been executed, that all necessary permits have been issued, that the drawings, specifications, and other documents are complete and that funding is available and will be disbursed according to the terms and conditions of the contract. With these in hand, the notice to proceed should be written by one party and acknowledged by the other and should contain in it a specific calendar date agreed upon by the parties which fixes the start of construction.

Other activities which involve preparing the project for field construction and for all of the prerequisite administrative processes are also especially important in integrated cost and schedule control.

## The Budget

The first of these is the conversion of the bid estimate into a production budget, one of the two primary performance models.

The bid estimate is not always prepared at the same level of detail or with the same level of refinement as required for a production budget. Integrated cost and schedule control imposes additional requirements relating to the preparation of the production budget. The activities to be included in the budget are determined by two separate requirements: first, the amount of detail required for cost control purposes; and second, the amount of detail required for the subsequent preparation of the schedule. The greater detail required by either is the level of detail which should be built into the budget.

A rather common error in early attempts at integrated cost and schedule control (particularly when using the critical path method for planning and scheduling) was to include more detail in the budget and schedule than required. The test for both relates to the level of *implementation* and *control* desired and the ways in which the work is to be monitored. Some experience with several projects where cost and schedule controls are integrated will quickly and clearly establish the requirements of the overseeing organization for the appropriate level of detail.

The questions to be asked in converting the bid estimate to the production budget and in anticipating the use of the list of budgeted activities for scheduling purposes are:

1. How is the job to be built?
2. What resources (cost types) are to be used in performing the work?
3. How will all of the activities be grouped into work packages?

The question of how the job is to be built will be addressed in terms of relationships between activities within the scheduling environment. On the assumption that budget and schedule are to be integrated, it is necessary to clearly have in mind how the job will be built in preparing the production budget since there will be a common list of activities for both budget and schedule.

The second question relates to the resources to be used. Preliminary decisions must be made concerning each activity: is it to be included as part of the subcontract; are the materials required to be purchased from a vendor; is labor provided by the contractor or is some other resource to be used?

Once the preliminary decisions on how the job will be built and what resources will be used have been made, all the activities constituting the project should be divided into work packages. The assignment of activities within a work package is a function of the style and practice of the industry within the local area. For example, within Construction Specification Institute Division 15, Mechanical, it may be the practice in a given jurisdiction to include all mechanical work in a single work package. In another jurisdiction, the practice may be to divide Division 15 into a series of specialities which may separately include plumbing, fire protection, ventilating, and heating and air conditioning. Whatever the initial decisions concerning work packages, there is, of course, flexibility in reassigning activities, dividing work packages previously agreed upon, or combining work packages for maximum convenience.

## Budget Sign-off

Since the production budget becomes a benchmark document for measuring cost performance, and since a number of work environments have a direct interest in the validity, content, and composition of the budget, a formal sign-off procedure for the production budget is recommended. Those groups who should approve the production budget include project management, field construction, contracts administration, financial and accounting, planning and scheduling, and management.

## The Schedule

Completion of the budget and its acknowledgment completes the preparation of the first of the two primary performance models. The second primary performance model is the schedule. While any scheduling method may be utilized, the critical path method (CPM) is the one preferred for integrated cost and schedule control since it deals with the construction of the project on the basis of activities and allows for the calculations of the time boundaries, which are controlled by how the project is to be built. Typically, the most difficult part of preparing for a CPM analysis of a project is generating the project's roster of activities. Since the need for the roster of activities has been anticipated and created in preparing the production budget (by asking such questions as to how the job is to be built and the resources are to be utilized), scheduling is made significantly easier.

The duration for activities must be established by any method appropriate to the scheduling process. The logic network diagram must be developed and the scheduling calculation made. The scheduling calculations will provide information on the time boundaries of all activities and events throughout the logic network diagram.

Experience clearly indicates that not all of the information generated by classic CPM analysis is useful for production control. In fact, when converting the critical path network to time scale, only two of all the time parameters generated (specifically, early start and duration) are required. It is strongly recommended for integrated cost and schedule control that all the other parameters generated by the CPM analysis (including the logic network) be filed away once the time scale network has been generated.

During the process of developing the schedule, additional refinements in the list of activities may occur, requiring that the budget be adjusted to reflect such changes. Less concern should be given to technical niceties in the scheduling process itself. Instead, the greater emphasis should be placed upon matching the schedule to the way in which the job will be built in the field, to making certain that every activity which is in the schedule is also in the budget, and that every activity in the budget is also in the schedule. To repeat, the common roster of activities is the basis for complete and total integration. This procedure must also be followed with any subsequent modification to either the schedule or budget (such as budget maintenance) which may be required as a result of a change in the work once the job is in construction.

## Schedule Sign-off

Similar to the production budget, the production schedule should be signed off as well. As a matter of fact, it is recommended that the sign-off procedures for both budget and schedule be accomplished simultaneously as soon as there is a consensus on both and a confirmation of an absolute match between the activities in the budget and the activities in the schedule. The sign-off procedure establishes both the budget and the schedule as the official primary performance models for the project which will be used in measuring project performance and in designing essential corrective action when such is required.

The kind of information now available on every activity within the project is worth noting. The budgeted value, the planned start and completion, relationship to other activities, the resources to be used in performance, and the work package in which it is likely to be included are all known about every activity in the project at the time these two primary performance models are completed and signed off. This makes an extraordinary amount of information available for subsequent project control. A tabular display of the kind of information available for each activity is shown in Exhibit 7-2. Also shown in Exhibit 7-2 is the activity information from the five derived performance models.

## DERIVED PERFORMANCE MODELS

The five derived performance models including the cost curve, the production curve, the Schedule of Values curve, the cash income curve and the cash requirements curve may all now be prepared. The preparation of each of these has been discussed in detail in Chapters 7 through 12.

### The Cost and Production Curves

The cost curve and the production curve may be directly plotted on the time scale network as previously discussed. The time scale network and budget are the only source data required for these two performance models.

### The Schedule of Values Curve

The Schedule of Values has associated with it an external approval requirement. The Schedule of Values should, therefore, be prepared as quickly as possible and submitted to the appropriate parties for review and comment.

Approval should be obtained as quickly as possible since the cash income curve may not be developed until the Schedule of Values has been prepared and preferably approved. From it, the Schedule of Values curve may be plotted using the methods described in Chapter 11.*

## Overview of Activity Information

The completion of the derived performance models adds several items of information to each activity. The budgeted cost and schedule for each activity is established by the primary performance models. The cost curve integrates budget and schedule into a single graphic display. The production value of each activity is shown on the production curve derived from the cost curve. The accounts receivable value of each activity is established by the Schedule of Values and is shown graphically on the Schedule of Values curve. The cash income for each activity is established by the cash income curve, and the cash requirements for each activity is established by the cash requirements curve, with both modified by certain contract terms and conditions of payment.

What is now known about each activity is the consequence of the planned performance of that activity from a time, cost, and cash point of view from its inception through its completion.

## SUBCONTRACTING AND PURCHASING

The summarization of this planned activity data by trade groupings or work packages provides a new dimension of information. The work package summaries are of particular interest and take on new meaning as the subcontracting and purchasing activities are addressed.

The assigning of every activity within the project into a work package has several beneficial results. The first is that the work packages provide an initial convenient summary of the resources to be used in performing those activities within the work package. A second and equally useful application of the work packages is in the preparation of requests for proposal (RFP), or bid packages, to the subcontractor and principal vendor sources. The bid package with its list of activities provides an additional description of the scope of the work on which the proposal is to be based, and also establishes some advance information on the breakdown required of the suc-

---

* The Schedule of Values can be appended to the contract or prepared for later approval depending on when source data is available. The controlling contract requirement is its approval, which may come as late as the first requisition for payment and as a part of the approval of that requisition.

cessful bidder for the payment schedule. The grouping of activities by work packages also greatly facilitates the competitive and comparative analysis required as part of any formal purchasing procedures.

## BUYING THE JOB

The next major group of actions required are those dealing with the letting of all subcontracts, issuing of all principal vendor purchase orders, and issuing of all supplier purchase orders, generally known as "buying the job."

Before a detailed examination of the purchasing or buying processes, a conceptual framework should be established. When a general contractor performs very little of the work on-site with his own resources, he is inseparably dependent on the performance of his subcontractors and principal vendors for producing the product required by the contract. When "lowest price" becomes the major criterion for buying, some of the most important aspects of setting up a job for construction may be lost. A different view of the purchasing or buying process should be developed and used when lining up subcontractors and vendors. This concept is that the general contractor, in buying out a job, is in reality selecting the team upon whom he must rely for quality of performance and for customer satisfaction. The "selecting the team" concept will change the order of importance of items of consideration and negotiation and cause these to be viewed differently as compared with cost being the primary consideration.

### A Suggested Buying Strategy

When the purchasing process is viewed as selecting a team of subcontractors and vendors to perform the work required for the project, six steps should be addressed in the order listed:

1. Clear the agenda on all issues of past performance on previous work.
2. Review the scope of the work in detail including the contract documents and the list of activities included in each work package.
3. Review the time schedules within which the work must be performed, using the time scale network.
4. Review the conditions under which the work must be performed in the field environment.
5. Review in detail the terms and conditions of contract, especially those related to payment, release of retainages, holdbacks, and back charges.
6. Negotiate the contract sum only after steps 1 through 5 have been agreed upon.

As a prerequisite to entering into a new contract, the performance on previous contracts should be reviewed. Any open agenda concerning poor performance (or a perception of poor performance) in terms of quality or time should be addressed and firm criteria established for the proposed work. This review also presents an opportunity for affirmation where past performance and past working relationships have been satisfactory.

A thorough review of the scope of the work, including *all* drawings, specifications, and other technical data, is essential and central to selecting a subcontractor or vendor.

Time constraints for the performance of the work are especially important because they create demands for organizational resources, both personnel and capital. Since the project schedule will have been prepared prior to subcontract negotiations, this information is instantly available to communicate the planned start and completion of every activity within the work package.

The conditions under which the work in the field must be performed are vitally important to subcontractors and principal vendors and may include cost factors significant enough to increase or decrease the cost of the entire project. When special sequencing of the work will be required or known adverse conditions may be anticipated, the methods of coping with these conditions should be discussed in detail and agreed upon.

The four items just discussed—past performance, scope, schedule, and conditions—are those most likely to be overlooked in subcontract negotiations since the parties may assume an understanding which does not exist. The only sure way to minimize the potential for disputes is to deal with these issues in the subcontract negotiation process.

Contractual terms and conditions themselves must be discussed in detail. When the terms and conditions of the prime contract are integrated into the subcontract requirements, these may affect the ability of the subcontractor or vendor to perform and must therefore be reviewed. Of particular importance are those terms and conditions which relate to both the amount of payment and the timeliness of payment. When retainages are to be held, these must be noted. When payment is keyed to owner funding to the general contractor, such stipulation should be reviewed and clearly understood. Provisions for holdbacks, retainages, and similar items need to be properly understood and documented; requirements for sharing of common expenses such as hoisting and/or trash removal need to be spelled out; provisions for back charges and remedial procedures when work is not performed consistent with time constraints or the qualitative requirements should be clearly communicated and agreed upon.

To summarize, the terms and conditions of contract frequently contain subtleties which affect the funding of the work of subcontractors and principal vendors. When these are properly understood and communicated, situations can be avoided which might otherwise lead to poor relationships and affect job performance.

## The Traditional Problem

Traditionally, the *amount* to be paid for the work package is usually the first item considered in subcontract and principal vendor negotiations. This is unfortunate since the five other items reviewed above—past work, scope, time, job conditions, and terms of execution—clearly affect the cost of the work to the subcontractor or principal vendor and, consequently, may affect either positively or negatively the value of the work and the amount agreed upon for the work package. Indeed, when such crucial items are left to "a common understanding," it may be subsequently discovered that each party to the contract had an understanding but that these understandings were not the same.

Negotiations of all subcontracts and principal vendor items required for a project should be accomplished as quickly and expeditiously as possible. All subcontracts should be prepared and executed promptly following the execution of the prime contract. Principal vendor items also should be negotiated as soon as possible after the execution of the contract and purchase orders immediately issued and signed.

Due consideration must be given to administrative time requirements. Lead time must be allowed from the commitment to a subcontractor or principal vendor to the full execution of the documents required to finalize the transaction. Lead time must also be allowed for ordering of material and equipment from mills and factories. Preparation of the shop drawings, technical data, and erection drawings necessary to execute the project, and their approval by the architect(s) and engineers, the owner, and, in some cases, by public agencies, require adequate time allowances as well.

## Administrative Tasks

The administrative tasks required to place product at the site of the work when required by the schedule falls outside of the context of this reference. A recommended procedure, however, is that immediately upon completion of the production schedule and its sign-off, every event which signifies the start of field work such as the structural steel, roof systems, mechanical and electrical equipment and so on, should be identified. Using these identification milestones for having material and equipment on the job, it is possible through a reverse sequencing diagram to create a technical administrative schedule for the project. This process can be carried back from the need for material on the job to include appropriate allowances for a shipment to the site from the site of fabrication, fabrication of the material, acquisition of material from mills, approval of shop and erection drawings, preparation of those documents, execution of the subcontract or purchase documents, and the contract negotiations resulting in the award of the item. With time allowances made for each of these activities, the administrative schedule for both purchasing and preparation of prerequisite materials can be clearly established.

Each organization will have its own methods of dealing with such administrative tasks. Suffice it to say in this context the assumption is that the necessary administrative purchasing activities will be carried out in time frames which will allow the field production schedule to proceed as planned.

### Prerequisites to Start of Construction

Beyond the simple execution of the contract documents, there are several other prerequisites to commencing construction in the field. These include issuance of the building permit by agencies having jurisdiction, confirmation that funding for the project is in place and will be paid according to the terms and conditions of the contract, and, of course, completion of the technical description of the project in the form of drawings and specifications. When it is confirmed that these four items are, in fact, accomplished, the fifth prerequisite to starting construction is a notice to proceed with the work. The items already discussed may be diligently pursued in anticipation of the notice to proceed, with the caveat concerning all subcontractor and principal vendor items that the agreements reached are subject to the notice to proceed.

### JOB START-UP

Also in anticipation of the notice to proceed, another group of activities prerequisite to job start-up can be accomplished. The recommended procedure is that a production authorization meeting be held just prior to the start of construction. The purpose of this meeting is to inform all parties of interest that the job is about to commence, to allow for a complete review of the contract documents, the budget, the schedule, all subcontracts and purchase orders issued through the date of the meeting, and similar items.

In most construction environments, the production authorization meeting is held when the primary responsibility for producing the job changes from project management to field construction. With this change of lead responsibility also goes the transfer of all documentation for the project generated up to that point. Among other things, certain job site purchasing requirements should be clearly identified for budget and purchasing control. An important and often overlooked item in setting up the job is recording the actual start date. This date is of contractural significance particularly when some issues subsequently develop with respect to the performance of the project.

The first set of instructions and the first group of activities should be identified for immediate action. The field organization should be set up immediately following the production authorization meeting along with all necessary temporary facilities and utilities. These may include a field office, storage trailers, telephones, temporary sanitary facilities, hoisting equipment, temporary power, and other required items.

## Cost Influence and Job Phases

In the contract award and preconstruction phase, another important issue needs to be considered. Exhibit 15-2 presents graphically the relationships which exist between the percentage of opportunity to influence cost as a function of job time and job phases.

**Exhibit 15-2 Opportunity to Influence Cost as a Function of Job Time and Phases**

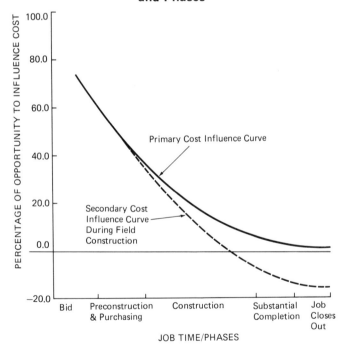

A number of historical studies support the information presented. Clearly the maximum opportunity to influence the cost of the project is during the bid phase and the preconstruction and purchasing phase. As the project moves into the construction phase, the preconstruction and purchasing phases will have fixed the cost on most of the items on which the cost can be fixed. The items remaining open as the project moves into the construction phase tend to be those items which are typically more difficult to control, such as random material supplier purchases, equipment rental, and on-site labor.

The secondary cost influence curve shows the effect of these items on which costs *cannot* be fixed. Unfortunately, the trends for these items become more negative as the project proceeds since the consequences of the bid phase and the preconstruction and purchasing phases of the project will have left only the more difficult items open; the logical consequence is that any lack of control on these non-fixed cost items will become progressively more negative with the progression of time. What is clearly indicated is that schedule control on nonfixed items is an essential part of cost control.

## SUMMARY

Activities which involve the award of the contract and those which prepare the job for actual construction are crucial to integrated cost and schedule control. These activities continue to build on the informational system established in the bidding processes.

In this chapter, the important relationships between primary and secondary or derived performance models in preparing the job for construction have been discussed; the methods of preparing these models however, have been covered in detail in earlier chapters.

The primary and derived performance models also create the basis upon which actual job performance will be measured. These bench marks of performance allow the utilization of simple performance ratios for providing management information.

The purchasing or buying of a job is in reality the selection of the team of subcontractors, principal vendors, and suppliers, on whom the general contractor must depend for job performance and for quality of product. The successful project requires that the entire construction team work together effectively and efficiently. The terms and conditions of contract along with schedule and price must be properly negotiated and established in order for the team to work effectively.

The physical start of work in the field environment should signify the change of lead responsibility from project management to field construction. This assumes that all necessary preparatory work leading up to the start of construction in the field has been accomplished and that all job related information has been communicated to field construction. The actual start of field construction should be keyed to the accomplishment of the five prerequisites to the start of construction: execution of the contract, acquiring the necessary building permits, availability of all technical documents including drawings and specifications, confirmation that funding is available and will be disbursed according to the terms and conditions of the contract, and, finally, in confirmation of the four previous items, the issuance of the notice to proceed.

On the assumption that the preconstruction activities have been thoroughly accomplished, Chapter 16 will consider the construction phase of the project.

# 16

# THE CONSTRUCTION PHASE
# OF THE PROJECT

The construction phase of the project implements and places into action all of the planning and scheduling which has been previously discussed. The construction activities required to complete a job and the orderly flow of work associated with those activities should be clearly shown on the construction schedule for the project. The construction schedule is, of course, one of the primary performance models. The construction schedule also provides one of the bases for measuring cost performance. There are a number of nonconstruction activities which must also be completed coincident with actual construction and which tend to take the form of a series of cycles and milestones. These cycles govern the orderly flow of construction in the field and give direction to the other administrative processes required for the proper and effective management of the construction job.

## JOB CYCLES

Job milestones are signficant isolated events signifying the completion of certain activities and the commencement of others. These vary according to the interests of the organization and the importance placed on each. Job cycles, however, are of particular interest and take several different forms, actually, points of view covering the same ground. Three kinds of job cycles are worth examining:

1. The production cycle.
2. Time cycles (weekly and monthly).
3. The informational cycle.

Each of these provides information on the flow of a job from a particular point of view and also unique data and information concerning overall project performance.

## The Production Cycle

The first of the job related cycles is the production cycle, which deals with physical construction and preparation for physical construction in the field environment. While its primary purpose is to facilitate field construction, the production cycle is keyed to certain nonproduction demands for data and information, such as payroll (weekly), job cost (weekly or monthly), accounts receivable (monthly), and accounts payable (monthly). Production tends to flow in a weekly/monthly cycle in response to these external informational demands although actual construction is usually keyed to job milestones, for example, foundations in place or building watertight. As a rule, production cycles are repetitive from job to job rather than within each job.

## Job Time Cycles

The second kind of job cycle is time related and includes certain weekly and/or monthly processes. The weekly/monthly cycles are primarily *data oriented*. The weekly time cycle may deal primarily with payroll information and payroll preparation. However, since payroll requirements dictate a weekly cycle, many other informational requirements for the project such as the planning horizon for turnaround documents and weekly updates of schedule may also be keyed to a weekly cycle.

The second time cycle is the monthly cycle, which is keyed to three major nonproduction processes. The first of these monthly nonproduction processes is accounts receivable (A/R); the second is accounts payable (A/P). These two monthly process cycles are very closely related to overall cash management of the project. In addition, most formal accounting processes and procedures are also tied to monthly accounting periods. These weekly/monthly cycles are repetitive by week and by month within the context of a single project.

## The Informational and Control Cycle

The third kind of time cycle concerns the informational processes associated with job site performance. This informational cycle becomes the basis for monitoring in progress job performance, for generating useful information with respect to job performance, for noting exceptions in performance and for designing corrective action when this is required.

## WEEKLY JOB CYCLES

Exhibit 16-1 is a flow diagram showing the weekly performance and data capture job cycle. Of interest is that a single activity in the flow diagram, daily/weekly activity production, is the only production activity shown.

**Exhibit 16-1. Weekly Job Performance and Data Capture (top);**

**Exhibit 16-2. Weekly Job Monitoring and Performance Evaluation**

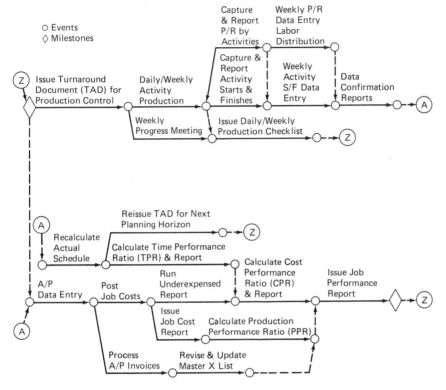

Exhibit 16-2 shows another form of weekly job cycle which deals with job monitoring and performance evaluation. This is another form of the weekly job cycle flow diagram which deals primarily with the informational processes related to job performance. These are especially important and relate directly to the third kind of job related cycle, the informational cycle.

## The Informational Cycle

The informational cycle shown in Figure 16-2 would appear to be more optional than mandatory. There is, however, a portion of the informational cycle which is essential to integrated cost and schedule control. It provides information which relates to the performance of individual activities, work packages, phases of the work and for the job in its entirety. The informational cycle consists of a number of major components:

1. Activity or work package performance.
2. Data capture on performance.
3. Analysis comparing actual performance to planned performance.
4. Exception reporting on activities falling outside of specified parameters.
5. Design of corrective action.
6. Implementation of corrective action.

This cycle is constantly repeated throughout the job at the activity level, work package level, and total project level. All activities must undergo the performance data capture and analysis portions of the cycle for integrated cost and schedule control. Those activities which are performing within allowed tolerances require no further action. However, those activities on which performance falls outside acceptable parameters or variances become performance exceptions. These require both the design and the implementation of corrective action. Once this is accomplished, the cycle repeats with performance, data capture, and analysis. Where the exceptional status is relieved, no further action is required. Where the exceptional status continues, the redesign of corrective action and reimplementation is again required.

### Weekly Job Performance and Administration

The weekly job performance and administrative cycle may vary widely from organization to organization. What is recommended is that, since all performance models will have been generated prior to the start of field construction, the information from the performance models be used as the basis for the weekly job cycle. The entire weekly and monthly job cycle processes may be greatly simplified when the information already prepared on expected performance can be used as a basis for production control and when the same information may also be used as the basis for data capture.

### The Turnaround Document

The process of supplying information to the field on expected performance and using the same instrument for data capture is referred to as a "turnaround document." These turnaround documents significantly reduce the on-site clerical load for job cost coding and provide the basis for simple ratio calculations for performance. Exhibit 16-3 shows a sample turnaround document.

## Exhibit 16-3. Sample Turnaround Document

```
          JUNE  6, 1985              GENERAL CONSTRUCTION CO., INC.              MONDAY 11:20 AM
                                  Actual Schedule sorted by Durations
PS075     JOB 9010  CONSTRUCTION MANAGEMENT SAMPLE      1 FLRS   75000 SQ FT              ESTIMATE  PAGE 04
          Daily Production Activity Check List & Time Report                     Report No  :   1
          Period Covered   0 - 3/ 1/85 to   88 - 7/ 1/85            Work Day: 71  Report Date: 06/07/85
=====================================================================================================

   Activity        Description         Comment Early Start  Completion Dur. Actual Actual Percent  Req   Pay   Hold Close
Vendor Number/ Name                                                         Start  Compl  Complete Quant Adjust Pay  Out
                                                                             A d j u s t   R e a s o n / C o m m e n t s

   3320 (9    1) CAST-IN-PLACE FLOOR SLAB        06/14/85  07/05/85  15____:_____:____:_____:_____:____:_____

-----------------------------------------------------------------------------------------------------
   3333 (9    1) SITE CAST-IN-PLACE CONCRETE     06/27/85  07/18/85  15____:_____:____:_____:_____:____:_____
   437 3365-1
-----------------------------------------------------------------------------------------------------
   4200 (9    1) UNIT MASONRY       ALL MASONR *04/19/85  06/07/65  35____:_____:____:_____:_____:____:_____
   273 2356-1 GOOD MASONRY, INC.    P.O.# 5185
-----------------------------------------------------------------------------------------------------
   4200 (9    2) UNIT MASONRY       MASY FILL   04/19/85  06/07/85  21____:_____:____:_____:_____:____:_____
   273 2356-1 GOOD MASONRY, INC.    P.O.# 5187
-----------------------------------------------------------------------------------------------------
   5120 (9    1) STRUCTURAL STEEL   ALL STRUCT  06/21/85  07/12/85  10____:_____:____:_____:_____:____:_____
   997 7211-1 STURDY STEEL CO., INC. P.O.# 5193
-----------------------------------------------------------------------------------------------------
   7150 (9    1) DAMPPROOFING                    05/19/85  05/20/85   1____:_____:____:_____:_____:____:_____

-----------------------------------------------------------------------------------------------------
   7206 (9    1) CAULKING           CNTRL JNTS *05/10/85  05/27/85 * 13____:_____:____:_____:_____:____:_____
   273 2356-1 WATERPROOFERS, INC.   P.O.# 5206
-----------------------------------------------------------------------------------------------------
   8100 (1    1) METAL DOORS & FRAMES  INSTALL  *05/10/85  05/10/85   0____:_____:____:_____:_____:____:_____
            EMP #   EMP NAME       HRS :EMP #    EMP NAME    HRS :EMP #    EMP NAME    HRS :EMP #    EMP NAME    HRS
MONDAY   :                            :                        :                        :
TUESDAY  :                            :                        :                        :
WEDNESDAY:                            :                        :                        :
THURSDAY :                            :                        :                        :
FRIDAY   :                            :                        :                        :
SATURDAY :                            :                        :                        :
-----------------------------------------------------------------------------------------------------
   8100 (2    1) METAL DOORS & FRAMES  MATERIAL *05/10 85  05/10/85 *  0____:_____:____:_____:_____:____:_____

-----------------------------------------------------------------------------------------------------
   15400 (9   1) PLUMBING SYSTEMS   UNDER SLAB *03/29/85  04/12/85 *  4____:_____:____:_____:_____:____:_____
   587 0013-1 GREAT CONNECTIONS, LTD.  P.O.# 5212
-----------------------------------------------------------------------------------------------------
   16111 (9   1) CONDUITS           UNDER SLAB *04/01/85  04/04/85 *  1____:_____:____:_____:_____:____:_____
   384 1053-1 POWER PLUS, INC.      P.O.# 5216
-----------------------------------------------------------------------------------------------------
   16400 (9   1) ELECTRICAL SERV.& DISTRIBUTION 06/01/85  06/08/85   5____:_____:____:_____:_____:____:_____
   384 1053-1 POWER PLUS, INC.      P.O.# 5216
-----------------------------------------------------------------------------------------------------
   81110 (4   1) Builders Risk Insurance  BY OWNER 06/01/85  06/01/85   0____:_____:____:_____:_____:____:_____
                                                                        CONFIDENTIAL
-----------------------------------------------------------------------------------------------------
WEATHER:_____ :_REMARKS:_____
TEMP___:_____:
FACTORS_CAUSING_DELAYS:_____:
_____:_BY:_____DATE:_____
```

The turnaround document can be issued on a weekly basis. The time period covered in the turnaround document is referred to as the "planning horizon." All activities which are—to be started, to be completed, or to be worked on during the planning horizon—should be on the turnaround document.

## The Turnaround Document as a Production Management and Control Tool

The turnaround document should be used for daily and weekly assignment of resources for activity production. Some degree of efficiency is introduced by the use of the turnaround document for production management in assigning work force to specific activities and for capturing data on both the start and completion of activities. Where percentages of completion are part of the reporting system, the percentage completion data may be captured at the same time. Data capture, particularly the start and completion of activities, must be accurate and complete to form the basis of integrated cost and schedule control. The weekly data capture on activity start and finish, and, where appropriate, on percentage completion, becomes the basis for weekly updates of schedule. The process of weekly update of schedule will be discussed in some detail.

## Payroll Data Capture

Parallel to the capture and reporting of activity starts and finishes and the preparation for weekly update of schedule are the capture and reporting of payroll information by activity for direct payroll items. The weekly payroll data is entered through some process which provides for labor cost distribution into job cost. A part of weekly processing is confirmation that the data on the reports is, in fact, correct and verifiable.

There is a wide variety of methods by which the turnaround document may be generated. A computer generated turnaround document is shown in Exhibit 16-4. A careful review of this sample turnaround document indicates that it contains all data with respect to planned performance of each activity. This information includes the appropriate job cost coding, the cost type, the level of indenture, the description of the item, the subcontractor and vendor responsible for the work, the telephone number, the expected start, the expected completion, the purchase order or subcontract number, and the planned duration of the activity. All of this information is provided by the primary and derived performance models previously discussed and by subcontract control. This represents the maximum information available on expected activity performance if we keep in mind that the control of

## Exhibit 16-4 Sample of Computer Generated Turnaround Document with Actual Performance Data

```
        JUNE  6, 1985              GENERAL CONSTRUCTION CO., INC.                    MONDAY 11:20 AM
                              Actual Schedule sorted by Durations
PS075    JOB 9010  CONSTRUCTION MANAGEMENT SAMPLE     1 FLRS   75000 SQ FT                ESTIMATE  PAGE 04
        Daily Production Activity Check List & Time Report                            Report No  :  1
        Period Covered   0 - 3/ 1/85 to   88 -  7/ 1/85              Work Day: 71     Report Date: 06/07/85
=================================================================================================================

 Activity        Description           Comment Early Start  Completion Dur. Actual Actual Percent  Req  Pay  Hold Close
Vendor Number/ Name                                                         Start Compl Complete Quant Adjust Pay  Out
                                                                              A d j u s t   R e a s o n / C o m m e n t s

  3320 (9    1) CAST-IN-PLACE FLOOR SLAB          06/14/85   07/05/85  15____|_____|_____|_____|_____|_____|_____

  3350 (9    1) SITE CAST-IN-PLACE CONCRETE       06/27/85   07/18/85  15____|_____|_____|_____|_____|_____|_____
  437 3365-1

  4200 (9    1) UNIT MASONRY            ALL MASONR *04/19/85  06/07/85  35_05/02_|_____|_____|_95_|_____|_____|_____
  273 2356-1 GOOD MASONRY, INC.         P.O. # 5185

  4200 (9    2) UNIT MASONRY            MASY FILL  04/19/85  06/07/85  21_04/29_|_05/30_|_100_|_____|_____|_____|_____
  273 2356-1 GOOD MASONRY, INC.         P.O. # 5187

  5120 (9    1) STRUCTURAL STEEL        ALL STRUCT 06/21/85  07/12/85  10____|_____|_____|_____|_____|_____|_____
  997 7211-1 STURDY STEEL CO., INC.     P.O. # 5193

  7150 (9    1) DAMPPROOFING                       05/19/85   05/20/85  1_05/19_|_05/20_|_100_|_____|_____|_____|_____

  7206 (9    1) CAULKING                CNTRL JNTS *05/10/85 05/27/85  * 13_05/10_|_05/27_|_100_|_____|_____|_____|_____
  273 2356-1 WATERPROOFERS, INC.        P.O. # 5206

  8100 (1    1) METAL DOORS & FRAMES    INSTALL    *05/10/85 05/10/85  0_05/10_|_____|_45_|_____|_____|_____|_____
            EMP #    EMP NAME     HRS |EMP #   EMP NAME    HRS |EMP #   EMP NAME    HRS |EMP #   EMP NAME     HRS
MONDAY   :                          |                    |                    |
TUESDAY  :                          |                    |                    |
WEDNESDAY:                          |                    |                    |
THURSDAY :                          |                    |                    |
FRIDAY   :                          |                    |                    |
SATURDAY :                          |                    |                    |

  8100 (2    1) METAL DOORS & FRAMES    MATERIAL   *05/10 85 05/10/85  * 0_05/10_|_05/10_|_100_|_____|_____|_____|_____

 15400 (9    1) PLUMBING SYSTEMS        UNDER SLAB *03/29/85 04/12/85  * 4_04/02_|_04/13_|_100_|_____|_____|_____|_____
  587 0013-1 GREAT CONNECTIONS, LTD.    P.O. # 5212

 16111 (9    1) CONDUITS                UNDER SLAB *04/01/85 04/04/85  * 1_04/01_|_04/04_|_100_|_____|_____|_____|_____
  384 1053-1 POWER PLUS, INC.           P.O. # 5216

 16400 (9    1) ELECTRICAL SERV.& DISTRIBUTION    06/01/85   06/08/85  5____|_____|_____|_____|_____|_____|_____
  384 1053-1 POWER PLUS, INC.           P.O. # 5216

 81110 (4    1) Builders Risk Insurance  BY OWNER  06/01/85  06/01/85  0____|_____|_____|_____|_____|_____|_____
                                                                                        CONFIDENTIAL
----------------------------------------------------------------------------------------------------------------
WEATHER:_____|_REMARKS:_____
TEMP___:_____|
FACTORS_CAUSING_DELAYS:_____
_____|_BY:_____DATE:_____
```

the project depends on the control of the performance of each individual activity.*

## The Weekly Progress Meeting

Any serious attempt at integrated cost and schedule control will include in the weekly cycle a weekly progress meeting. While the agenda for the weekly progress meeting is an open one, the first item on the agenda is a schedule review beginning with the previous week's production compared to the schedule and proceeding to the following week's planned production including workforce and resource requirements for meeting or improving the levels of production being experienced.

## Progress Meeting Minutes

The processes for reporting discussions and actions of the weekly progress meeting are especially important. There is a propensity for writing long scenarios covering progress meetings. In fact, the informational processes flowing from the weekly progress meeting should be directed toward identifying problems and required actions, assigning a specific responsibility for those actions, and for assigning a specific date by which the actions are to be completed.

Some convenient way of distributing the information to all interested parties is also essential as is the need for collecting information on the completion of those items. A recommended procedure is to use a daily/weekly project action checklist; a sample is shown in Exhibit 16-5. The key to the utilization of the project action checklist as the reporting vehicle for weekly progress meetings is to limit the data entry to simple direct identification of action items, leaving the full documentation of items involving disagreements and disputes to other reporting mechanisms.

## Weekly Schedule Updates

The idea has previously been developed that the relationship between activities in a construction project fall into two categories, physical constraints and planner's choices. Since the majority interactivity relationships fall into the area of planner's choices, it is not at all unusual that the flow of the work at the job site will not precisely match the original plan and schedule. It

---

* It should be apparent that a relatively large amount of clerical work is required for the manual preparation of turnaround documents. These processes are, of course, greatly facilitated when a computer is available and the software applications accommodate weekly cycle requirements.

### Exhibit 16-5. Project Action Checklist

PROJECT ACTION CHECKLIST

OWNER: _Jones Realty_      PAGE: _1_ of _1_

PROJECT NAME: _Jones Office_      JOB NO.: _8510_

DISTRIBUTION: _Owner, Archt., Contractor_      REPORT NO.: _2_

_all Subs and Vendors_      DATE ISSUED: _10/15/85_

ATTENDED BY: _JF, JM, BL, AZ, ED, HT_

| ITEM NO. | DATE ISSUED | DESCRIPTION | ACTION REQ'D. BY | DATE | COMPLETED | |
|---|---|---|---|---|---|---|
| 1 | 10/8 | Plumbing rough-in | Plumber | 10/15 | 10/14 | — |
| 2 | 10/8 | Elect. rough-in | Electrician | 10/15 | | Expect 10/20 |
| 3 | 10/8 | Partition layout | Drywaller | 10/15 | | Expect 10/20 |
| 4 | 10/15 | Set stairs | Misc. Iron | 10/22 | | |
| 5 | 10/15 | Set door frames | Carpenter | 10/22 | | |

is, therefore, especially important that the schedule (or at least a portion of it within the planning horizon) be recalculated routinely on a weekly basis. The information or list of activities requiring attention for the planning horizon period must be based on what is actually happening on the job site rather than on some historic schedule which may have been prepared months earlier and which, as time proceeds, becomes progressively less related to the actual performance of the work.

## The Time Performance Ratio

After the schedule has been recalculated, the turnaround document for the next planning horizon may be prepared. The time performance ratios (TPR) may also be calculated and the report issued after the schedule has been recalculated. One method of calculating the time performance ratio (TPR) as a measure of production follows:

Time performance ratio (TPR)
     = actual duration of all activities completed during the week
         ÷ planned duration of all activities completed during the week

This time performance ratio calculation may be used for individual activities, for all activities completed during the time period, for all activities within a work package, or for all activities to date. The weekly summary of activities completed provides direct information on whether the appropriate resources are being applied to complete the activities within the allowed time period. The summary by work package also provides direct information on how a particular subcontractor or vendor is performing. The time summary for all activities completed to date provides an additional measure of overall job production.

## Weekly Job Costing Data Sources

The weekly job cycle includes a number of other job performance related administrative tasks, including accounts payable data entry, the posting of job costs, and the completion of under-expensed reports. The job cost programs are driven by the two sources of job costs, payroll and accounts payable. Payroll labor distribution updates the job costs from a payroll point of view. The accounts payable data entry updates job costs from an accounts payable point of view. Since the budget and schedule are fully integrated with all subcontractor and principal vendor data, it is possible to confirm or verify accounts payable directly through the performance data from the turnaround document. When the accounts payable and labor distribution data entries are completed, it is possible to issue the job cost report, that is, the formal accounting record of job cost. The job cost report is current, however, only to the extent of data entry for both labor and accounts payable. Invoices for accounts payable data entry may unfortunately lag behind actual performance of the work by four to six weeks.

## The Under-expensed Report

It is also possible to generate an under-expense report (that is, a report listing all activities started, completed, or completed by some percentage based on the turnaround document) and to establish the accounts payable and job cost value of those items from the accounts payable payment schedule for each subcontractor and vendor responsible for performing the work. This process results in job costing information beyond the total of the formal accounting job costs by adding in the under-expense report which will be as current as the turnaround data entry.

## The Production Performance Ratio

The completion of the job cost report allows the calculation of a production performance ratio. The production performance ratio (PPR) is calculated by the following formula:

Production performance ratio (PPR)

= actual value of all work in place during the weekly period
÷ planned value of all work planned to be completed during the period.

Note that both are expressed in terms of production value.

The cost performance ratio (CPR) may also be calculated by using the formula indicated below:

Cost performance ratio (CPR)

actual cost of all activities completed during the period
÷ planned cost of all activities completed during the period.

Note that the planned and actual activities list must match exactly.

## PPR and CPR Differences

It is important to note the specific difference between the production performance ratio and the cost performance ratio. In the case of the production performance ratio, there is no concern for the *composition* of the activities completed compared to the activities planned, since what is being measured is the *total* value of product in place, actual compared to planned. In the case of the cost performance ratio, however, the composition of activities, actual and planned, must be an absolute dead match in order for the cost performance ratio to be valid; that is, only those activities included in the planned data are to be included in the actual data. Only in this way can valid information based on calculating the cost performance ratios be derived.

## Changes in Scope of the Work

Another major weekly function is to revise and update the Master X List (Exhibit 18-2) to provide a proper basis for the accounting for changes in the scope of work, reassignment of work, and back charges so that these may be properly entered into the performance model and accounted for in all the performance ratios.

## Interpretation of Performance Ratios

With the completion of these activities, a job performance report may be issued and should consist of the time performance ratio, the production performance ratio, and the cost performance ratio. In each case, actual performance matching planned performance should result in a performance ratio of 1.0. In the case of the *time* performance ratio, superior performance (when actual time is less than planned time) will result in a time performance ratio less than one, whereas poor performance will result in a time performance ratio greater than one. In the case of the *cost* performance ratio, superior performance (when actual cost is less than planned cost) will also result in a cost performance ratio of less than one, whereas poor performance will result in a cost performance ratio of greater than one. In the case of *production* performance, however, superior performance (when more product is actually placed than planned to be placed) will result in a production performance ratio *higher* than one, and poor performance will result in a production performance ratio *less* than one. These three ratios taken together provide information on overall job performance and performance trends. The weekly job performance report becomes a useful business management tool for looking at ongoing trends in individual job performance, for making comparisons between jobs, and for evaluating overall organizational performance.

## THE MONTHLY JOB CYCLE

The weekly job cycle continues throughout the duration of the project. The information generated weekly, however, results in the completion of a monthly job cycle. The *monthly* job cycle includes, among other things, generation of the monthly accounts receivable requisition, month-end production verification, preparation of the formal requisition, processing of accounts payable, and a complete series of month-end job status reports. Each of these is essential in the ongoing flow of the project and particularly important in overall cash management.

A typical monthly job cycle flow diagram is provided in Exhibit 16-6. Here two separate sub networks are indicated. The first of these deals with the accounts receivable process and the related final processing and paying of accounts payable. The second deals with the monthly job monitoring and performance evaluation, which is similar to the weekly job monitoring and performance evaluation but updated to a monthly status.

### Accounts Receivable Verification

The integration of cost and schedule control and the integration of the data capture processes through the turnaround document provide all the data required for the preparation of accounts receivable. Based on a specific requisition cut-off date either established by the contract or agreed to by the parties, the data entry cut-off for preparing the accounts receivable requisition is also established. Since data entry is maintained currently through the weekly

**Exhibit 16-6. Monthly Job Cycle Flow Diagram**

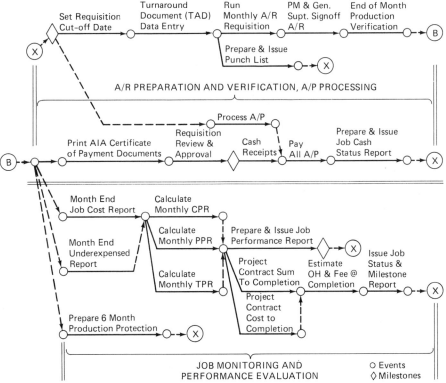

cycles, only some minor updating to a cut-off date is required in preparation for running the monthly accounts receivable requisition.

The information required to run the monthly accounts receivable requisition includes data on activities which have been started, completed, or completed to a certain percentage and the scheduled value of each of those activities. The turnaround document uniquely provides the required performance information. Chapter 11 describes the preparation of the Schedule of Values, which establishes the accounts receivable value for each activity as well. This is indicated in the derived performance data in Exhibit 7-2. With a scheduled value for each and every activity along with the status of its performance, the accounts receivable requisition can be automatically generated. Two separate verifications are required: one for the project manager and general superintendent, and a second serving as an independent end-of-the-month production verification.

Note that the systems integration allows a direct preparation of the certificate of payment or other required payment documents, which in turn allow the review and approval of their funding.

Simultaneously, the accounts payable data entry processes are kept current. The performance verification relating to each activity used in the preparation of accounts receivable is the identical data used for accounts payable, the result being an exact match of both accounts payable and accounts receivable. Upon receipt of cash funding for the requisition, all accounts payable may be paid. With both the cash income and cash requirements confirmed, a monthly job cash status report can be issued.

## Job Monitoring and Performance Evaluation

Once the end-of-month production verification has been accomplished, the end-of-month job monitoring and performance evaluation reports can also be prepared. Note that the performance data captured in the turnaround document for each activity is used whenever other downstream processes require performance information on activities. This is an excellent example of the effect of single data capture where the turnaround document data is used to automatically prepare the accounts receivable requisition, to process the accounts payable invoices, and simultaneously, to provide all the raw data necessary for the weekly and monthly job monitoring and job performance evaluations.

## Month-end Job Cost Reports and Under-expensed Reports

Two month-end reports concerning job cost are of vital interest to management. The first of these, the formal job cost report, is an accounting report and is based upon the current data entry of accounts payable and payroll. The major weakness of this report is that it is normally out of phase with actual production by several weeks. Typically, job performance and the accounts receivable requisition are billed through the end of a production month. The accounts payable invoices, however, may not be received until after the end of the production month, this delay creating a disparity within formal accounting records relating to accounts receivable and accounts payable.

In integrated cost and schedule control systems, however, it is possible through the month-end under-expensed report to generate current, accurate and valid information on the accounts payable value of activities completed but not yet billed by the subcontractor and principal vendor base. The same performance information from the turnaround document is fully integrated with the payments schedule for each subcontractor and vendor through subcontract control. The identical processes establish the current value of items which have not yet been expensed, the source of the term "under-expensed"; that is, the cost has been incurred and the value established prior to receipt of the accounts payable invoice.

## Month-end Performance Evaluation

The information available through the turnaround document and through the other performance data allows the monthly cost performance ratio, the monthly production performance ratio, and the monthly time performance ratio to be calculated. These should result in the preparation and issuance of the monthly job performance report, the second month-end report of great interest to management. Monthly job performance reports may be of interest, but, like job cost reports, require additional processing to provide the kind of information which management requires. The comparison of actual performance to planned performance gives some indication of trends resulting from historic job performance. What is needed is a way to project the ultimate consequence of the job given historic performance.

## Projections to Completion

Valid information on historic performance provides a framework which allows the projections of the contract sum to completion and of the contract cost to completion. These are accomplished by using historic performance as a performance model which in turn is used to forecast the ultimate outcome of the project. The specific methods of projecting contract time and contract amounts to completion will be presented in detail in Chapter 28. With a projection of contract sum to completion and a projection of contract cost to completion, it is possible to estimate overhead and fee at completion. It is also possible to issue a job status report and a milestone report.

## MONTH-END JOB REPORTS

There are five major status reports which may result from the completion of the monthly job cycle. Three of these—the job cash status report, the job performance report, and the job status and milestone report—have already been discussed. In addition, as a separate informational process using turn-around document data, a *punch list report* should be prepared and issued.

The punch list (see Exhibit 16-8) is qualitative in nature, whereas the other reports deal with quantitative issues. The punch list report gives some indication of the quality control capabilities of the organization. It also provides early visibility to problems which may be experienced as a job nears substantial and final completion.

The fifth of the reports is the six months production forecast (Chapter 13). Monthly updates of the six months production forecast, which takes into account historic performance on all projects in work, present meaningful management information not available through any other source. The six months production projection should be included routinely as a part of the monthly job cycle and may include a cash flow analysis.

## Cash Flow Analysis

From a business management point of view, one of the most useful information reports which may be generated as part of the monthly job cycle is the six months cash flow analysis for each job. With the monthly update of schedule and historic and actual receipts and disbursements, it is possible to project receipts and disbursements through the duration of the job. A detailed discussion of these methods of projections is included in Chapter 13.

The cash flow analysis is prepared in much the same manner as the production forecast beginning with the most recent historic month, a projection for the current month resulting from the end of month update of schedule, and a projection of four additional months or to the end of the job (whichever provides the most useful information). When this information is prepared for each job, summary data for the cash flow analysis of all work in progress may be quickly and conveniently developed. Since the format and source data are the same for both the production forecast and the cash flow analysis, a single report may be issued providing both sets of information. The procedures for developing the six months projection and cash analysis report are presented in Chapter 29.

## SUBSTANTIAL COMPLETION

One of the most significant milestones in the life of a project occurs at substantial completion. Substantial completion is particularly interesting because it is a clear indicator that the project is moving toward actual completion and there are certain contractual and economic consequences resulting from the issuance of a certificate of substantial completion. In addition to the normal monthly cycle of items which occur concurrently with substantial completion, there are a number of specialized kinds of items which are associated specifically with substantial completion, shown in Exhibit 16-7.

The first of these is a complete owner-architect-contractor inspection of the project. This inspection usually results in a listing of those items which have been completed but on which corrective work is required or of some other condition which exists but is inconsistent with the contract requirements.

**Exhibit 16-7. Substantial Completion Activities (in Addition to Monthly Cycle Items)**

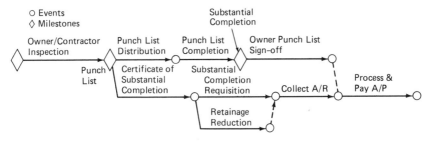

## The Punch List

This list of activities, commonly called a "punch list" in the trade, is distributed for immediate action and completion. An appropriate punch list form is shown in Exhibit 16-8.

It is worth noting that the punch list form illustrated is similar to the project action checklist in that it does not take the form of a written scenario. It, rather, focuses on specific items to be completed, identification of the responsible party, contract requirement for completion, with space allowed for the recording of that completion.

### Exhibit 16-8. Sample Punch List Form

PUNCH LIST

NAME _Jones Realty_  PROJECT _Jones office_  LIST NO. _1_

ADDRESS _1234 Main_  DATE ISSUED _10/22/85_

_____  PREPARED BY _J. Smith_  PAGE _1_ OF _1_

TELEPHONE _555-1625_

| Item | Area | Corrective Work Required | Craft | Contractor | Notice | Complete |
|------|------|--------------------------|-------|------------|--------|----------|
| 1 | 1ST floor | Air condit. balance | Hvac | Mech Inc. | 10/22 | |
| 2 | " | Adjust doors | Carp. | Contractor | 10/22 | |
| 3 | " | Touch up paint | Painter | Brush Inc. | 10/22 | |
| 4 | " | Stretch carpet | Carpet | Carpet Inc. | 10/22 | |
| 5 | " | Clean windows | Cleaner | Janitorial Serv. | 10/22 | |

## Certificate of Substantial Completion

The list of items to be completed becomes a part of the certificate of substantial completion. Assuming that that list is not extensive, two actions result. The first of these is assigning a value to the punch list which becomes the basis for any holdback against the contract sum. The second is that any retainage held during the job is released and paid. The result is that the only amount withheld at substantial completion is the value of the items not yet completed or requiring correction. A substantial completion requisition is prepared along with the appropriate certificate of payment which allows the payment of the entire contract sum except for the value of the work contained on the punch list. Accounts payable may be processed as the cash is received from the substantial completion requisition. Upon completion of the punch list items, which list has been acknowledged by the parties to the contract, the project may move toward final completion and job closeout. The time between substantial completion and final completion on a job which has been properly constructed with appropriate quality control processes and which has been performed in a proper and expeditious manner and which has been handled with administrative efficiency, should not be more than 30 days. The activities required to move a project from substantial completion to final completion and closeout are discussed in detail in Chapter 17.

## SUMMARY

All of the activities which have been previously discussed lead up to the construction phase of the project. The purpose of all the preparatory work is to establish a valid framework for firm control of both cost and schedule of the project. Many of the opportunities for cost control will have been achieved or lost prior to the commencement of construction. Purchasing represents a major area of cost control where most of the work on the job is performed by subcontractors.

The actual physical construction seems almost anticlimactic. It is true, however, that a properly planned, scheduled, budgeted, and administered project should result in good cost and schedule control and in earning the planned overhead and fee.

The flow diagrams presented provide an administrative and project management framework for the project. The integration of cost and schedule control and the associated data capture procedures allow all the necessary administrative tasks to be carried out quickly and efficiently. Single data capture on activity performance provides the data and information required for all related accounting, administrative, project management, field, and overall management informational processes.

# 17
# JOB CLOSEOUT

One of the continuing significant organizational weaknesses in construction companies is contract administration. The project flow diagram dealing with final completion and closeout, developed in this chapter, makes apparent that the effectiveness of the closeout procedures and, inversely, the difficulty associated with these procedures will be in direct proportion to the effectiveness of contract administration from the inception of the job through its completion. If the job contract records have been maintained in a current status (including maintenance of a master X list, processing of all items on the Master X list toward final resolution, and accurate documentation of back charges and holdbacks), the administrative closeout of the job becomes significantly less difficult and less time consuming. Where this is *not* the case, difficulties will be encountered, the time and effort involved in job closeout greatly extended, and, in all probability, job profitability reduced.

Chapter 18 will develop the full issues and procedures concerning contract administration; here the focus is on the administration of closeout.

## THE FINAL COMPLETION AND JOB CLOSEOUT FLOW DIAGRAM

This diagram contains two separate sub networks. The first of these, Exhibit 17-1, shows the activities associated with the administrative and accounting close-out for the project.

This stage of the project deals with such general items as final inspections, establishing final contract amounts, final requisition, collecting accounts receivable, warranties, guarantees, and similar items. Under normal circumstances, the accounting and administrative closeout of a project might be considered the end of a project. In integrated cost and schedule control processes, however, an abundance of historical information on job performance, an important contribution to the organization's informational data base, is available and conveniently accessible at the time the accounting and administrative closeout is being accomplished. Exhibit 17-2, then, shows the flow diagram associated with the final job analysis and evaluation.

## Exhibit 17-1. Accounting and Administrative Closeout

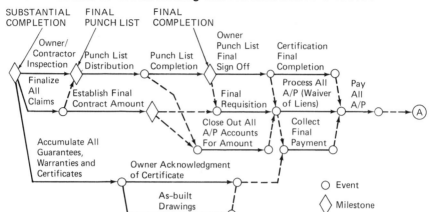

Final job analysis and evaluation activities provide the final job cost report, the calculation of a variety of performance ratios, the formal calculation of final overhead and fee, and a parametric analysis of the project. In addition, the overhead burden to be carried by the job may be calculated based on overall organizational production. This information allows the calculation of the net fee contribution to the organization by the specific project, which

## Exhibit 17-2. Final Job Analysis and Evaluation

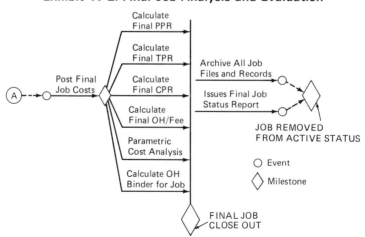

concludes the final job closeout. The remaining activities are the archiving of all job files and records and the issuing of final job status report along with its removal from active status.

## The Final Punch List

Substantial completion is the first significant milestone leading to the final completion and closeout of the job. A listing of all the activities requiring completion or corrective work, called a *punch list,* survives substantial completion and requires attention prior to final completion. The first step in initiating the final completion process is a reinspection of the project by the owner, the designer, and the contractor to determine whether the items set forth at substantial completion have been completed and there are any remaining items requiring attention. The inspection and the resultant punch list are prerequisite to final completion and tend to be more detailed and specific than the punch list prepared at substantial completion. This is because the completion of the substantial completion items allows a more detailed review of the remaining corrective actions which might be required.

It is apparent that the number of items included on the final punch list are inversely proportional to the quality control exerted by the general contractor during the construction of the entire project. What is recommended, then, is a weekly/monthly punchout of the job by the general contractor so that corrective actions are maintained in a *current* status. When this is done, few items will appear on any final punch list.

## Finalization of Claims

Coincident with the final inspection of the job site is the finalization of all claims associated with the job. There are two separate groups of claims: the first involves the relationship between the owner and contractor; the second involves the contractor with subcontractors and principal vendors. Since there are significant interrelations between both groups, all claims should be dealt with at the same time so that when the job is closed out it may be done with finality. All claims from subcontractors and principal vendors with respect to a job must be processed by the general contractor so that those relating to the owner/contractor agreement may also be properly presented and brought to a satisfactory conclusion.

Not all claims a subcontractor has with the general contractor affects the owner/contractor agreement. Claims made by one subcontractor may result in backcharges to another subcontractor. These must also be dealt with and finalized.

## Current Administration of Claims

It is, of course, strongly recommended that charges and back charges to subcontractors and principal vendors be currently maintained. When appropriate, supplementary purchase orders must be issued to formalize these charges. The processes for handling claims to the owner and charges and backcharges to subcontractors and vendors are dealt with in detail in Chapter 18, on contract administration.

## Final Punch List and Sign-Off

The recommended form of punch list has been discussed in Chapter 16, relating to substantial completion. The principal purpose of the punch list is not to create a historic scenario on the circumstances leading up to an item being included on the list (though important for the archives). Rather, it is to identify clearly each item, to identify the party responsible, to establish a date for the item's completion, to distribute the information to the parties of interest, to record the item's completion, to collect data on actual performance, and to confirm the item's completion in the form of a sign-off. The use of the punch list form allows a distribution directly to all parties who must take action in order to complete the list.

When the punch list is completed, a reinspection may be scheduled to obtain acknowledgement from both the owner and the designer that the project is, in fact, complete. It is crucial that this acknowledgement be *in writing.* When a contract does not specifically provide for a certificate of final completion, the general contractor should initiate a certification that the project is completed and the contract fully performed.

## Claims and Waivers

The resolution of all outstanding claims between the owner and contractor is a part of the completion of the punch list since items included on the punch list may be the subject of claims. It is therefore important that all claims outstanding at the time of final completion by fully documented and that these be brought to a conclusion by the appropriate processes of contract. The earlier the claims are addressed and presented, the greater the probability of resolving these so that open claims will not stand in the way of closing out a job which is otherwise completed.

Once the punch list is completed, all changes negotiated, and the final contract sum established, the final requisition for payment may be initiated and presented to the owner. Some contracts will require the execution of mechanics' releases of liens from the general contractor, all subcontractors, principal vendors, and suppliers. These should be obtained as early in the closeout process as is practical so that they may be presented when required by the contract documents. Even where the contract does not require that release of liens be obtained for all monies disbursed, this procedure should be followed nonetheless.

## Final Payments

Assuming that all claims have been resolved with the owner and that all claims, holdbacks, and backcharges have been finally negotiated with the subcontractors, principal vendors, and suppliers, all accounts payable may be processed and prepared for final payment upon receipt of funds from the owner. Prompt processing is especially important since the information needed to update most job cost programs or procedures is supplied by both payroll and accounts payable. The processing of accounts payable will also reconfirm that all open claims with respect to the project between the contractor and the subcontractors, principal vendors, and suppliers have been resolved. Any items remaining open or coming to light at this stage must be resolved as quickly as possible.

Parallel with processing the final requisition is closing out all accounts payable for the amount to be paid when final payment is received. This closeout procedure is important and frequently requires the participation of a wide variety of people within the organization, including project management, field construction, administration, and accounting. It should be noted that these accounts payable closeout procedures require an effort inversely proportional to the currency with which all accounts payable transactions, including claims, have been maintained.

## Warranties, Guarantees, and Certificates

In preparation for final completion and job closeout, many contracts require that certain documents concerning guarantees, warranties, certificates, and similar technical data, be presented as a prerequisite for final payment. The accumulation, preparation, and presentation of project related documentation is an important part of job finalization and closeout even when not specifically required by the contract. For example, the contractor should provide to the owner any warranties, guarantees, or certificates relating to that project as well as a complete set of "as-built" drawings for future reference at the time of closeout. As-built drawings are especially important when changes have been made from the working drawings during field construction or when change orders have been introduced into the project but not documented through revisions to the working drawings. This kind of documentation should be routinely a part of the job closeout procedure whether or not specifically required by the contract.

## The Importance of Timeliness and Good Administration

Two issues will control the efficiency with which a job is certified for final completion and the contract closed out. The first of these is the quality of the ongoing administrative processes while the project is in construction. The second relates to the amount of effort expended during the period between substantial completion and final completion. A special effort will be required to maintain interest in the project once it has reached substantial completion and before it has been finalized. When these items are tended to on a current basis, the closeout of the job is accomplished quickly and efficiently. Where such is not the case, a job otherwise satisfactory and profitable may deteriorate into an unhappy relationship between contractor and owner, significantly reducing the goodwill generated during the course of the job and reducing the probability of repeat work.

The collection of funds constituting final payment for the project is especially important. Collection of the funds on a timely basis allows every aspect of the job to be closed out quickly and efficiently. The final payment from owner to contractor constituting the holdback for unfinished or for corrective work may not be required for the funding of accounts payable. The general contractor may, therefore, elect to make final payment to all subcontractors and vendors at the time the payment for substantial completion is received, holding back only the value of items which are on the punch list.

It may be necessary to close out the contracts of subcontractors and principal vendors where they are unable, unwilling, or uninterested in performing the final work required to complete the project. In this case, the subcontractor or principal vendor item may be closed out with appropriate deductions for uncompleted or uncorrected work, with that work being completed by others. In this connection, an adequate personnel resource is always essential to insure the timely completion of work.

## Final Job Costs and Evaluation

Completion and payout of the accounts payable transactions along with the final labor costs allow the finalization of job cost, an important milestone relating to the final completion of the project. It further allows the calculation of a variety of performance ratios and certain other analytical routines which provide data and information for future work.

The production performance ratio should be calculated as an indication of the diligence with which the project has been pursued to a completion.

Of course, when the job is finally completed, the PPR will equal a nominal 1.0. The time performance ratio gives a clear indication of the time required for the various activities and establishes a valid statistical basis for estimating durations of activities. This kind of information is especially useful when durations are being established for new projects. And, as always, a primary interest is in the cost performance ratio, which indicates where the project stands at final completion with respect to the budget. With all the necessary adjustments having been made in the contract sum and that final contract sum agreed upon, the final overhead and fee may be calculated. All of the foregoing items relate to measuring the performance of the project from a historic point of view.

The use of historic performance data in establishing costs and schedules for future work is especially vital to an organization. There are two sets of information which may be generated for future reference. The first of these is total unit cost pricing based on the actual cost of all activities involved in the project. The generation of unit cost information increases the validity of new estimates in the context of integrated cost and schedule control. Within this context, it is important that the organization be able to generate independently its own quantifications and cost estimates (fair value) using historic unit prices without reference to marketplace pricing.

The generation of unit prices through a post job analysis of cost is especially important. These unit prices are, of course, available for future estimating and provide better information on the ability of individual companies to perform certain kinds of work as compared with standard pricing systems.

## PARAMETRIC COST DATA AND PRICING

Parametric methods of estimating allow pricing of a proposed job long before drawings and specifications are available, that is, when only the basic job concept and size are known. One of the primary values of parametric pricing is that it allows the generation of valid costing information for a future project based on historic job costing. In such a case, similar historic projects serve as the basis for the parametric cost data.

A major requirement for generating parametric pricing is redistributing historic estimates from the standard 16 division trade breakdown to a functional breakdown of the project. The functional breakdown of the completed project by major elements such as site work, foundations, superstructure, exterior closure, and similar items allows information to be prepared on a proposed new project while taking into account variables unique to the new estimate. In some cruder methods, the square footage of a building may be taken as the primary variable in parametric pricing and a cost per square foot may be applied to the square footage of the new project as the basis for generating cost information. Except when all of the elements in a project vary in direct proportion to the variance in, for example, square footage, some distortion will result from using such a single parameter for generating parametric pricing.

Each project is made up of a number of functional items. One summary listing (The Uniform Cost Index, UCI) includes:

1. 1000 Foundations
2. 2000 Substructure
3. 3000 Superstructure
4. 4000 Exterior closure
5. 5000 Roofing
6. 6000 Interior construction
7. 7000 Conveying system
8. 8000 Mechanical
9. 9000 Electrical
10. 10000 General conditions, overhead, and fee
11. 11000 Equipment
12. 12000 Site work

Each of these bears an arithmetic relationship with certain basic parameters of any project. For example, the exterior closure does not vary directly with the square footage of the building but rather in proportion to the quantity of exterior included in the original estimate. Using the square foot quantities for the exterior closure from the historic estimate and the proposed new estimate to construct the variable factor generates a much more valid parametric estimate of the exterior closure. Also, site work is more nearly associated with the area of the site rather than the floor area of the building placed on the site. The site area calculations for similar completed projects determine the variable factor for site work.

## OVERHEAD PRORATION

The last of the calculations recommended earlier to be made when a job is completed is to prorate company overhead to that (and every other) project. This requires a series of computations which are best done on a monthly basis. The percentage of total value of monthly company production each job's production represents is calculated. This percentage is then multiplied by the total overhead for the company during that period. The result of this calculation is the dollar amount of overhead to be charged to that job for that production period. When these calculations are made for each month, the overhead chargeable to that job as a function of total production may be calculated.

When the overhead is calculated by the process described, the percentage it represents of production cost should also be calculated. These calculations will result in historic data concerning the overhead percentage for the organization's operation. This information is essential in determing bidding strategies and the relative cost of overhead as a percentage of production.

As the volume of business of an organization *decreases* over time, the percentage for overhead on projects in work increases. Within limits, the corollary is true. That is, as a company's volume *increases* over time, the percentage for overhead tends to decrease, until saturation of organizational resources is reached. At this point, inefficiencies in handling a given volume of work exceeding organizational capacity occur.

Additional caution is required in using actual value of production as a basis for prorating overhead costs. Poorly performing jobs will be charged a disproportionately lower overhead cost while higher performing jobs will be penalized by carrying a higher portion of organizational overhead.

Two procedures are available to make overhead distribution more equitable. First, planned production may be used rather than actual production in determining the pro rata distribution of overhead.* The alternative involves a penalty system in which the production performance ratio is used to modify the overhead distribution in favor of the high production projects. The first, however, is perhaps the simplest and fairest.

## Individual Job Fee or Profit

These contingencies and limitations having been noted, the calculation of the percentage for overhead for each individual job provides instantly for the isolation of the fee portion of the spread (sum of overhead and fee) in a project, clearly indicating the contribution to profitability each project makes.

The (time) variable related to the duration of a project is a more significant factor than the dollar volume of the project. That is, the controlling variable is the *rate* of product placement, which in turn controls the duration of the project. This is why *production control* (as distinguished from *profitability*) is the focus of the integrated cost and schedule control process.

## Completing the Closeout

There are two last items which should be attended to as part of the closeout of the project. The first of these is to archive all job files and records to

---

* For example, suppose planned production is $100,000 with an overhead of 10%, or $10,000. Actual production is $50,000, with an overhead of $5,000 at the 10% rate. In this case, allocate the *planned* overhead charge, or $10,000, against the job.

some inactive medium, which simply means removing all files from active status. Where the job is processed through a computer, these active files are removed from the on line storage capacity to allow for additional on line processing capacity. Second, and perhaps the most interesting from a management point of view, is the issuance of a final job status report. This report may include as much information about the job and its performance as the information system can provide.

The significant value of the final status report is that it sets permanently in the record for future reference the *actual* final status of the project as compared to the expected final status *perceived* or *projected* at anytime by those involved in handling the project. There is a tendency among production personnel to focus on the performance of the job during the period of maximum productivity. Typically, however, the perceived final results of a project will deteriorate from the time of maximum production through the final completion and closeout. The amount of attrition may be significantly reduced by diligent attention to the necessary contract administrative details, to be fully discussed in the next chapter.

## SUMMARY

The final completion and job closeout involve accounting and administrative action, final job cost analysis and evaluation as well. Efficient contract administrative procedures throughout the life of the job greatly facilitate final completion and job closeout.

The efficient handling of the administrative aspects of job closeout significantly enhance the financial and accounting closeout. Administrative prerequisites to job closeout frequently stand in the way of collecting the final payment, however. Final payment is usually related to cash requirements necessary for the final payment of all accounts payable.

Final completion and job closeout should include a significant effort related to final job analysis and evaluation. These latter two reports provide substantial data and information on historic performance which may be used in planning and preparing for future project performance. They provide a basis for independent fair evaluation of projects proposed for future construction, which in turn allows necessary and proper evaluation of the marketplace and consequent adjustment of historic data for future jobs. The final completion and closeout of each job should receive the same preparation, diligence, effort, and commitment of resources to their accomplishment as do the bidding process to acquire the work and the planning, scheduling, and budgeting to set the job up for production.

# 18

# CONTRACT ADMINISTRATION

Changes to construction contracts are a part of contract construction. Disputes frequently result from changes in the scope of the work, changed conditions, or the effect of changes on the work. Change related disputes can, however, be minimized through efficient administrative procedures.

There is no direct relationship between the size or cost of the work and the number of changes which may be anticipated. There are, however, certain similarities in the number of changes in specific types of work.

Large engineering and institutional projects on which extensive planning is done and where the scope of the work is fully defined have the fewest changes in relation to size. Speculative commercial construction such as shopping centers, where construction often commences before the scope of the work is fully defined, has by far the greatest number of changes during construction. In commercial work, it is not unusual to have one change in the work for every $20,000 of the original contract sum. A shopping center project costing $5,000,000 might expect as many as 250 changes in the work during its completion.

Multiple changes on several projects being processed simultaneously make efficient administrative procedures essential to total project control. These are inseparably a part of integrated cost and schedule control.

## THE SYSTEM FOR MANAGING CHANGE

The system for managing changes in the work presented is a logic system. It has been developed and refined as a result of extensive use in the construction industry for more than 30 years. It can handle projects with 10 changes or 1,000 changes with equal efficiency. In short, it is sufficiently flexible to adapt its basic control elements to most construction contract environments.

### Basic Control Elements

The basic control elements of the system are listed below in the order in which they come into existence.

1. The working contract file.
2. The master X number list.
3. The master X file.
4. The X number.
5. The X file.
6. The work authorization.
7. The stop order.
8. The proposal for change.
9. The change order file.
10. The change order.
11. The production authorization.

Each of these basic control elements has a specific function in the firm administrative control of the construction project. It is assumed that the contract documents have been properly prepared and executed, and are available.

## Control Elements Explained

The control elements for contract administration and their organization and use within the context of a unified system are essential to integrated cost and schedule control. Each is explained in some detail:

**The Working Contract File.** The executed original contract documents should be placed in a secure area for safekeeping. To preserve the integrity of the original contract documents, a separate working contract file folder is used to maintain a copy of the executed contract, the notice to proceed, and other contracturally relevant items. The copy of the contract and all other items except change orders are fastened to the righthand leaf with the contract on top. The lefthand leaf is used for change orders with the latest on top. When the contract file is laid open, a copy of the executed contract is on the right and the latest change order is on the left. The latest change order provides the complete current status of the contract, the adjusted contract sum, the adjustment to contract time, and changed terms and conditions. Exhibit 18-1 shows the contract file.

**The Master X Number List.** The master X system is first and foremost a memory system. From its origin, "X" has referred to *extra* or *claim*. While this definition still applies, it is expanded to include any contemplated change or modification to the work or to the contract or its terms and conditions. Every contemplated change or modification is immediately placed on the master X number list and is thereby assigned an X number. This procedure dilligently followed brings each item within administrative control and creates a record of the sequence of events relating to each contemplated change.

**Exhibit 18-1. Working Contract File**

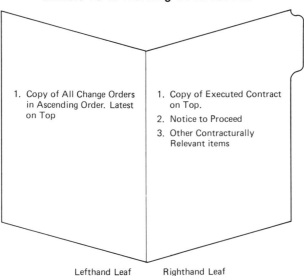

| 1. Copy of All Change Orders in Ascending Order. Latest on Top | 1. Copy of Executed Contract on Top. |
|---|---|
| | 2. Notice to Proceed |
| | 3. Other Contracturally Relevant items |

Lefthand Leaf          Righthand Leaf

Items are placed on the list as they come to light. No value judgment is made concerning the probabilty of that item subsequently becoming a change order.

The X numbers must be assigned from the master X number list only. As many subordinate copies of the list may be maintained as desired but there can be *only one master X number list.* At the beginning of each project, responsibility should be assigned for maintaining the master X number list. It makes no practical difference who maintains the master list provided two absolute rules are followed:

1. There must be no value judgment or debate concerning the addition of any item to the master X number list. The mere fact that one of the parties to the contract feels an item should be listed is sufficient. The normal contract administrative procedures then will operate to subsequently validate the item or clear it from the list of changes.
2. There must be only one master list maintained by one party to the contract from which all X numbers are assigned.

In its evolution over the years, the master X number list has been expanded to record the processing of a contemplated change or modification to the contract from the time it is assigned an X number through its final contractural dispostion. The processing procedures will be described later in detail.

Exhibit 18-2 shows the master X number list form. The purpose and function of each column is worth noting. The "File" column contains check spaces to indicate that the X file for each item has been prepared. The X number list is used to assign X numbers; these are prenumbered. The "Date" space is used to record the day on which each item is added to the master X number list. The "Description of Claim" space is for a very brief identifying description of a claim, three or four words at most. Columns one through four, then, are used for the initial data entry for each change or X number as it occurs.

The remaining columns track the processing of the change through its final deposition. "Date of Notice" is for recording the date upon which notice of claim as required by the contract is given to the parties to the contract; "Date Work Authorization" records the date a work authorization is issued

## Exhibit 18-2. Master X List

MASTER X NUMBER LIST

Job Name_____ Job No._____    First Notice Rec'd By_____
Location_____    To Be Sent To_____
Owner_____    Cost Data Rec'd By_____
Contract Date_____    Apprvl. To Be Issued By_____
Notice to Proceed Date_____    Continuing Notice Rec'd By_____

DON'T FORGET; THIS MASTER "X" LIST IS A MEMORY SYSTEM,
IF IN DOUBT WRITE IT DOWN. DON'T FORGET

| X Number | File | Date | Description of Claim | Date of Notice | Date Work Authoriz. | Date Stop Work Order | Proposal | | Prod. Auth. Change Order | | | | | Notes |
| --- | --- | --- | --- | --- | --- | --- | --- | --- | --- | --- | --- | --- | --- | --- |
| | | | | | | | Date | Amount | Cond. | Final | # | Date | Amt. | |
| 1 | | | | | | | | | | | | | | |
| 2 | | | | | | | | | | | | | | |
| 3 | | | | | | | | | | | | | | |
| 4 | | | | | | | | | | | | | | |
| 5 | | | | | | | | | | | | | | |
| 6 | | | | | | | | | | | | | | |
| 7 | | | | | | | | | | | | | | |
| 8 | | | | | | | | | | | | | | |

Initial Data Capture — Required Notices and Authorization — Proposal for Change Information — Internal Authorization — Change Order Data or Final Deposition as allowed by the contract when prior to the execution of a change order; "Date Stop Work Order," the date of any stop work order issued in connection with that item; "Proposal, Date, Amount," the date and amount of the original and subsequent proposals for change for that item; "Production Authorization, Conditional, Final," the dates on which the production authorization(s) are issued; and "Change Order Number, Date, Amount" records change order numbers, the date on which the last party executes the change order, and the amount. The last column is blank and is used for various notes, cross-references or other information.

The notice requirements for claims section at the upper right of the form holds certain contractural information needed in the processing of claims. "First Notice Required By" records the contractural requirement, usually in days, that notice of a claim must be given; "To Be Sent To," the name of the party to the contract to whom notice is to be sent; "Cost Data Required By," the contractual requirement, again in days, that stipulates the time period allowed for making monetary claims; "Approval to Be Issued By," the name of the party to the contract who will issue or grant approval of proposals for change; and "Continuing Notice Required By," notes any contractural requirement which requires more than one notice of claim.

**The Master X File.** The master X file (Exhibit 18-3) contains the master X number list securely fastened to the righthand leaf. On larger work, the

### Exhibit 18-3. The Master X File

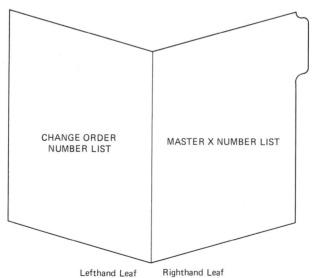

CHANGE ORDER
NUMBER LIST

MASTER X NUMBER LIST

Lefthand Leaf     Righthand Leaf

## Exhibit 18-4. Change Order Number List
## (Showing C.O. Numbers Assigned)

CHANGE ORDER NUMBER LIST (for the purpose
of knowing what change order numbers
have been assigned)

CROSS OUT NUMBER AS IT IS ASSIGNED

| C.O. | ~~1~~ | ~~26~~ | 51 | 76 |
|---|---|---|---|---|
|  | ~~2~~ | ~~27~~ | 52 | 77 |
|  | ~~3~~ | 28 | 53 | 78 |
|  | ~~4~~ | 29 | 54 | 79 |
|  | ~~5~~ | 30 | 55 | 80 |
|  | ~~6~~ | 31 | 56 | 81 |
|  | ~~7~~ | 32 | 57 | 82 |
|  | ~~8~~ | 33 | 58 | 83 |

master X number list may actually consist of a number of pages, which are fastened in chronological order with the latest or higher X numbers on top. The lefthand leaf is used for a consecutive change order number list (Exhibit 18-4) to facilitate assigning change order numbers. As each change order number is assigned, that number is marked. This convenience is particularly helpful on larger jobs.

**The X Number.**  The X number is a numerical identification assigned to each contemplated change or modification. It has no contractural significance or status of itself. It is placed on an X file for orderly filing and data retrieval pending further processing of the claim.

**The X File.**  A separate X file (Exhibit 18-5) is prepared for each X number when a contemplated change or modification is assigned an X number. All data concerning that change, correspondence, notice of intent to file claim, cost estimates, bids, and so on, are maintained in the X file and fastened to the righthand leaf of the file. When the proposal for change is written, the X file copy is attached on the right on top of all supporting data.

The lefthand leaf is used for attaching all related authorizing documents such as work authorizations, purchase orders, notices to proceed, production authorizations and, finally, the change order. When the X number results in a change order, a copy of the change order is attached on the left on top of all other authorizing documents. The change order number is assigned

**Exhibit 18-5. Typical X File**

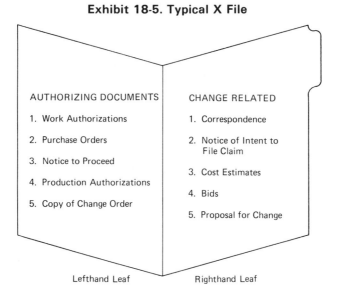

AUTHORIZING DOCUMENTS

1. Work Authorizations

2. Purchase Orders

3. Notice to Proceed

4. Production Authorizations

5. Copy of Change Order

CHANGE RELATED

1. Correspondence

2. Notice of Intent to File Claim

3. Cost Estimates

4. Bids

5. Proposal for Change

Lefthand Leaf          Righthand Leaf

to the X file and the X file then becomes the change order file. When the change order file is laid open, the (latest) proposal for change for that X number is on the right. The change order is on the left, giving an immediate visual comparison between proposed and authorized.

The *X file tab* (Exhibit 18-6) should be set up to provide information required at a glance. Color coding is recommended. Space is needed for job number, X number, future change order numbers, and "Finalization" note. Descriptor lines should include the X number description using exact wording as in the master X file, the name of the party to whom the claim is to be submitted, the name of the job, and the job location.

The *change order finalization checklist* is attached to the face of the X file immediately below the tab, as shown in Figure 18-6. This checklist and its use is described in detail under Change Order File.

**The Work Authorization.**  The ultimate or final owner/contractor authorization for any change is a *change order.* Circumstances on occasion require that changes be implemented *prior* to the issuance of a change order. The *work authorization* is used for this purpose and is an interim measure pending the issuance of a change order. The work authorization may be made by letter or by using standard forms and is always superseded by a change order.

**Exhibit 18-6. X File Tab and Change Order Finalization Order**

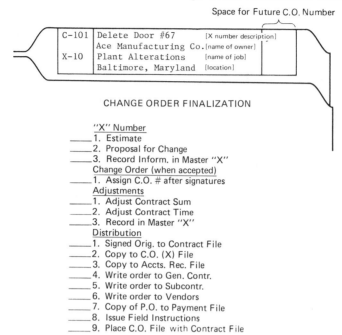

Space for Future C.O. Number

| C-101 | Delete Door #67 | [X number description] |
| | Ace Manufacturing Co. | [name of owner] |
| X-10 | Plant Alterations | [name of job] |
| | Baltimore, Maryland | [location] |

CHANGE ORDER FINALIZATION

"X" Number
_____ 1. Estimate
_____ 2. Proposal for Change
_____ 3. Record Inform. in Master "X"
Change Order (when accepted)
_____ 1. Assign C.O. # after signatures
Adjustments
_____ 1. Adjust Contract Sum
_____ 2. Adjust Contract Time
_____ 3. Record in Master "X"
Distribution
_____ 1. Signed Orig. to Contract File
_____ 2. Copy to C.O. (X) File
_____ 3. Copy to Accts. Rec. File
_____ 4. Write order to Gen. Contr.
_____ 5. Write order to Subcontr.
_____ 6. Write order to Vendors
_____ 7. Copy of P.O. to Payment File
_____ 8. Issue Field Instructions
_____ 9. Place C.O. File with Contract File

**The Stop Order.**  The stop order is similar to the work authorization except that it allows the owner to interrupt the flow of work. It may be issued verbally, by letter, or on a standard form. When issued verbally, it must be always confirmed in writing immediately by the contractor since it constitutes the basis for a delay claim.

**The Proposal for Change.**  The proposal for change (Exhibit 18-7) is made in writing as a formal submission from the contractor to the owner for a change in cost, in time, in conditions of work at the site, or in contract terms and conditions. This can be done by letter or by using any of a number of standard forms available. Care should be exercised, however, so that *proposals for change* are not confused with *change orders;* this will be further explained under the section dealing with change orders. The proposal for change form may also be used to give formal notice of intent to file a claim.

**The Change Order File.**  Within the framework of this system, the X file containing all the supporting data for a given change becomes the change

## Exhibit 18-7. Proposal for Change

A B C Construction Co.
1234 Fifth Avenue
City, State

PROPOSAL FOR CHANGE:

Ace Manufacturing Co.
City, State

JOB NO. __101__ X __15__
ESTIMATE NO. __9101__
DATE: __Dec. 10, 1985__

ATTENTION: Plant Manager

We propose to effect the following changes or extensions to our original agreement and if accepted, you agree to increase/decrease your payments to us in the lump sum of:

Five hundred fifty dollars
_____

($  550.00  )
_____

and to extend the contract time by ____10____ calendar days.

SCOPE OF WORK:

| | | |
|---|---|---|
| 1. | Delete door #67 | $ 750.00 |
| 2. | Add masonry and finishes | 250.00 |
| | Net deduct | 500.00 |
| 3. | Credit overhead and fee | 50.00 |
| | Total credit | $ 550.00 |

This change does/does not represent and change the working conditions at the job site.

All terms and conditions of the original agreement including terms of payment, apply hereto with equal force and effect.

Your signature below accepting this proposal effects a formal Change Order and authorizes us to proceed with such changes of extensions to the work.

May we have your immediate acceptance or rejection on this proposal to minimize delays.

ACCEPTED:                    PROPOSED:

THRU CO # _____ FOR:_____ FOR:_____

CO #_____ BY:_____ BY:_____

CO #_____ TITLE:_____ TITLE:_____

DATE:_____ C.O. NO.____DATE_____

ADJUSTED CONTRACT SUM INCLUDING THIS CHANGE ORDER WILL BE:

_____

($            )
_____

order file upon issuance of the change order. The *change order number* is contracturally significant and once it is assigned it takes precedence over the X number. To facilitate the completion of all administrative details required in the conversion of an X number to a change order, a change order finalization checklist is attached to the front of the X file below the tab. Each item on the list is checked off as it is completed. When all are complete, the administrative processing of the change order is complete.

The completion of all checklist items is noted by drawing a diagonal line through the checklist, placing a capital "F" in a circle on the righthand file tab removing the file from among the X numbers and refiling it numerically according to the change order number with the working contract file. Exhibit 18-8 shows a finalized change order file.

**The Change Order.** There are a number of different meanings given to the term change order in the industry. For its use with this system, a change order is defined as follows:

A change order is a written legally binding modification to the contract between the owner and contractor setting forth its effect on the contract

**Exhibit 18-8. Checklist and Finalized Change Order File**

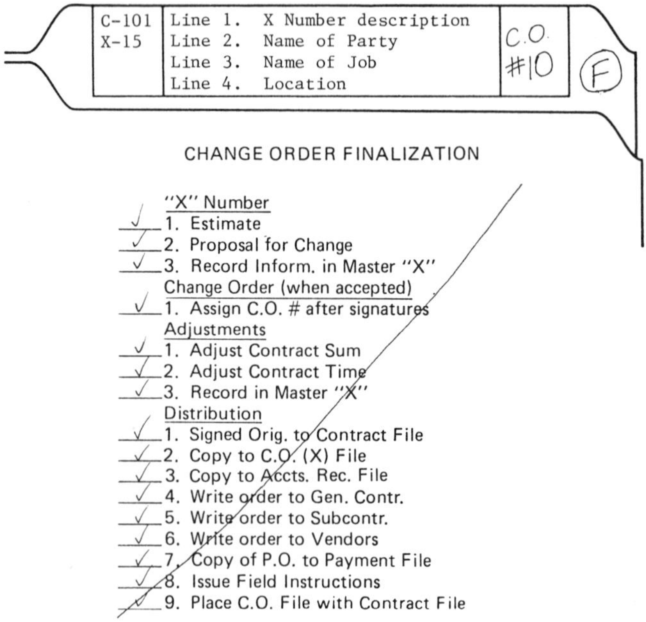

sum, the contract time, the contract terms and conditions, and/or the condition of work at the site; and is signed by the parties to the contract.

Within this definition, the change order must set forth its effect on time, cost, working conditions, and terms of the contract.

The change order's effect on cost is a lump sum amount and contains an adjusted contract sum (except where the original contract sum is other than a lump sum). It adjusts the contract time, stipulates whether any of the terms and conditions of the contract are modified and describes any changes in the field working conditions. It must be signed (executed) by the parties to the contract.

Five prerequisites to issuing a change order are a more stringent administrative constraint than required by some of the widely used standard forms of contract. Each prerequisite, however, has its specific purpose. Each is designed to separate points of agreement from points of dispute and in forcing the timely resolution of all changes or modifications to the contract. The five prerequisites to the change order are explained:

*Time.* The effect of a change on contract time must be determined. Making it a prerequisite to the issuance of a change order expedites that determination and documents the time consideration in a valid, legally binding way, especially important in contracts containing penalty and bonus provisions.

*Amount.* The cost of a change must be determined. It is usually easier for the parties to agree on the value of a change *before* the work is performed. Lump sum amounts in change orders permit the current updating of the contract sum. Even where the basic contract is other than lump sum, the accumulated lump sum of change orders can and should be maintained.

*Terms.* Every change order must state the effect it has on the terms of the contract. If there are none, the change order should so state.

*Changed Working Conditions.* When the change order is the result of changed working conditions, this change must be described.

*Signatures.* Finally, a "change order" is not a change order until signed by the parties to the contract. Only *after* signatures are affixed can the change order number be assigned.

The prerequisites for issuing a change order are necessary because change orders have economic and legal consequences. Ideally, no change or modification should be placed into production until a change order for that work has been issued. When practical considerations or emergencies may intervene, other procedures, such as the work authorization given by the owner are available to allow the work to proceed pending the issuance of the change order. These expediencies, however, must not be permitted to compromise the prerequisites and procedures for the issuance of change orders.

## Exhibit 18-9. Executed Change Order

A B C Construction Co.
1234 Fifth Avenue
City, State

PROPOSAL FOR CHANGE:

Ace Manufacturing Co.
City, State

JOB NO. __101__ X __15__
ESTIMATE NO. __9101__
DATE: __Dec. 10, 1985__

ATTENTION: Plant Manager

We propose to effect the following changes or extensions to our original agreement and if accepted, you agree to increase/decrease your payments to us in the lump sum of:

Five hundred fifty dollars
_____

_____ ($    550.00    )

and to extend the contract time by _____10_____ calendar days.

SCOPE OF WORK:

|   |   |   |
|---|---|---|
| 1. | Delete door #67 | $ 750.00 |
| 2. | Add masonry and finishes | 250.00 |
|   | Net deduct | 500.00 |
| 3. | Credit overhead and fee | 50.00 |
|   | Total credit | $ 550.00 |

This change does/does not represent and change the working conditions at the job site.

All terms and conditions of the original agreement including terms of payment, apply hereto with equal force and effect.

Your signature below accepting this proposal effects a formal Change Order and authorizes us to proceed with such changes of extensions to the work.

May we have your immediate acceptance or rejection on this proposal to minimize delays.

ACCEPTED:                     PROPOSED:

THRU CO # __4__ |150,380.00 FOR: Ace Manufacturing Co. FOR: ABC Construction Co.

CO # __5__ | (550.00) BY: *James Jones* BY: *John Smith*

CO # __5__ |149,830.00 TITLE: Plant Mgr. TITLE: Project Mgr.

DATE: 12/18/85 _____ C.O. NO. __5__ DATE 12/15/85

ADJUSTED CONTRACT SUM INCLUDING THIS CHANGE ORDER WILL BE:

One hundred forty-nine thousand Eight hundred thirty dollars
_____

_____ ($ 149,830.00)  )

Exhibit 18-9 shows a proposal for change which has been processed into a valid change order and assigned a change order number.

**The Production Authorization.** All of the basic control elements thus far discussed have been essentially in the area of owner/contractor relationships. Internal control and communication within the contractor's organization is equally important and the production authorization is used for these purposes.

The production authorization is the contractor's internal communication tool and is in two versions. The first of these is the *conditional* production authorization (see Exhibit 18-11, p. 282). It is essentially the same as the production authorization except for its different color and its "conditional" limitation. The second is the (final) production authorization (Exhibit 18-10), which is used only after the change order for the work has been issued.

### Exhibit 18-10. (Final) Production Authorization

**(ABC)** CONSTRUCTION CO.

DISTRIBUTION
- ☒ Construction Manager
- ☒ Field
- ☒ Accounting
- ☒ Contract Administration
- ☐ _____
- ☐ _____

PRODUCTION AUTHORIZATION

JOB NAME: Century Office Project    PRODUCTION AUTHORIZATION NO.: 001
LOCATION: Metropolitan State    DATE ISSUED: 2/10/86    X NO.: —
OWNER NAME: Office Developers Inc.    JOB NO.: 0686    C.O. NO.: —
OWNER ADDRESS: 1234 Main Street    AUTHORITY: CONDITIONAL ☐    FINAL ☒
Midtown, State    CONTACT PERSON: John Jones    PHONE: 321-9876

Description of Work and/or Actions Required:

Constuction office building complex in accordance
with contract documents.

All work and/or actions covered by this Production Authorization shall be reported on Daily Reports and Daily Time Sheets as JOB NO.: 0686    X NO.: —    C.O. No.: —

| OWNER AUTH. | CREDIT APPR. | ADMINISTRATIVE APPR. | PLACED IN PROD. |
|---|---|---|---|
| BY: Contract | BY: A. Smith | BY: J. Miller | BY: A. Schultz |
| TITLE: | TITLE: Controller | TITLE: V.Pres. | TITLE: Chief of OP. |
| DATE: 2/15/86 | DATE: 2/16/86 | DATE: 7/20/86 | DATE: 2/27/86 |

## Exhibit 18-11. Conditional Production Authorization

DISTRIBUTION
☒ Construction Manager
☒ Field
☒ Accounting
☒ Contract Administration
☐ _____
☐ _____

(ABC) CONSTRUCTION CO.

PRODUCTION AUTHORIZATION

JOB NAME: Century Office Project          PRODUCTION AUTHORIZATION NO.: 015
LOCATION: Metropolitan State          DATE ISSUED: 3/17/86          X NO.: 21
OWNER NAME: Office Developers Inc.          JOB NO.: 0686          C.O. NO.: Pending
OWNER ADDRESS: 1234 Main Street          AUTHORITY: CONDITIONAL ☒          FINAL ☐
          Midtown, State          CONTACT PERSON: John Jones          PHONE: 321-9876

Description of Work and/or Actions Required:

    Hold foundations for Building #3 pending owner decision
    on increasing building size by adding two additional bays.

All work and/or actions covered by this Production Authorization shall be reported on Daily Reports and Daily
Time Sheets as JOB NO.: _____ X NO.: _____ C.O. No.: _____

| OWNER AUTH. | CREDIT APPR. | ADMINISTRATIVE APPR. | PLACED IN PROD. |
|---|---|---|---|
| BY: Verbal J. Jones | BY: A Smith | BY: J. Miller | BY: R Schultz |
| TITLE: Project Manager | TITLE: Controller | TITLE: V. Pres. | TITLE: Chief of OP. |
| DATE: 3/17/86 | DATE: 3/17/86 | DATE: 3/19/85 | DATE: 3/9/85 |

The production authorization contains all of the information contained in the change order along with other instructions to field personnel which are deemed necessary to the implementation of the change order.

The conditional production authorization is used for all other change related communications. Examples include holding work in areas of a pending change, allowing work on a change to proceed based on a work authorization pending the issuance of a change order, or for merely informing the field personnel that a change is under consideration. The conditional production authorization is an interim document and must always be superseded by a (final) production authorization following the issuance of the change order. Conditional production authorizations are always assigned the related X number. Production authorizations are always assigned the related change order number and cross-referenced to the X number.

## FILES AND FILING PROCEDURES

Several filing and clerical comments are appropriate before proceeding with an examination of the system in operation.

Two separate file drawers or sections of files should be used for maximum efficiency.

One section is used to retain the master X file followed by all X files in numerical sequence according to the X number. During the course of a job, this file section may be quite active and should therefore be readily accessible to the personnel responsible for maintaining the master X file and for processing the X numbers through final disposition.

The second section contains the working contract file followed by all change order files in numerical sequence according to change order numbers, in turn followed by all void and finalized X files also in numerical sequence according to X numbers. Except for the working contract file, the files placed in this section are fully processed and are maintained for record and reference purposes only. Also, since this section contains confidential contractural data, it should be maintained in a more private and secure location.

As the X numbers are processed, each X file becomes either a change order file or a void and finalized X file. In both cases, these are transferred from the X file section to the contract file section when processing is complete. This procedure has two purposes. The first is to continually clear the more active X file section of unnecessary files. Second, since the X files in the X file section at any time represent the open or unresolved X numbers, a quick review of these files will indicate the number of open items. Such reviews are especially convenient during the most active periods of a job and when the work is nearing completion.

File folder tabs should be color coded with one color for X/change order files, another for the master X file, and a third for the contract file. Recording this information on the master X list neatly by hand will prove quite convenient and expeditious. Using regular staples to attach materials to the X/change order file folders will substantially reduce the bulk of those files; it is also less expensive and faster. The change order finalization checklist can be typed and duplicated and then glued or taped to the front of each file folder directly below the file tab. If large quantities of X/change order files are required, the checklist may be printed directly on the file folder.

## THE SYSTEM IN OPERATION

The purposes of administrative procedures are control and efficiency. The wide range of variables which are possible in changes or modifications to construction contracts make flexibility and adaptability of equal importance.

The following case studies are presented to examine the system in operation under a considerable range of situations.

GROUP I. Changes in Scope of Work by Owner

*Case 1.* The owner is considering the addition of a small item which has no effect on any work presently in progress.

*Case 2.* The owner is considering a substantial change which will affect work currently being performed in terms of cost and time factors.

*Case 3.* The owner finds it necessary to make an immediate change in the work affecting work already in progress as well as work to be shortly commenced with time a major factor.

*Case 4.* The owner finds it necessary to make a change but cannot agree with the contractor on the cost of the work or the time required for its completion.

GROUP II. Changes in Conditions of the Work

*Case 5.* Some unexpected condition such as unsuitable soil bearing, rock, or obstructions is discovered in the field and requires a change in the work to compensate for these conditions.

*Case 6.* An emergency such as an accident at the site of the work develops, requiring immediate action to maintain or restore safe conditions.

GROUP III. Contractural Claims

*Case 7.* Claims by either party to the contract may be made even though no actual change in the work is involved.

*Case 8.* The voided and finalized X number.

## Group I. Changes in Scope of Work by Owner

The preceding eight cases do not represent *all* the possible changes to a construction contract. They do serve to illustrate the features of the system and show how the basic control elements can be applied in every other situation. These illustrations assume that the contractor is maintaining the master X list.

**Case 1.**  Case 1 represents the simplest kind of change, a change which is totally at the option of the owner and has no effect on other parts of the work. The following steps are involved in processing the change:

1. The owner decides to examine the possibility of making a change in the contract.
2. The owner enlists the aid of the designer to quantify and qualify that change.
3. The data concerning the proposed change is forwarded to the contractor, requesting a proposal for change.
4. The contractor enters a brief description (three or four words at most) in the next open space on master X number list and assigns the next X number to that item.
5. An X file with the color coded tab is prepared as described in Exhibit 18-6. The change order finalization checklist is attached to the front of the file immediately below the tab. When the X file is prepared, a checkmark is placed in the X file column on the master X number list to avoid file duplication. Descriptive data concerning that change is stapled to the righthand side of the file.*
6. The contractor prepares a cost estimate for the change using sub prices, material prices, and, on occasion, his or her own estimates. To the cost, the contractor adds the percentages allowed by the contract for general conditions, payroll insurances and taxes, and overhead and fee.
7. The contractor prepares the proposal for change with the appropriate number of copies and forwards these to the owner for review and action. To expedite review, all appropriate supporting data is attached.**
8. The owner reviews the proposal for change, finds it acceptable, signs it, and returns the appropriate number of copies to the contractor for final processing.
9. Upon receiving the executed proposal for change, the contractor adjusts the contract sum in the space provided, adjusts the contract time, and assigns the change order number. One copy of the change order is returned to the owner. The change order is thus signed, sealed, and delivered.
10. The change order number is assigned to the X file in the space provided (see Exhibit 18-8); the X file thus becomes the change order file. The formation about the change order is recorded in the master X file.

---

* Since only a proposal has been requested, no entry in the "Date of Notice" column, the "Date of Work Authorization" column, or "Date of Stop Work Authorization" column is required.
** Supporting data should be attached to the proposal for change especially where large numbers of changes are to be processed. This procedure substantially reduces the review process and the time and effort required to clarify details of the proposal.

11. The signed original of the change order is placed on the lefthand leaf of the working contract file. A copy is also placed on the lefthand leaf of the change order file.
12. Appropriate accounting records of the change order are made and orders are issued to the appropriate subs and vendors.
13. The production authorization is issued to instruct the contractor's staff and field personnel on the actions required pursuant to the change order. A copy of the production authorization is placed on the lefthand leaf of the change order file under the copy of the change order.
14. Several other actions are required by the contractor to complete the change order finalization checklist on the front of the change order file. A checkmark indicates the completion of each finalization item. When all are completed, the processing of the change order is complete or final, indicated by a diagonal line through the completed checklist, with a capital "F" in a circle on the righthand end of the tab. At this point, the file is removed from the X file and refiled with the working contract file according to the change order number. (See Exhibit 18-8).

These explanations may appear somewhat tedious. However, except for the clerical details, the steps listed are required to complete a change order transaction in any event. The clerical details will not be repeated in the subsequent cases; only the necessary adjustments will be noted.

## Case 2. A Contemplated Change Affecting Work in Progress.

Case 2 differs from case 1 in that action is required to discontinue work which may be affected by the change. The procedures to be followed are the same as case 1 with two major exceptions. The owner must issue a stop work order (preferably in writing) stipulating exactly what work is to be held and the duration of the hold. The contractor must file a notice of claim for the record. Where the stop work order is given verbally by the owner, it should be immediately confirmed in writing by the contractor. Additional clerical steps are now required as follows:

1. The date of the stop work order should be immediately recorded on the master X list on the line of the appropriate X number. The data used should be the actual date on which the stop work order becomes effective.
2. Since a stop work order will affect contract time and could affect cost or terms, notice of claim should be given as required by the contract. The notice of claim can be included in the confirmation of the stop work order or in acknowledgement thereof. In either case, the date

on which the notice required by the contract is made is also recorded on the master X number list.

As noted above, the remaining procedures outlined for case 1 apply.

**Case 3. Work Authorized Pending a Change Order.** Case 3 presents a situation in which work must begin on a change without waiting for estimates, proposal for change, negotiations or change orders. While this situation has greater potential for disputes than waiting for a change order, it can nonetheless be handled by using a work authorization. The work authorization, originated by the owner preferably in writing, should contain the following:

1. An adequate description of the change to be undertaken and its effect on other work.
2. Expected effects on contract time and completion date.
3. A statement concerning how the costs are to be determined, including which of the methods of computation set forth in the contract will apply.
4. The terms, including how progress payments will be made pending the issuance of a change order.

When the owner's work authorization is verbally conveyed, it should be immediately confirmed in writing by the contractor. The date of the work authorization should be recorded in the master X file. The work authorization remains in effect until the change order is issued. Every effort should be made to process the proposal for change to completion without delay. The importance and value of resolving open items on a timely basis cannot be overemphasized.

The work authorization is the owner's direction to the contractor to proceed with certain work. The contractor must relay these instructions to the office and field staff. Since the change order is prerequisite to issuing a (final) production authorization, a procedural difficulty is encountered. This difficulty is overcome with a color coded conditional production authorization which provides the necessary instructions to proceed with the change. The conditional production authorization is an interim document and must always be superseded by a (final) production authorization as soon as the change order is issued.

Both the work authorization (*owner's* document) and the conditional production authorization (*contractor's* document) are temporary expedients. These should not be used except in situations in which prerequisites to a change order and production authorization cannot be met on a timely basis.

**Case 4. Work Must Proceed, But Price, Terms or Time Are Not Agreed Upon.** Case 4 presents yet another possibility concerning a proposed change in the work. It begins as case 1 and proceeds through step 7 (p. 285). The proposal for change and all supporting data is submitted to the owner for review. The owner reviews the proposal for change but disagrees with something in the proposal. After one or more attempts to negotiate the differences fail to produce an agreement, action must be taken to implement the change. Most standard contract forms provide for this contingency. The AIA General Conditions are particularly useful in this situation. It allows the owner the right to instruct the contractor to proceed with the change in the work on a time and material basis with stipulated percentages for overhead and fee. Cost records are kept by the contractor and regularly acknowledged by the owner. Immediately upon completion of the work, an accounting of cost is made. This procedure is known as a *force account.*

The work authorization is to be used to set forth the owner's instructions and option under the contract to proceed with the work. The primary difference between this and case 3 is that an accounting of cost (rather than an *estimate*) is made when the work is completed and becomes the basis for the proposal for change. The remaining procedural steps are the same as in case 1.

Cases 1 through 4 are changes that originate with the *owner* and come about as the result of some change in the owner's needs. These cases serve to illustrate the flexibility of the system in formally handling any change which the owner may find either desirable or necessary.

### Group II. Changed Conditions of the Work

Case 5 and case 6 are changes resulting from conditions beyond the control of *either* the owner or the contractor.

**Case 5. Non-emergency Unexpected Conditions.** Case 5 involves the discovery of unexpected conditions in the field. Upon this discovery, the contractor places the item on the master X list (following all previously discussed procedures) and notifies the owner of the discovery usually in the form of a notice of claim requesting instructions from the owner on an appropriate strategy to remedy the problem. Assuming no emergency is involved, this strategy becomes the description of the scope of the work. Any of the previously discussed procedural applications can now be used in processing the change. In all probability, a combination of case 2 and 3 procedures would apply. What is important, however, is to follow the appropriate procedures and to record all actions taken through the issuance of the change order.

**Case 6. Emergency Action on Unexpected Conditions.**  Case 6 differs from case 5 in that an emergency exists which requires immediate action to maintain or restore safe conditions. Most contracts allow the contractor to act unilaterally in emergencies. Such action is required even if the contract is silent on the issue. In emergencies, corrective action comes first but good administrative procedures should shortly follow.

In this case, two X numbers are preferred. The first X number is limited to the unilateral emergency action taken by the contractor to stabilize a dangerous situation. The second X number is used for the non-emergency corrective or remedial work which can be jointly determined by owner and contractor. This allows the system to function within practical considerations without compromising administrative control.

## Group III. Contractural Claims

Group III claims deal with *contract issues:* terms and conditions, or errors and omissions.

**Case 7. Other Claims.**  Case 7 covers the general category of claims other than those related to changes in the work or other modifications. While some claims originate with the owner with respect to penalties for delays in the work and similar items, most of such claims originate with the contractor. These are frequently related to errors and omissions in the description of the work or ambiguities in the contract documents. A request for added time and compensation related to conditions beyond the contractor's control may also result in a claim. The absence of claims in these categories is indirectly a measure of the clarity of the contract documents.

The procedures for processing non-scope claims are the same as the procedures for processing changes in the scope of work. The same diligence in following procedures is required for these claims as for other changes in the work. However, since the resolution of a claim may not require action at the site of the work, it may receive only secondary attention. Instead, all open items, changes, and claims alike should be treated as administrative emergencies to be resolved aggressively.

## Case 8. The Voided and Finalized X Number

In case 8 the owner finds the proposal for change unacceptable (on a nonessential or optional item) or decides not to proceed with the change. When the proposal is rejected, a finalization procedure is followed to clear the X file from the active X file roster. A line is drawn through the change order finalization checklist; "Void" is written on the file tab in the space provided

### Exhibit 18-12. The Voided X File

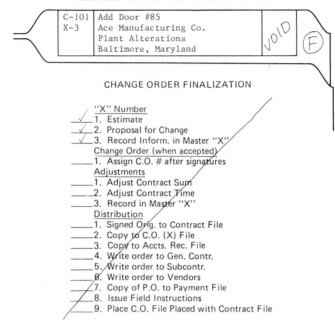

for the change order number; an encircled "F" is placed on the righthand end of the file tab; and appropriate notation is made in the master X file. Exhibit 18-12 shows the voided X file for an X number *not* resulting in a change order.

The finalized void X file is removed from the X files and refiled with the working contract file behind the change orders in numerical order according to X numbers.

## SUMMARY

One of the greatest weaknesses throughout the construction industry is poor and ineffective contract administration. Many owners and contractors both hold the notion that there should be no changes, modifications, disagreements, or disputes during the performance of a well prepared construction contract. As a result, when such develop, as they surely will, the contracting parties are confronted with the unexpected task of handling changes and resolving disputes and disagreements which delay the fulfillment of the owner's needs and delay compensation to the contractor, to the detriment of both.

Expensive and time consuming litigation is often the final recourse and all of the parties to the contract are the loosers. It is a sad commentary on the perils of construction as a creative activity when this occurs.

The purpose of the preceding comments is to focus attention on the risk of disappointment which is inherent (though by no means inevitable) in construction. A proper concept of the dynamics of construction activity and a realistic philosophy of interpersonal relationships and firm efficient administrative procedures will minimize these risks. Proper recognition of these concepts will give the parties to the contract confidence that their interests are being protected. With these also will come the freedom to enjoy the creative activity of construction.

I. Dynamics of Construction
    1. The job will be finished.
    2. There will be changes, modifications, and claims.
    3. There will be disputes and disagreements.
    4. Changes can be accommodated.
    5. Disputes can be resolved.
II. Philosophy of Interpersonal Relationships
    1. The assertion of a claim is not an attack on personal or corporate integrity.
    2. Disputes and disagreements should not flow from or result in bitterness and animosity.
    3. A solution is available to every problem.
III. Firm and Efficient Administrative Procedures
    1. Firm procedures are necessary to command respect and compliance.
    2. Procedures need to be sufficiently flexible to adapt to every situation.
    3. Efficient administrative procedures reduce the work load and enlist cooperation and usage.

The added purpose of this summary is to point out the indispensability of good administrative procedures. These should be agreed upon in advance and become an integral part of the terms and conditions of the contract. The administrative procedures must be in place before the work begins. Certainly the economic benefits to all parties should motivate the use of the procedures set forth in this chapter.

Efficient administration has its own reward: a satisfaction that comes from executing any difficult or complex assignment well. There is simply no substitute for the assurance of firm control in the construction process.

The greatest reward of all comes with the satisfactory closeout of the job and the completion of the contract when all of the changes, modifications, and claims are finalized. Now all parties to the construction contract can enjoy the solid result of their efforts and the satisfaction that comes only with the creative activity of converting an owner's need for construction to a reality.

# 19
# CASH MANAGEMENT INFORMATION AND STRATEGIES

The issues of cash flow and cash management are given little attention in most business, management, professional, and accounting references. There seems to be an assumption that cash is available, that credit is available to routinely handle cash deficits, and that there is sufficient working capital to accommodate organizational growth and development. These assumptions may not be valid for the construction industry.

## COMPARISON OF MANUFACTURING AND CONSTRUCTION

A typical *manufacturing* enterprise builds its business volume and its production based on a relatively large number of small items which it produces and sells. Poor performance of any one item represents a relatively small percentage of the total work in progress and of the business volume of that organization. These relatively small factors do not normally represent significant cash management problems.

With an examination of the typical production of manufacturing as compared to construction, these significant differences emerge. The business volume of a typical *construction* company on the other hand is characterized by a relatively *few* specialized contracts, each of which represents a relatively large amount of its volume. These contracts have varying cash requirements associated with them, some of which may have a considerable and immediate impact on the company's total cash flow.

## CONSTRUCTION CASH FLOW FACTORS

Among the factors to be taken into account in construction is the need for cash during the preconstruction and mobilization phase of the contract. Certain activities require funding prior to the first payment from the owner. Payroll must be funded on a weekly basis. Certain supplier items must be paid prior to the receipt of payment. Time phased differences between cash requirements and cash income are directly controlled by the anticipated receipt of cash resulting from job performance.

Another significant factor affecting cash flow relates to the different terms and conditions of payment which may be included in the general contract as compared to the terms and conditions of payment and the cash funding demands related to the resources used in job performance. The most significant of these concerns funding of labor, supplier, and principal vendor items. No retainage provisions are customarily included in these agreements. The typical owner/contractor agreement, however, requires a retainage (usually 10%) which is universally held from the value of all work in place for any given production period. This disparity automatically and routinely creates a cash shortfall for each project in construction. The cumulative effect of this cash shortfall tends to result in a serious cash flow problem for the contractor even when it is known in advance and its consequence anticipated.

Chapter 13 has presented several examples of the effect of typical contract terms and conditions of payment on cash flow and net cash position. Exhibit 13-9 shows the cumulative effect of these on five typical jobs covering a 12 calendar month period.

## THE NEED FOR PLANNED CASH MANAGEMENT STRATEGIES

Each of the items just discussed requires a much more significant role for cash management in the construction industry than in other industries. Effective cash management requires accurate, valid, and contemporary information on what is likely to happen to the firm's cash on hand so that appropriate strategies may be developed. When a cash shortfall is indicated, some specific plan must be designed to accommodate the shortfall. When the cash flow information indicates that excess cash will be generated, appropriate investment strategies should be developed to maximize income. These sources of cash related information and how this information may be used in developing and maintaining a consistent set of cash management strategies will be further explored.

### Sources of Information

The overall processes of integrated cost and schedule control which have been developed thus far provide an abundance of information concerning cash income, cash requirements, and net cash (or the delta), all related to individual projects. When these are summarized on a cumulative basis, the same information is generated for the entire business enterprise (See Exhibit 13-9). From a *cash* management point of view, certain of the performance models are of significantly greater interest than others and are, consequently, much more useful in cash flow analysis.

## The Cost Curve

Among the five derived performance models, the first two provide very little information on cash flow or little helfpul to cash management. The cost curve provides very interesting information on the rate at which *costs* are incurred and may indirectly measure the performance of the job from a cost point of view. The cost curve, however, presents very little information to management concerning *cash* flow.

## The Production Curve

The production curve, which is closely associated to the cost curve (in that it is the cost curve plus the overhead and fee), similarly provides little information for cash management purposes. In fact, under certain circumstances, the high levels of production which are considered positive can significantly increase cash flow *deficits* if the accelerated production is not closely tied to the cash funding it requires. For example, the delivery of major equipment to a project several months in advance of job requirements would create especially steep costs and production curve slopes when any equipment is received at the job site.

Two possible negative impacts with respect to cash flow are worth examining:

The first occurs when no provision has been made in the contract for payment for stored materials. In this case, the contractor may be called upon to pay for the materials (and/or equipment) delivered earlier but with no hope of recovering cash income for those payments until the materials are placed in the regular monthly requisition.

Another more subtle but equally important impact relates to the differences in terms and conditions of payment for the general contractor and for principal vendors. When the terms of payment to principal vendors are net, five days after receipt of payment by the general contractor the total value of the material is due and payable to that principal vendor. Considering that the normal payment to the general contractor has been reduced from the scheduled value of the work by a 10% retainage, the amount of funding available to the general contractor against the stored materials item is equal approximately to the amount of the principal vendor's invoice less 10%. This difference creates an immediate cash shortage and requires some method of funding or accommodating the 10% shortfall. Thus even with a desirable increase in production, that production may result in an undesirable trend with regard to cash flow and cash management.

## The Schedule of Values, Cash Income, and Cash Requirements Curves

The Schedule of Values curve, the cash income curve and the cash requirements curve, are the three derived performance models which provide the greatest amount of information with respect to cash flow. These are consequently the most useful in developing cash management strategies. The procedures for preparing both graphic and tabular versions of each have been presented in detail in Chapters 11 and 12.

The first of the performance models to be considered as part of cash management is the Schedule of Values. Most general contracts provide for a payment according to a "schedule of values" which is used exclusively for payments of the owner to the contractor. The Schedule of Values assigns a value to each group of activities or work packages within the project for payment purposes. The amount of detail to be included in the Schedule of Values is a subject of negotiation, but once established, it becomes the basis for payments.

The value assigned each work package is of particular interest from a cash management point of view. The preparation of the Schedule of Values should take into account the initial effects of job start up, especially those associated with the mobilization of the project; the funding of a series of initial project costs such as performance and payment bonds, insurance coverages, building permits; and, perhaps, the acquisition of the temporary on-site plant and equipment. With a list of work packages to be included in the Schedule of Values and the dollar values assigned to each, payment will subsequently be made on the basis of the percentage of each of these work packages which has been completed during the billing period.

## Schedule of Values Weighting

In distributing the overhead, fee, and other transparent items within the Schedule of Values, contractor commonly weight this distribution in favor of those work packages to be performed in the early states of a project. This procedure, which is not universally accepted, does allow for early generation of cash from the early work packages in the project schedule to account for early cash demands.

Another common practice, among owners and lending institutions, is to limit payment to the contractor on any work package to not more than the amount to be disbursed to the subcontractor performing the work. When this practice is tolerated in lump sum contracts in violation of the terms and conditions of payment, the lump sum contract is in effect converted into a job cost accounting contract where payment is based on the lesser of the scheduled value less the retainage versus the amount billed by the subcontractor.

Except in cases in which a specific contract requirement limits payment to the general contractor to the amount to be paid to the subcontractor or vendor, no basis for job cost accounting should be considered or allowed as part of the payments process. Where, however, an accounting with respect to job cost is a consideration within the payments process, the payment process must recognize the differences in payments and the funding requirements incurred by the general contractor. These differences will include the weekly payment of payroll along with payroll's related insurances and taxes, the need to pay and discount supplier bills, the net payment and retainages with respect to principal vendors, and the payments to be made to subcontractors.

It is not at all unusual (in the absence of cost related payment provisions in the contract) for an owner to attempt to control payment against a Schedule of Values (which is designed specifically for payment purposes) on the basis of actual cost rather than percentage of work completed without due consideration to the differences in payment requirements between the general contractor and his subcontractors, principal vendors, and supplier, and his own work force. The important issue with respect to the Schedule of Values is that it be established within the terms and provisions of the contract and that it be faithfully administered according to those same terms and conditions.

The Schedule of Values, however, merely provides the basis for calculating the amount to be paid to the general contractor by the owner from time to time. It does not provide information on expected cash income.

## Contract Controls of Cash Income

Information with respect to cash income is derived from the income curve. The income curve uses the information presented in the Schedule of Values, and while this deals with the value to be assigned for work which is completed, other information is required in order to determine both when the payments will be made and the amount to be paid. The contract provides this information.

There are three items in the terms and conditions of payments which need to be established:

The first of these is the *time period* for which they are to be made. Typically, a production period relating to the equivalent of a calendar month will be established. This may in fact be a calendar month. Or it could be a different time period, such as the 26th of the previous month to the 25th of the current month. This time period is used for measuring product in place.

The second concerns *when* payment for that work may be expected and is also established by the terms and conditions of the contract. The expected pay date may be established in one of two ways: the first stipulates a number of calendar days after receipt of the requisition for payment; the second establishes a specific day (such as the tenth of the month) for when payment may be expected. In either case, clear information is given concerning the production time period for which payment will be made and the day of the month payment is to be expected.

The third item to be established is the *amount* to be paid, which is determined by the scheduled value of the work in place and other provisions with respect to retainages and/or holdbacks.

## Schedule of Values, Cash Income, and The Question of Owner's Retainage

Typically, a contract between a general contractor and an owner may provide for payment equal to the total value of the work placed during the production period less a retainage (usually 10%) as a contingency against defective work.

There has been considerable discussion in recent years concerning the validity of an owner retainage as a guarantee against the contractor's performance of the work. The primary argument against retainages is that by the time payment is received by the contractor some of the work included in the payment will have been in place for as long as six weeks. If the production perdiod cut-off is the 25th of the month and payment is to be made by the 10th of the following month, there is an additional 15 days of work which will be in place but not included in the current payment, that is, the payment received on the 10th of the month.

Another argument against retainages is that given the reality of negative cash flow during the first half of a job, a further penalty in the form of an owner retainage tends to encourage the *front end loading* of Schedules of Values and requisitions as an offset against these retainages.

With the scheduled values times percent of work complete determining how much is to be paid, and contract terms the date payment is due from the owner, the amount of cash to be received and when it is expected to be received is clearly established. This information is instantly available from the cash income curve and related tabular data presented in Chapter 12.

## Cash Requirements

Similarly, but in a somewhat more complex environment, the demands for cash or cash requirements may be predicted from the cash requirements curve. There are actually five different kinds of demands for cash which can be anticipated and predicted, all based upon the information set forth in the terms and conditions of payment in the subcontract and purchase documents and other payment requirements:

1. Items which require cash on demand such as building permits and bond fees.
2. Weekly funding of payroll and payment of insurances, taxes, and withholdings associated with payroll.
3. Supplier items which must be paid within the terms of payment to earn the discounts available.
4. Principal vendor items for which, while payment may be tied to the receipt of cash from an owner, the issue of retainages affects the amount of cash required.
5. Subcontractors for whom terms and conditions of payment are similar to those in the general contract and payment is tied to the receipt of payment.

When time of payment is considered, all demands for cash may be conveniently summarized into four payment groups, discussed in Chapter 12 and shown in Exhibit 12-1, p. 151.

## CASH ANALYSIS

The major issues of both cash income and cash requirements are understanding the kinds of data available and extracting the information required so that it can be presented in a form useful for cash management purposes.

It is again worth reemphasizing that when the plan, schedule and budget is in place for a project, when the terms and conditions of payment for the general contract are known and when the terms and conditions of payment to subcontractors, principal vendors and suppliers are known, the planned or projected cash income and cash requirements are firmly established. These are not a matter of independently generating information on cash flow; they are a matter of extracting the data which is already firmly fixed and established by the preceding factors. The necessary information required to develop effective cash management strategies is available as a result of the integrated cost and schedule control methods presented in Chapters 11, 12 and 13.

## THE DIFFERENCE BETWEEN CASH FLOW AND PROFITABILITY

Another significant issue in dealing with effective cash management strategies is the need to recognize the difference between cash flow and profitability.

*Profitability* is the essential long term objective of any business enterprise. It is the basis upon which organizational growth and development is established. In the ongoing day to day management of a business, however, profitability tends to be an abstract issue to be dealt with in the quarterly, semiannual, or annual financial statements.

From a management point of view, the more immediate concern is *cash flow*. Cash flow and cash availability must be the constant concern of management. Payroll must be met on a weekly basis. The payment for payroll insurances, taxes and withholdings must be made when required in order to avoid serious legal consequences. Payment on a timely basis to earn discounts is an essential part of long range profitability. Paying subcontractors and vendors when and in the amounts to which they are entitled is an important part of a growing and maturing business enterprise. There is a dependency of the general contractor on the peformance of subcontractors and principal vendors, who must be paid if they are to perform. And inevitably there are unanticipated but immediate on-demand cash requirements, thus the need for some flexibility in cash management to fund them.

A common example of the difference between cash flow and profitability is worth considering. The most profitable job which has been completed for six months and but for which the funds are still not collected or collectable will not contribute to the need for cash to meet the payroll on Friday. To meet this and similar situations, cash flow and effective cash management strategies associated with cash flow must be a constant and primary concern of management.

## CASH INFORMATION FOR MANAGEMENT

Current valid information with respect to cash is essential management information. There are three components of the information required related to cash flow: These are (1) cash income and when it is expected to be received, (2) cash requirements and when those requirements must be paid, and (3) the delta or increment between the first two. Keep in mind that these discussions relate to planned cash income, planned cash requirements, and the resulting delta—clearly the direct result of how the job was planned, scheduled and budgeted and how the terms of payment in the contract, subcontracts and purchase orders were negotiated. Because of this cause/effect relationship, it is imperative that the planning, scheduling, budgeting, and contract negotiations be carried out with great care and in the full knowledge of their effect on cash.

Planned performance is, of course, clearly distinguished from actual performance. It is possible to predict with a high degree of accuracy the planned cash consequence of each project when both the primary and derived perfor-

**Exhibit 19-1. Composite Cash Income Curve and Cash Requirements Curve**

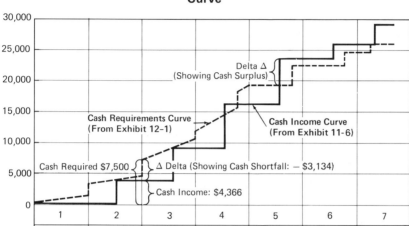

mance models have been created. Exhibit 19-1 combines the cash income curve from Exhibit 11-6 and the cash requirements curve from Exhibit 12-1.

When the cash income curve and the cash requirements curve are shown together as in Exhibit 19-1 or have been made on transparencies of equal scale, a simple comparison of these two curves allows management to read the planned cash position of the job at any point. For example, the planned cash position immediately after the end of production period 2 is approximately $3,100. The exact planned cash position may be calculated from the tabular data.

Net cash after 2nd period
    = cash income (Exhibit 11-6) − cash requirements (Exhibit 12-2)
                         = $4,366 − $7,500 = −$3,134

Either the monthly or cumulative curves may be used for these comparisons.

Exhibit 19-2 shows the combined monthly cash income and cash requirements curves.

Within an individual project, tabular data on cash position should be generated at the level at which cash will be controlled. This may be weekly on a project with large quantities of direct labor. More than likely, an analysis associated with the production period and payment cycle on the income side will provide information in sufficient detail. In plotting cash requirements on a monthly basis to integrate a number of projects into an overall summary, it may be necessary to maintain accurate data at the weekly level because

**Exhibit 19-2. Combined Monthly Cash Income and Cash Requirements Curves\***

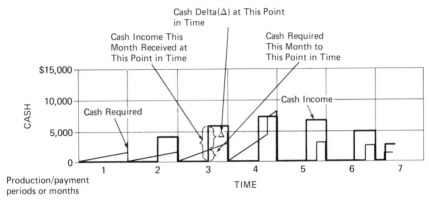

of time or phase differences between projects which significantly vary overall cash requirements. The simplest form of tabulation, however, is a receipts/disbursements summary for the production interval. An example of the kind of summary required is provided in Exhibit 13-9.

## Cash Flow Information

When summarizing income through receipts and cash requirements through disbursements, it is possible to relate overall cash to the total cash flow. This process also provides a convenient way of summarizing net cash requirements, also illustrated in Exhibit 13-9. Monthly net cash shows the cash or delta resulting from each production period individually. Cumulative net cash shows the net cash flow over the report period and assumes a zero beginning balance. These cumulative figures therefore clearly predict the expected cash result during the report period and are an exceptional indicator of cash trends. When a positive or negative beginning balance is added to the first or historic month and projected through the report period, valid cash position, cash flow, and cash trend information are provided.

For each accounting period, the planned receipts and disbursements are totaled. Where the receipts exceed the disbursements, the net positive cash is shown under the receipts column. When disbursements exceed receipts, the net negative cash is shown in the disbursements column. This method

---

\* Note that comparative data and delta are valid only within each month and are not cumulative. See Exhibit 19-1 for cumulative data.

of summarization allows a quick and convenient calculation of the delta on both the monthly and cumulative basis. It is especially interesting from a cash management point of view to have valid information conerning the cash position of a project if the planned schedule and budget for the project were actually accomplished. The cash income curve and cash requirements curve are the basis for making these projections. These cash analytical methods have been fully discussed and developed in Chapter 13.

## CASH POSITION UPDATES AND CASH PROJECTIONS

Prior to the commencement of work on a project, the information from the cash income and cash requirements curves is tabulated on a monthly basis. The income table provides information on cash to be generated by each month of production. The cash requirements table provides information on the cash requirements resulting from that production. The monthly receipts and disbursements schedules may be extended throughout the life of the job. The cash projection is updated for each month's production.

The monthly update has two components. The first is a review and comparison of the previous month to determine how planned cash receipts and disbursements compare to actual receipts and disbursements. Second, on the basis of actual experience, the projection is revised. Once revised, the projection is also extended for the current month and for the remaining months of the project. On individual projects, it is desirable to extend the projections throughout the life of the project. This creates significant information on how the project is progressing and allows opportunity to adjust performance when a need for such is indicated by a review of the historic data.

Concerning the historic data, it is possible to make several simple ratio calculations as an indicator of job cash performance. The first of these is the cash income ratio (CIR). This calculation is made as follows:

Cash income ratio = cash income actual/cash income planned

Where the cash ratio is one, the income is as planned. Where it is greater than one, the income is greater than planned. Where it is less than one, this is an indication that the cash income being generated by the job is falling behind the planned performance. A similar ratio calculation may be made for cash requirements:

Cash requirements ratio (CRR) = cash required actual/cash required planned

Care must be exercised in drawing conclusions about the cash requirements ratio since many variables will affect the results. The question of composition of actual and planned data is the same as the cost performance ratio.

Since the primary and derived performance models, including the cash income curve and cash requirements curve, are automatically updated with the weekly/monthly updates of schedule, current actual data is always available. It is important to note that cash income, cash requirements, and the net cash are not only based on the planned schedule and budget but are also directly derived from these primary performance models.

### Project Management Responsibilities

The current maintenance of planned schedule and budget and the updating of these to reflect historic job performance is, therefore, an essential part of the informational processes associated with integrated cost and schedule control as these relate to individual projects. The individual project and the responsibility for its cash management are a primary managerial responsibility. Both historic performance and projections of the project through completion from a cash point of view with monthly receipts and disbursements, cash delta, and accumulated cash throughout the project are especially important for project management purposes. All updates and projections should be first completed at the individual job level. These completed analyses are then summarized at the organizational level.

## COMPANY SUMMARIES

From a company point of view, a summary of the cash projections and analysis including all work in progress provides exceptional contemporary management information concerning the cash position of the organization. Due to the dynamic nature of an all-projects analysis, the routine management information processes should be built around a six month cash analysis. Longer periods of analyses may be generated when these seem important for particular purposes such as the completion of a quarter or the completion of a fiscal or calendar year. The six month projection, which includes the first month as the most recently completed historic month, the current month for which projections will be most valid, and four subsequent projected months, is useful. The validity of the data on an all-projects summary reduces as time is extended in the projection beyond the current month. Therefore, the summary of all projects in work should be generally limited to six months.

### The Six Month Projection

The six month projection of all projects in work with the calculation of the monthly delta and assuming a zero cash balance at the beginning of the historic month provides valid information on ongoing performance. A number of additional factors can be added into the summary to provide new and interesting information. First is adding the general overhead expenses to each month under cash disbursements. When the delta is calculated and the cumulative cash position is summarized, a clear picture emerges relating to net cash gain or loss resulting from the total business enterprise. Note that this analysis still assumes a zero cash balance at the beginning of the first historic month since this provides a direct read of information on contemporary performance.

The six month projection provides significant information beyond cash projections and cash analyses. It gives a clear indication of expected levels of product placement or production even though these analyses are made primarily from a cash point of view. Of course, separate production projections may be made (fully developed in Chapter 13). When clerical resources are limited, the cash analysis and projection provides the more meaningful and critical management information.

In an ongoing construction operation where the production volume is more or less constant, the collection of retainages will tend to add to the cash position, while the withholding of new retainages will conversely tend to reduce the cash position. These offsetting factors tend to balance out over time so that analysis on a continuing basis unincumbered by retainage considerations does provide valid information with respect to production. What is available is a clear indication of production trends with a five month planning horizon including the current month (six months including the previous or historic month).

## The Effect of Significant Changes in Production

When significant increases or decreases in production levels are forecast, there is sufficient visibility to these changes to plan and implement business and cash management strategies to cope with their effect. From a business management point of view, if a growth in monthly production is clearly forecast, it may be necessary to plan for and implement organizational growth to properly handle the increased levels of production. When decreasing production is indicated, the planning horizon should be sufficient to allow extra sales and marketing efforts to acquire new negotiated work or to increase the levels of competitive bidding in order to maintain appropriate business volume.

The most critical of these is the effect of varying production on cash position. Exhibit 13-11, Multiple Job Cash Analysis, clearly indicates that rising production will tend to consume cash while falling production and completion of contracts in work (related to a drop in business volume) will tend to produce cash from the release of earned retainages which have been withheld. Exhibit 19-3 shows changes in cash position plotted against changes in production based the data from Exhibits 13-6 and 13-11.

**Exhibit 19-3. Change in Cash Position as a Function of Change in Production**

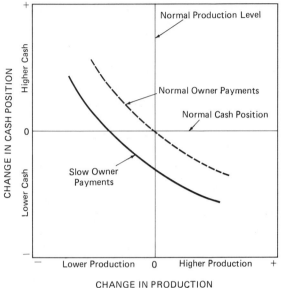

CHANGE IN CASH POSITION

Higher Cash

Lower Cash

0

Normal Production Level

Normal Owner Payments

Normal Cash Position

Slow Owner Payments

Lower Production    0    Higher Production    +

CHANGE IN PRODUCTION

The effect of varying production on cash positions is based on several assumptions, the most important of which is that there is no significant variation between planned performance and actual performance. Nonetheless, varying demands for cash may be safely predicted, stimulating the design and implementation of effective cash management strategies before the need arises.

As a separate issue, most construction organizations have projects being developed or negotiated which, while not presently in construction, will make some contribution to future production. Through the probability analyses discussed previously, it is possible to reasonably estimate the contributions these future jobs will make to the business and production of the organization within the planning horizon.

## PRODUCTION FORECASTING AND ITS CONSEQUENT EFFECT ON CASH DEMANDS AS A CASH MANAGEMENT TOOL

A separate analysis is needed to properly forecast production. Chapter 13 has presented the methods and procedures of production forecasting in detail. Exhibit 13-4 is a sample of a production forecast. In addition to jobs in construction, other jobs to be included in production forecasting are selected by the chosen threshold of probability (0.9 is recommended). These are analyzed to project the production cost and net income for those jobs in the sales and marketing pipeline at or above the selected probability. They may be added to the production forecast for work in progress to obtain a complete forecast of production, cost, and planned net income for the planning horizon. The *production* forecast and the planned net income associated with it are *not* to be confused with the *cash* analysis and the net cash delta even though they appear to be similar; cash analysis is far more complex. While the production forecast will directly and appropriately show an immediate increase in net income as a result of increasing production, the cash analysis will just as surely show a predictable reduction in net cash during the period production is increasing, and until production and consequently net cash stabilizes at some new plateau.

In adding future projects to the analysis, clear recognition should be given to the difference between (1) contracts in work and maintaining cost and schedule control as well as profitability on those contracts, and (2) anticipated new work. Contracts in work, the cash flow these generate, and the profitability resulting is what is happening now and provides the resources for current management. Projected new work anticipates what might happen in the future. With the addition of new projects at an appropriate probability level, a complete analysis is now available forecasting the consequence of business activity for the organization through the selected planning horizon.

## BEGINNING BALANCES AND CASH MANAGEMENT

There is still another increment of information which may be added to either the cash analysis or the production forecast. In all of the previous studies, the assumption has been a zero cash balance at the beginning of the historic month shown on the analysis. This recommendation comes from a desire to isolate the cash performance of the organization in a contemporary time frame. The beginning balance may now be added at the beginning of the (first) historic month and the effect of adding the cash balance may be then projected throughout the planning horizon. The result of these calculations is a net cash delta (or a new net income in production forecasting) for the organization. Where the cumulative data is generated, it is possible to provide the anticipated cash position of the organization monthly through the planning horizon.

## THE CASH ANALYSIS AND CASH MANAGEMENT STRATEGIES

In order to complete the cash analysis, the effect of any credit accommodations which must be repaid during the planning period may be included in the analysis. The information now available allows management to review its cash position and to design specific ways of coping with cash management problems and obversely, the opportunities presented.

When a cash deficit is indicated, a plan may be developed for the utilization of credit accommodations or other cash sources for funding cash shortfall. Cash management strategies for dealing with a cash shortfall may not always involve the use of credit. Controlling cash requirements may be accomplished by balancing cash receipts and disbursements, by deferring capital expenditures, or by deferring accounts payable to more closely match cash availability. Nonetheless, developing credit accommodations with one's banker or lending institution is a part of business maturity; using such accommodations as lines of credit enhances the company's banking relationships and preserves liquidity.

## Deferring Accounts Payable

Deferring accounts payable should be one of the last resorts in the cash management strategy arsenal. When it is considered appropriate, the details of the strategy must be carefully planned in full consideration of the cash analysis data. It should be reduced to writing and agreed upon within the organization. The specific implementation including scheduled pay out of all accommodations should be discussed with the principal accounts in the accounts payable roster. It is especially important to provide complete information to the subcontractors, principal vendors, and suppliers involved so that these resources will not be unreasonably inconvenienced through lack of information. Since the deferring of accounts payable is, in reality, a credit accommodation, there may be cost associated with using this credit source for balancing cash position.

An example of a complete six month cash analysis including all cash management strategies suggested is provided in Exhibit 19-4. With sufficient current information on cash position, a number of cash management strategies are available to management; several of the more obvious have already been discussed.

The budget and schedule are the primary performance models and as such fix the consequence of the derived performance models. The former may, therefore, be used, because of their relationship to the Schedule of Values, cash income, and cash requirements to manage cash effectively. For example, skewing of the Schedule of Values may be weighted in favor of early work packages to increase positive cash flow, a method discussed earlier.

Another strategy is to make sure that the items throughout the project are carefully balanced between those with high cash demands and those with low cash demands. Some of the factors affecting cash demands which may be minimized by project strategy are the amount of direct labor to be

# Exhibit 19-4. Six Month Cash Analysis and Cash Management Strategies (Data from Exhibit 13-9)

| Production period | 1 | | 2 | | 3 | | 4 | | 5 | | 6 | | 7 | |
|---|---|---|---|---|---|---|---|---|---|---|---|---|---|---|
| ($ in $1,000.00) | RECEIPTS | DISBURS. | RECEIPTS | DISBURS. | RECEIPTS | DISBURS. | RECEIPTS | DISBURS. | RECEIPTS | DISBURS. | RECEIPTS | DISBURS. | RECEIPTS | DISBURS. |
| Job # 2 (1,000  10%) | -0- | 31.9 | 90.0 | 151.1 | 270.0 | 277.8 | 315.0 | 235.2 | 117.0 | 94.0 | 63.0 | 54.8 | 745.0 | 63.8 |
| 5.0000  3 (5,000  10%) | -0- | 159.5 | 450.0 | 755.5 | 1,350.0 | 1,389.0 | 1,575.0 | 1,176.0 | 585.0 | 470.0 | 315.0 | 274.0 | 725.0 | 319.0 |
| 3.0555  4 (3,000  8%) | -0- | -0- | -0- | -0- | -0- | 97.5 | 270.0 | 461.7 | 810.0 | 848.8 | 945.0 | 718.7 | 351.0 | 287.2 |
| 2.5943  5 (2,500  6%) | -0- | -0- | -0- | -0- | -0- | -0- | -0- | 82.8 | 225.0 | 392.0 | 675.0 | 720.7 | 787.5 | 610.2 |
| 8.3808  6 (8,000  5%) | -0- | -0- | -0- | -0- | -0- | -0- | -0- | -0- | -0- | 267.3 | 720.0 | 1,266.3 | 2,160.0 | 2,328.2 |
| Monthly totals: | -0- | 191.4 | 540.0 | 906.6 | 1,620.0 | 1,764.3 | 2,160.0 | 1,955.7 | 1,737.0 | 2,072.1 | 2,718.0 | 3,034.5 | 4,168.5 | 3,608.4 |
| Monthly cash before O.H. | -0- | 191.4 | | 366.6 | | 144.3 | 204.3 | | | 335.1 | | 316.5 | 560.1 | |
| Less overhead | -0- | 100.0 | | 100.0 | | 100.0 | | 100.0 | | 100.0 | | 100.0 | | 100.0 |
| Monthly net cash | -0- | 291.4 | | 466.6 | | 244.3 | 104.3 | | | 435.1 | | 416.5 | 460.1 | |
| Cumulative totals | -0- | 191.4 | 540.0 | 1,098.0 | 2,160.0 | 2,862.3 | 4,320.0 | 4,818.0 | 6,057.0 | 6,890.1 | 8,775.0 | 9,924.6 | 12,943.5 | 13,533.0 |
| Cum. cash before O.H. | -0- | 191.4 | 540.0 | 558.0 | | 702.3 | | 498.0 | | 833.1 | | 1,149.6 | | 589.5 |
| Cumulative overhead | -0- | 100.0 | | 200.0 | | 300.0 | | 400.0 | | 500.0 | | 600.0 | | 700.0 |
| Cumulative net cash | -0- | 291.4 | | 758.0 | | 1,002.3 | | 898.0 | | 1,333.1 | | 1,749.6 | | 1,289.6 |
| **Funding sources** | | | | | | | | | | | | | | |
| Working capital | 291.4 | — | 500* | | 500 | | 500 | | 500 | | 500 | | 500 | |
| Line of credit | — | 291.4 | 258 | | 502.3 | | 398 | | 833.1 | | 1,249.6 | | 789.6 | |
| | 291.4 | 291.4 | 758 | 758 | 1,002.3 | 1,002.3 | 898 | 898 | 1,333.1 | 1,333.1 | 1,749.6 | 1,749.6 | 1,289.6 | 1,289.6 |

Cash management assumptions:

1. Initial capitalization    $ 750,000
2. Line of credit    $1,500,000
3. Minimum cash reserve    $ 250,000
4. Net minimum balance =    $ 250,000*

* Therefore $500,000 is maximum working capital contribution to current cash requirements.

utilized, the amount of material to be bought from suppliers which require immediate full funding and the amount of principal vendor items in the project. In each of the foregoing cases, retainages are held on the cash income side but these items must be funded in full on the cash requirements side. Management policy and operating style may dictate how these items can be handled to reduce cash deficits. For example, subcontracting all labor, material, equipment, and services closely ties cash requirements to cash income with regard to both the amount to be paid and when payment of retainages and holdbacks must be made. Subcontractors are normally subject to the same retainage and holdback provisions as the contractor and are usually paid after the contractor is paid by the owner.

## Income Producing Surplus Cash

Monthly cash projections and cash analyses may reveal the generation of excess cash during the planning horizon. Contemporary investment opportunities and rates of return make it essential that all cash on hand in excess of immediate cash requirements be placed in an interest earning environment. Such environments are available to handle funds on a daily or longer term basis. Even when cash will be available for investment for as short a period as one day, it must be an ongoing and routine part of the investment strategy of the company to invest any cash available. Over a full year, a significant contribution may be made to the overall profitability of the organization resulting solely from the proper, immediate, and continuous investment of available cash on a day to day basis.

## The Use of Credit

The dynamic nature of the construction industry and the ongoing production and operation of the business enterprise will result in frequent cash shortfalls. Some cash management strategies have been discussed as alternatives to the use of credit accommodations with lending institutions. There will be times, however, when short term credit will be required to fund short term cash deficits. It is therefore essential that each business organization establish and develop mature banking relationships which include lines of credit commensurate with the business volume of the organization.

Credit accommodations will most likely be generated with the commercial bank being utilized by the organization, and other sources as well. The terms and conditions of each line of credit should be negotiated to be consistent with the organization's requirements and used in accordance with the agreed upon conditions. Most credit accommodations will involve a time period during which the credit accommodation is not utilized. The "out time period" must be carefully included in the cash management strategies of the organization and provisions for repayment of any outstanding portion of the line of credit built into the cash management plan.

The information generated on cash position from the processes of integrated cost and schedule control should provide the basis for a balanced and manageable cash management plan. The plan should be firmly developed and reduced to writing, and carefully monitored by management. The credit accommodations with lending institutions should be used as the primary method of balancing cash within the context of the cash management strategy. This also enhances relations with one's banker, while preserving liquidity.

## Short Term Cash Demands and Working Capital

Another advantage coming out of a detailed cash management strategy as a result of cash analysis and cash flow projections is the ability to project *working* capital requirements. It has already been noted that cash income is normally tied to some monthly payment date but that cash requirements are *spread out* over the whole month and well into the following month associated with that income. This means that analyzing *total* cash receipts and disbursements on a monthly basis will *not* account for all of varying cash *demands* within that time period. It is therefore necessary to carefully analyse weekly cash demands over a period of not less than three months to determine cash requirements on a weekly basis as a function of total production and to generate the working capital requirements for the organization. It is preferable to make six month cash analyses paralleling the six month production projections, noting that validity decreases with time, especially beyond three months. To those projections of working capital requirements relating to cash demands must be added those larger working capital requirements associated with cash reserves, bonding capacity, and similar items.

## SUMMARY

Earlier chapters have presented a number of detailed techniques for accumulating cash management data and information. This chapter sets in the context of business management those earlier data and informational processes. A major procedure in this reference consists of structuring data from the smallest detail to the large summaries which address management information. These procedures are no less true for cash management and cash analysis.

The validity of management information summaries for cash analysis is predicated upon the validity of the cash analysis made at the individual job level where both cost and schedule control are to be exerted. Examples in this and earlier chapters illustrate single job information and multi-job summaries. These summaries also are presented for production periods or monthly periods, and the cumulative data is presented as well. The performance models are always presented in both tabular and graphic form.

A central recurrent need in construction business management is current, valid information on cash income and cash requirements. With this information on hand, cash management strategies may be designed and developed. The sources of information required for cash analysis and cash management are those primary and derived performance models discussed earlier. The Schedule of Values, cash income, and cash requirements curves, along with the terms and conditions of payment associated with the contract and subcontracts, provide the essential information for cash management and cash flow projections. In the final analysis, however, the terms and conditions of payment for the contract and for the subcontracts and purchase orders control both the amount of income and the timeliness of income, and cash requirements as well.

The techniques of cash analysis have been thoroughly developed in Chapter 13. In this chapter, the application of this data to cash management strategies is seen to depend on the maintenance of current data at the individual job level. But a significant difference between cash flow and profitability is also noted. While continuing profitability is essential to organizational growth and development, cash flow is a much more immediate and urgent consideration.

Company or organizational summaries are best presented in a six months projection which displays information for each monthly period separately as receipts and disbursements. Receipts and disbursements are derived from the earlier performance models.

The differences between cash analysis and projections and production forecasting analysis are noted. Also noted are the significant differences in the way the information generated by each is utilized. Production planning is useful in forecasting production and in designing sales and marketing strategies to maintain an appropriate level of production. Cash analysis, on the other hand, is designed to maintain adequate cash flow appropriate to cash requirements.

The impact of varying levels of production on cash flow are revealed and discussed. Subject to certain assumptions, when business volume and production increases, cash is consumed; when business volume and production decrease, cash is produced.

Part II has disclosed a number of weapons in the cash management arsenal which an organization has at its disposal. Many of these are related less to the specific issues of cash income and cash disbursements and more to good solid ethical business practices. They include:

1. Adequate liquid capital appropriate to the planned production volume.
2. Diligent production methods and quality control procedures.
3. Good and consistent contract administration.
4. Accurate and timely accounts receivable billings.
5. Aggressive collection of accounts receivable.
6. A complete and accurate cash analysis capability.
7. On-demand credit accommodation at the maximum level available.
8. Good subcontractor and vendor credit.
9. Well managed processing and payment of accounts payable.
10. Overall good business management and controlled growth.

These may not exactly look like the ten commandments of cash management, but the consistent adherence to all will result in both effective cash management and successful business management.

# III
# Construction and Production

Part III, Construction and Production, covering Chapters 20 through 25, addresses the issues of construction and production. Since Parts I and II deal essentially with preparing a complete project game plan, Chapters 20, 21, and 22 focus on implementing the game plan and making it work. Chapter 20 deals with the construction team and the relationships which are involved. Chapter 21 deals with the administrative prerequisites to starting construction and Chapter 22 more specifically with the game plan and its effective utilization.

These three chapters attempt to make clear that there is neither magic nor mystery involved in integrated cost and schedule control. What is stressed is the importance of thorough planning and scheduling, of setting out expected performance in advance and of using the information on expected performance as a basis for maximizing production in the field.

Chapters 23, 24 and 25 address the question, "How are we actually doing compared to how we had expected to do?" Chapter 23 deals with production management utilizing the project game plan previously developed. Because of the importance of communications, Chapter 24 is devoted to the methods of communicating the game plan to the field environment and the associated methods of obtaining information on actual performance from the field environment. The turnaround document, as it is called, is conceptually developed and examples presented.

Chapter 25 deals extensively with performance measurements, specifically, the processes by which raw data on performance is converted to useful management information. The issues of integrated cost and schedule control are clearly dependent on the timeliness, the quality, and the validity of information on actual performance. Methods of measuring performance are therefore an important part of integrated cost and schedule control.

# 20
# THE CONSTRUCTION TEAM AND ITS WORKING RELATIONSHIPS

In the previous chapters, a wide variety of general concepts have been presented for integrated cost and schedule control. These concepts have been expanded and articulated in the form of specific recommendations, which in turn have been further developed as specific processes, procedures, and techniques which result in a complete and clearly articulated project strategy. From this strategy, the methods of extracting performance models have been further discussed and developed. These methods include, among other things, the various time, production, cost, and cash related performance models which are of high interest, if not essential, to systems integration.

The first 19 chapters have dealt essentially with planning the project to a point where the complete strategy and statement of expectations have been developed and presented for all interested parties. With this groundwork in place, it is appropriate to turn to the more specific issues of construction and production and of dealing with the general issue of the game plan as related to implementing the approved project strategy. Since the resources which will be required for the project have been established and at least some concept of how those resources are to be utilized in building the project has been set, attention can now be focused on selecting the construction team and marshalling the necessary resources for construction.

## THE CONSTRUCTION TEAM

In a rather general sense, there are three functional entities involved in the design, development, and construction of a project, that is, which make up the construction team. First, there must be an owner with specific needs and the resources to meet those needs. Second, a designer is required who can articulate those needs in a technically competent way within the limitation of the owner's resources. And there must be a constructor to articulate a project strategy with respect to time and cost and to manage the construction endeavor through to its successful completion. A good working relationship— including efficient communications—between these three functional entities is essential to a well executed project.

## The Owner

The owner, preferably within the contract itself, should designate a representative who has a clearly defined responsibility and authority to act on behalf of the owner in all matters concerning the project. This person should be the primary channel of communication among the owner, the designer, and constructor. The limitations to be imposed on the authority and responsibility of the owner's representative should be exactly set out in the terms and conditions of the contract to avoid confusion and misdirected communications—especially important when issues of dispute or disagreement arise during the course of construction, as they surely will.

## The Designer

Similarly, the *designer's* representative should be identified for the project. Where the designer has one representative for the construction at the job site and a different representative for design, this should be clearly established prior to the start of construction. Since the designer's representatives will be intensely involved in the technical administration of the project (including processes related to review and approval of shop drawings and other submittals, review, and processing of requisitions for payment; and the issuing of various certificates which may be required by the project), it is important that these processes and procedures be clearly described in writing prior to the commencement of the work.

## The Constructor

The *constructor's* representative is the third member of the team with overall responsibility for the project. Regardless of the organizational systems, functionally the general contractor's (constructor's) representative is a project manager who is responsible for the project from its earliest conceptual stage through its final delivery.

## Team Communications

All communications concerning the project should be directed through each of the three representatives responsible for bringing to bear resources which may be required to address a specific problem, issue, or opportunity.

## Quality of Communications

The quality of communications between the owner, the designer, and the constructor, and the freedom with which communications may be carried out is a key ingredient in the success of a project. Good rapport between these three representatives is the key to the overall enjoyment of the project by the participants. Any construction project should be a creative, productive, and enjoyable experience happily shared by three functional entities each of which brings its own unique training, skills, experience, knowledge and expertise to a project. This results in a productive and satisfactory expenditure of effort and a satisfactory end result.

## Two Kinds of Communication Systems

The owner and the designer each will have particular organizational structure and requirements as these relate to a specific project. These vary widely and, apart from the communications interface, are not considered in detail. The communication interface, however, is crucial. The most effective communication among the three team members is free and open as shown in Exhibit 20-1.

This diagram indicates a freedom to communicate with any other member of the construction team, *keeping the third party properly informed.* A more limited project communications system is shown in Exhibit 20-2.

This communications system involves the designer functioning as a conduit and clearing agent of all communications between the constructor (general contractor) and the owner. Although in common usage, it is more limited and may not afford the best communication for effective project completion. There may also be serious legal and contractural implications associated with this communications diagram. In most construction contracts, the contract is between the owner and the general contractor. There is usually a *separate* agreement between the owner and the designer. The designer is not normally a party to the owner/contractor agreement even though the designer may be named in that agreement with certain specific agency functions.

**Exhibit 20-1. Diagram of Free and Open Construction Communications**

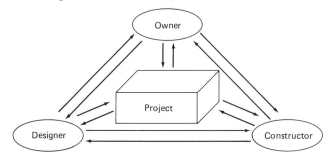

**Exhibit 20-2. Diagram of Traditional Construction Communications**

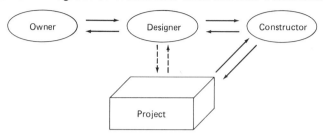

When there is freedom of communication between contracting parties, the potential for misunderstanding and misinterpretation is significantly decreased. Indeed, the very nature of the construction enterprise and its dynamic characteristics in all likelihood will result in at least some technical disagreements between parties otherwise favorably disposed to each other. Maintenance of rapport and accommodation when these disagreements do develop is essential to their easy or convenient resolution.

## Dealing with Disputes

An industry mindset which tends to attribute disputes and disagreements to poorly prepared contract drawings, specifications, and poorly written contract documents. This industry perception is as unfortunate as it is untrue. The technical experts responsible for preparation of the contract documents are, in fact, skilled and competent in what they do. Nonetheless, honest disputes and disagreements are a part of any construction project.

When all is considered, the best approach seems to involve a mindset which *assumes* that disputes and disagreements will develop during the course of a project. This assumption allows (rather, requires) the parties to the contract to build into the contract appropriate procedures for amicably resolving such disagreements and disputes. Included should also be some procedures for final disposition should this become necessary. This agreement will remove the surprise element from evolving disagreements and allow their orderly resolution without adversely affecting either the completion of the project or the enjoyment of a creative construction activity.

## THE GENERAL CONTRACTOR'S ORGANIZATION

The general contractor's organization of the project and the assignment of people resources for the construction of the project are especially important.

## Project Management

Depending on the contractor's organizational structure, a person is usually designated early in the processes of negotiation to perform the role and function of project manager. This function includes overall responsibility for the project from the initial contacts and negotiated process through final completion and delivery of the product to the owner.

## Contract Administration

Closely associated with project management functions from the general contractor's point of view is contract administration. Contract administration deals with all the areas pertaining to the contract and, more specifically, with the identification of emerging changes relating to the contract and its performance and processing such changes administratively to their final disposition. The responsibility for contract administration should be identified by the general contractor. While an individual may be identified with contract administrative functions, the project manager is primarily responsible for seeing that these functions are carried out.

## Financial and Accounting

In addition to contract administration, there are, of course, the vital areas of financial and accounting. In most general contracting organizations, the financial and accounting function is separate. In the owner/designer/contractor interface, financial and accounting is responsible for generating the monthly requisition. Internally, financial and accounting is responsible for handling accounts payable, and accounts receivable, for maintaining job costing systems and procedures, and for providing management information relating to project performance.

These three—project management, administration, and accounting—together represent the contractor team for any project.

## The Field Organization

The field organization for the general contractor is a function of organizational style, job complexity, duration, and the extent to which the contractor's own resources will be utilized in the field for the performance and completion of certain activities. In a project on which most of the work will be performed by subcontractors, a field superintendent may be the extent of the general contractor's field staff. On a larger, more complex job, there may also be a clerk of the works or other clerical assistance for the superintendent. On larger projects and where the extensive utilization of the contractor's own resources is anticipated, the field organization may be expanded to include one or more assistant superintendents, field engineering, several craft supervisors, skilled workers, and unskilled labor. The project manager may also be resident in the field. The development of a project strategy should have clearly established the project game plan so that the requirements with respect to field organization are both well known and well established prior to the commencement of actual construction.

A chart showing both office and field organization is presented in Exhibit 20-3.

In addition to the organizational interface between owner, designer, and constructor, and the project management/office administrative and field relationships within the general contractor's organization, another major set of relationships must be established before the construction team is complete.

## SUBCONTRACTORS, PRINCIPAL VENDORS, AND SUPPLIERS

Three other groups with an intense interest who make an indispensable contribution to the construction of a job are subcontractors, principal vendors, and suppliers. How the general contractor nominates and selects each of these is a major factor (if not the *controlling* factor) in how well the project is performed and in the overall satisfaction of the parties who make up the construction team.

### Definition and Distinctions

The physical work in the field environment is performed either by the work force of the general contractor or by a subcontractor. The performance of either controls the progress of the work at the site and is the key to schedule control. The workforce of the general contractor is the employed craftpersons and laborers and requires no definition. The definition of a subcontractor is simply "one who performs physical work at the job site but who is not an employee of the general contractor."

The subcontractor is distinguished from though often confused with a principal vendor. A principal vendor is defined as "one who fabricates product to specific job requirements for integration into the work at the job site by either a subcontractor or by the general contractor's workforce." A principal vendor may include delivery of material to the job site and unloading. But a subcontractor puts work in its permanent place at the job site whereas a principal vendor does not.

A similar characteristic of principal vendors and subcontractors is as a general rule both perform their work on a lump sum basis. The form of

## Exhibit 20-3. Typical General Contractor's Office and Field Organizations

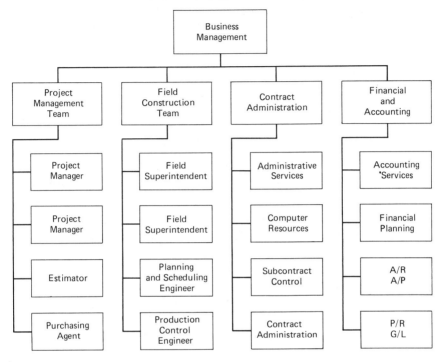

TYPICAL FUNCTIONAL GENERAL CONTRACTOR'S ORGANIZATION

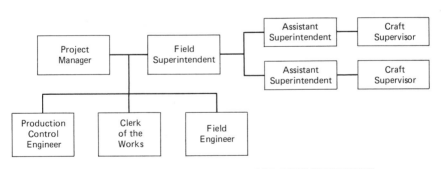

TYPICAL FIELD ORGANIZATION FOR LARGE, COMPLEX PROJECTS

agreement differs, however: a "subcontract" authorizes the subcontractor to do work; a "purchase order" authorizes a principal vendor to fabricate and deliver product.

The third group involved, the material suppliers, do not as a general rule work on a lump sum basis. Suppliers deliver product or materials to the job out of "normal stock" or deliver product "on call" usually at the request or direction of the field supervision.

The terms and conditions of payment to each group also differ. Subcontractors are usually paid a percentage of the value of work completed less a retainage. Principal vendors are normally paid the value of product delivered without a retainage but on occasion with a discount for timely payment. Payment of both subcontractors and principal vendors is usually tied to receipt of payment by the general contractor. Suppliers, on the other hand, are usually paid based on invoices usually supported by delivery tickets with some payment discount but with the time of payment unrelated to payment by the owner to the general contractor.

The tabulation in Exhibit 20-4 provides a more convenient summary of

**Exhibit 20–4. Major Groups Who Deliver Goods to the Job or Work in the Field Environment**

| GROUP | IDENTIFI-CATION | LOCATION OF WORK | RESOURCE COST TYPES | KIND OF AGREE-MENT | PAYMENT TERMS | CONTROL AND/OR SUPV. |
|---|---|---|---|---|---|---|
| I | Gen. contr. labor & equipment | On-site | 1, 3, 4 | Hourly employee | Weekly | Gen. contr. supv. |
| II | Material suppliers | Off-site | 2 | Unit price P.O. | 2% 10th prox. | On call by supv. |
| III | Subcontractors | On-site | 5, 7, 8, 9 | Sub-contract | Monthly after gen. contr. gets pd. less 10% retainage | Sub supv. |
| IV | Principal vendors | Off-site | 6 | Lump Sum P.O. | Monthly after g. c. gets paid; no retainage | Project mgt. or g. c. supv. |

the differences between the four major groups who deliver goods or work in the field environment.

What emerges from the foregoing is that actual placement of product in the field is performed by either subcontractor or general contractor work forces. The on-site production schedule involves those activities to be performed by these two entities. The critical relationship between the contractor and subcontractor workforces and the principal vendors and suppliers involves the administrative function with the task of insuring that principal vendor products and supplier materials arrive at the job on a time schedule consistent with the on-site production requirements.

## SELECTING THE GENERAL CONTRACTOR'S PRODUCTION TEAM

Now that the essential technical details of creating the production and administrative schedules have been discussed, what becomes important is to identify the process by which the general contractor's team of subcontractors, principal vendors, and suppliers is selected for a project. This process, commonly known as "buying the job," is in reality the nomination and selection of the general contractor's production team. The notion that selecting this team is "buying the job" creates a focus on the cost to be paid to subcontractors and suppliers—a much too narrow definition. It ignores those major considerations which go far beyond mere price to focus on nominating a team with the technical experience and capability of performing the work and whose members have a high probability of working together to produce a successful project.

### The Minimum Information Needed

It should be obvious that the team selection process cannot intelligently begin until certain minimum information concerning the time and cost expectations of the project have been prepared. The budget and schedule provide indispensable data on the expected performance of the project and must be an integral part of the selection process. Further, the team selection for a particular project should be done in the context of not only these two primary performance models but the derived performance models as well. The more information which can be presented to team nominees, the more effective the selection process will be and the more likely cooperation among the team members.

## BUYING, SELLING, AND NEGOTIATING

Selecting team members for the project, including subcontractors, principal vendors and suppliers, involves a carefully articulated mixture of buying, selling, and negotiating. The buying aspect of this process involves committing team members to a time/price/performance environment at the lowest reasonable cost. The selling portion of the process is convincing each prospective team member of the desirability of the project, the need for their cooperation, expertise, and participation as a team members and the benefits which will accrue to them as a result of becoming a member of the team for the project under consideration. A negotiating aspect of this process is accommodating authentic differences of opinion, need, and perception concerning the requirements of the project. All of these—buying, selling, and the negotiating—fit within a reasonably definitive negotiating process leading toward reaching an agreement concerning all the rights and responsibilities of each member of the team.

## Prerequisites to Buying a Work Package

There are a number of suggestions which will be helpful to the process of selecting team members and "buying" the project. On the assumption that the prerequisites to starting the job have been accomplished, an orderly series of negotiating steps may be undertaken. It is important here to re-emphasize the major prerequisites to buying: competitive pricing, comparative analysis, and proposed award below budget, coupled with an independent fair value evaluation of the purchase transaction.

Essential, in addition, is that the *scope of work* to be included in the work package be defined precisely and represented by the roster of activities from the budget and the schedule comprising the work package.

## Clearing the Old Agenda

The first step of negotiation should be a review of the past working relationships and performance when the parties have collaborated previously. If the working relationships have been positive and the performance of both parties to the new negotiation satisfactory, these relationships need to be reaffirmed. Where difficulties have been encountered relating to quality or timeliness of performance, timeliness of payment, or any other item, these items should be fully discussed and an understanding reached concerning the expected performance of both parties on the new project. Few things can be as disconcerting as an old agenda creeping into and negatively affecting a new project.

With clear expectations on the part of both parties, the overall specifics relating to the performance of the current project may be considered. In the negotiating process, it is assumed that the technical data used to describe the scope of the work and the activities to be included within the work package have been previously presented, thoroughly discussed, and are completely understood. It is also helpful when selection has proceeded to a point where one subcontractor, principal vendor, or supplier is the nominee to become a team member that they be told they are the prospective team member. Where possible, the specific motivations for the selection or nomination of the team member should also be presented.

## Terms and Conditions of Agreement

The second step of negotiation should be defining the terms and conditions of agreement under which the work is to be performed. There are usually four primary considerations affecting the terms and conditions of the contract.

The first of these is the contract requirements imposed on subcontractors, principal vendors, and suppliers as a direct result of the terms and conditions of the *prime contract,* between the general contractor and the owner. This information, of course, should have been previously included in the request for proposal information or the bid package and should, at this point, contain no surprises. With this information having been previously provided, including all supporting technical specifications, its coverage in the final negotiating session should be limited to a review of these items for clarity.

The second item associated with terms and conditions directly relates to the terms and conditions of the *agreement between the contractor and the nominee.* Here, too, this information should have been included and presented as part of the request for proposal. As noted in item one, the negotiable limitations imposed by the prime contract must be taken into account. Items which are clearly *not* negotiable should have been previously delineated. It is reasonable to expect that continued pursuit of the negotiating process to the point of nomination of a team member should be taken as evidence that these nonnegotiable terms and conditions have been understood and accepted by the nominee.

A third and closely related area are terms and conditions imposed by *external agencies.* These may include governmental agencies having jurisdiction over the work relating to codes, conditions of performance, and similar items which may be inducted into the technical requirements of the project by reference, inference, or common practice. Although a person presenting an offer of contract to perform work in a technical area regulated by a public agency is likely fully conversant with the agency's requirements and intends to comply with them, it is necessary to reconfirm this understanding.

A fourth area of terms and conditions relates to requirements by *third and fourth parties* with interest in the contract such as lending institutions, funding agencies, and similar parties of interest. These conditions may relate to timeliness of funding, the amount of funding in relationship to the value of work in place, prerequisites to funding, requirements for release of liens and waiver of lien rights, and various certificates to be presented as evidence of performance.

All four of these items relate to the terms and conditions affecting the performance of the work. Each should be reviewed by both contractor and nominee to make certain that there is understanding as well as agreement relating to the terms and conditions under which the work will be performed.

## Condition of Work at the Job Site

The third major area of negotiations to be considered are the conditions under which the work is to be performed at the job site. All special conditions relating to job site performance such as temporary facilities, utilities, requirements for hoisting, temporary power, temporary water, temporary sanitary facilities, clean up, and safety requirements should be reviewed in detail. Issues relating to accessibility into the site and limitations on such accessibility also need to be reviewed for clarity. Special expectations of performance requiring the team member nominee to provide a service for other team members, such as hoisting, lifting, or temporary power should have been included in the request for proposal package and should be reviewed at this time. Provisions with respect to extraordinary conditions such as unsuitable sub soil, rock, and hidden obstructions should also be reviewed. To the fullest practical extent, the *process* by which questions, disputes, and claims will be resolved should also be thoroughly discussed and agreed upon.

## The Time Schedule for Performance

The fourth major area of consideration in the negotiating process is the time schedule for the performance of the work. Here, a direct appeal should be made to the project schedule previously prepared for the performance of the work. Those items or activities contained within the work package should be clearly identified and then reviewed on the schedule in terms of calendar dates. Assuming that the primary performance models of budget and schedule have already been prepared, this technical information with respect to time performance should be instantly available.

Where necessary, and in particular, with respect to labor intensive trades like concrete, masonry, and carpentry, workforce leveling analyses should be made first to determine the average size and peak load workforce requirements and then to confirm that the nominee for such work has the people resource necessary to meet job requirements. These analyses may lead to the discovery that a sufficient work force to meet the project schedule may not be available and other performance strategies may need to be developed.

A work force leveling analysis will allow a specific plan and schedule to be developed for the performance of the work package.

Closely associated with the review of the project time schedule should be a discussion and understanding of the on-site preparation which the team member nominee may expect when called upon to place workforce and/or expend money in the performance of the work at the job site. When on-site preparation is properly carried out in the field, it assures the subcontractor that the workforce assigned to the project may arrive at the job site with few, if any, impediments to productive work. When this understanding is not complete or when actual performance varies from this expectation, the subcontractor's cost is increased as a result of low productivity. The work is consequently less profitable and some dissatisfaction with the working relationship may develop, leading to claims and disputes.

## Considering All Cost Consequences Before Negotiating Price

The four items just discussed as primary points of negotiation have serious cost consequences. It is therefore strongly recommended that final negotiation of the subcontract sum or purchase order amount be delayed until after these items have been thoroughly discussed, negotiated, and agreed upon. One of the primary concerns here is a propensity within the industry to negotiate price almost to the exclusion of these other, equally important considerations. Since the other (four) items suggested have serious cost implications, it is unrealistic and a waste of time to negotiate cost or price until the other items have been agreed upon. This approach certainly maintains the common interest of the parties, and it should lead ultimately to a common and mutually satisfactory understanding.

## Price and Final Negotiations

Now, it is appropriate to proceed with the negotiation of price and to confirm the selection of a team member nominee to the general contractor's production. To summarize, price negotiation is the last item in the series which has included clearing the old agenda relating to previous work, fixing terms and conditions of agreement, setting the work conditions at the job site, and establishing the expected schedule for the performance of the work.

Techniques of negotiation are a matter of individual preference and style. Prerequisites to buying a work package have been presented earlier and need not be repeated. Both fair value and market value have also been discussed. There are, finally, two basic objectives to be accomplished in buying a work package. First is to buy from the constructor's team member *exactly* what was sold to the owner. Second is to buy the work package at the *lowest reasonable cost* consistent with job requirements.

As suggested earlier, when the parties entering into negotiations have agreed that it is mutually desirable to work together on the project, the negotiating process is greatly facilitated, with a concomitant increase in the probability of coming to an agreement in the five major areas presented.

## THE TEAM'S NEED FOR CONTINUING GOOD ADMINISTRATION

The differences between the role and function of subcontractors, principal vendors, and suppliers has been discussed. Two additional issues with respect to performance now must be considered. The first of these is the actual time required for on-site placement of product. Closely associated with and frequently controlling the placement of product at the job site is the responsibility of both subcontractors and principal vendors to finish all of the administrative prerequisites which result in the timely delivery of material to the job site.

A firm system for handling preparation of shop drawings and other technical data, submission of this data for review and approval by the parties of interest including the prime contractor, timely completion of the review process, and the fabrication of the products is an essential part of the subcontractors' and principal vendors' administrative process. Where necessary, particularly on major components for the project, the administrative processes themselves and the related time schedules should be reviewed also by the general contractor. This review by all parties should confirm that the administrative processes will result in the material being delivered to the job site when required.

## BUY/SELL RELATIONSHIP AMONG TEAM MEMBERS

In most commercial construction there are significant buy/sell relationships within the context of the owner/contractor agreement as these relate to the agreements between the general contractor and subcontractors, principal vendors, and suppliers. The purchasing process should result in the general contractor buying from the subcontractors, principal vendors, and suppliers exactly what has been previously sold to the owner. The technical data comprising the scope of the work and the list of activities from the budget and schedule should be the same in all agreements. The list of the technical documents, drawings, specifications, addenda, and so on, should also be the same in all agreements. For example, the specific intention of the general contractor to purchase from the plumbing contractor "all of the plumbing" which has been sold to the owner is the primary criterion for this aspect of the scope definition and establishes a commonality of interest among all team members relating to the performance of the work. Where different definitions of scope other than those in the general contract are utilized in subcontracts or purchase orders; there is the distinct likelihood of ambiguity and disagreement.

## Simple Descriptions of the Scope of the Work

Descriptions of the scope of the work to be included within any work package should be clear, simple, and direct. There is a strong tendency within the industry to recite scope of work, terms and conditions, and other items in a redundant way rather than allowing the prime contract documents supported by the performance models to define the scope of the work. Every effort should be made to eliminate ambiguity and unnecessary restatement of the work.

## The Work Package Roster of Activities

Using a roster of activities to define a work package in addition to the drawings and specifications carries with it the concept of unity. This means that each work package must be considered an indivisible whole. The work package roster of activities taken directly from the budget and schedule should be used as the basis for the subcontractor and vendor payment schedule. All work required by the prime contact for the work package must be included in the activities listed for the work package. The work package is intended primarily to provide a basis for measuring performance and for payment. The work package roster of activities is *not* to be interpreted as a *limitation* on the scope of the work within the work package.

## Defining the Scope of the Work Package

The following recommendations are presented for use in defining the scope of work within a work package.

1. Use the drawings, specifications, and the prime contract as the primary exhibits for the scope of the work.
2. Avoid an unnecessary recitation of the scope of the work in a way that may create ambiguity.
3. In a subcontract and purchase order to a principal vendor, the scope of work should read as follows:

> The subcontractor (principal vendor) shall perform all ＿＿ work in accordance with the drawings and specifications enumerated in the prime contract except that ＿＿ as indicated on the drawings will be performed by others, there being no other exceptions.

The four specific points of definition are (1) the appeal to the prime contract, drawings, specifications, and other technical data; (2) the general reference to the trade or technical specialty included within the subcontract or purchase order; (3) a statement of specific exceptions where such exceptions may exist; and (4) (perhaps the most important) an exception closure statement. The exception closure statement is intended to make clear that the work package purchased is the exact work package previously sold to the owner; that only those specific exceptions which are set forth apply; and that there are no others. The description of the scope of the work should be kept as simple as possible, with the avoidance of a detailed re-recitation of the scope of work as given in other documents except where there is absolutely no alternative to create the degree of clarity which may be required. Most times, the definition given above, with appropriate attached exhibits, should suffice.

## SYSTEMS INTEGRATION

The issue of total system integration must remain constantly before the construction team members so that both cost and schedule control may be handled quickly and conveniently. The use of the roster of activities as part of the scope description serves a triple purpose. First, it provides the basis for system integration of both schedule and budget. Second, it provides the basis for periodic or progress payments to the subcontractors, principal vendors, and suppliers. Third, it serves as the basis (within the larger context of the entire project) for the accounts receivable and periodic requisitions to the owner.

## SUMMARY

The owner, designer, and constructor can effectively work in a cooperative team effort which results in a project meeting the owner's needs and within the owner's resource limitations. Properly structured and nurtured throughout the construction process, this team should not only produce the required product of construction but do so in an environment which allows each member to thoroughly enjoy the creative activity that is construction.

Full recognition should be given to the likelihood of disputes and disagreements developing during the building of any project. The construction team may creatively address these as opportunities by providing appropriate administrative procedures which will encourage the timely resolutions of disputes and disagreements. These should not be allowed to adversely affect the timely completion of the project, its cost, or the enjoyment of the team members.

The general contractor's organization and team require special attention. In addition to the contractor's own work force, a relatively large number of other producing entities consisting of subcontractors, principal vendors, and material suppliers are required to fabricate and deliver the necessary products and (in the case of subcontractors) to place these at the site of the work. The seemingly conflicting interest between buying at lowest cost and creating essential team relationships is resolved by placing the proper priorities in the selection of the team. The most basic of these priorities involves a full recognition of the kinds of contributions which each team member on the contractor's production team makes to the ongoing processes with due regard to the rights and responsibilities of each.

The proper selection of the construction team to set up appropriate working relationships is an essential part of integrated cost and schedule control. All of the information embodied in the planned performance and planned outcome of the project must be fully utilized within the team relationships to properly manage the project and to complete it within the time schedule and at or below the budgeted cost.

While disagreements may yet arise (and they most likely will) during the course of the project, good administrative procedures and properly negotiated documents well understood will allow the convenient and proper resolution of all disputes before they become disruptions.

# 21
# ADMINISTRATIVE PREREQUISITES TO STARTING CONSTRUCTION

The many aspects of preparing a job for construction have been discussed in detail. The project statement of expectations in the form of production budget and production schedule has been presented. The overall game plan for the project has also been reviewed. On the assumption that maximum production (rate of placement of product) is both desirable and beneficial for the parties to the contract, it is now appropriate to turn attention to administrative prerequisites to the starting of construction if maximum production is to be accomplished.

Administrative processes vary significantly among owners and general contractors. These processes range from pure reaction to highly controlled and finely tuned project guidance. Chapter 18 deals extensively with administrative procedures; these now should be considered in the context of actions to be accomplished prior to the expenditure of time, effort, resources, and dollars in the field environment. Good administrative processes facilitate the flow of essential information among all the parties of interest during the performance of the contract. Since disagreements are likely to develop in any undertaking as complex as construction, having administrative procedures in place to deal with these as they develop will significantly improve on the job production.

In addition to administrative procedures for managing change, there are several groups of prerequisites which should now be considered; most have been dealt with in detail through the series of flow diagrams and discussion in Chapter 15. The focus here is on those items as prerequisites to starting construction.

## THE FIVE PREREQUISITES TO CONSTRUCTION

The first group of prerequisites deal with the owner/contractor agreements. While these may vary somewhat from contract to contract, the five general prerequisites to construction are as follows:

1. The contract itself should be signed.
2. The drawings, specifications, and other technical data relating to the contract should be appropriately acknowledged.
3. All building and other permits required for the execution of the project should be in hand.
4. The sources of funding for the project should be confirmed.
5. With the four previous accomplished, the notice to proceed acknowledging the formal start of contract time should be issued.

### Owner/Contractor Agreement, Offer of Contract

The *execution* of the contract is an important procedural step. It is usually the culmination of a series of offers and counteroffers which ultimately result in the contract being signed, sealed, and delivered. Some understanding of these processes will be helpful in determining when this is accomplished. When one party to the contract prepares a contract and presents it to the other party, it is not a contract at all but an *offer* of contract. This is true whether the contract has been signed by the preparing party or not. A signature on the offer of contract by the preparing party will simply reduce the procedural steps by one to complete the contract execution. The offer of contract is extended on the condition that it is accepted *as presented.* The party receiving the offer of contract has several options which include accepting the contract as offered and signing it, modifying certain terms and conditions of the contract and then signing it, or rejecting the offer of contract and presenting a new offer of contract. Each of these has special implications.

**Contract Acceptance and Delivery.**   Assuming the party receiving the offer of contract is amenable to its terms, conditions, and provisions, the signature on the contract does *not* complete the process of *execution.* Not until the contract has been delivered to the person offering the contract has it been executed. Unless some provision in the bid documents provides otherwise, the person presenting the offer of contract may withdraw that offer at any time prior to its signature and its delivery. Once the contract is delivered, i.e. executed, it becomes an effective and binding legal document.

When the contract has not been signed by the *offering* party, it may be accepted and executed upon its return. In this case, the transaction does not become complete and binding until the contract is returned to the party *first signing.*

**Counteroffers.**   In the second situation presented, assume that the offer of contract is generally accepted but that certain terms and conditions need to be modified to satisfy the receiving party. The contract may be modified and, where appropriate, initialed and returned to the offering party. In this case, the contract documents being returned to the offering party does not constitute the delivery of the contract but rather the presentation of a counteroffer which must now be accepted by the original offering party. The exchange of offer and counteroffer may continue until both parties are satisfied with the content of the proposed contract, have signed and initialed the modifications, with the contract becoming a final and legal, binding document when it is in possession of both parties with no further counteroffers.

**A New Offer of Contract.** In the third case, a new draft or proposed contract represents a new offer of contract and constructively replaces the original offer of contract. It is possible for each party to present an offer of contract which may be different in content and substance. While it is normally considered that the second offer of contract supersedes the earlier offer of contract, this is not always the case. The exchange of contracts through the offer of contract and counteroffers may be significantly reduced during contract negotiations in which all the essential terms and conditions of contract are agreed upon prior to the preparation of the physical contract documents. Once agreement has been reached, the preparation of the contract documents, due to their legal and contracturally binding nature, should be prepared by or under the direction of competent legal counsel. The major issues to be addressed in the preparation of contracts are simplicity, clarity, mutuality, and freedom from ambiguity.

## Drawings, Specifications, and Technical Data

The second of the five prerequisites to commencing construction is the collection, formal acknowledgement, and distribution of drawings, specifications, and technical data. The best method is to provide for the signing and initialing of an appropriate number of sets of these documents, and to attach and distribute them with the contract itself. When a formal contract signing meeting is held, the acknowledgement signatures and initials on the drawings, specifications and technical data should be a part of the formal process. It is important to include all technical data which is contracturally relevant in the package of material. Technical data which frequently has contractural significance, such as test borings, may be overlooked in assembling the package of technical data unless extra attention is given to making sure that such is included.

## Building and Other Permits

The third prerequisite to starting construction as it relates to the owner/contractor agreement is to have all necessary building permits and all other related permits in hand. The related permits may include those for grading and/or storm water management, sediment control, plumbing, electrical installations, and such other specialized permits relating to licensure which may be required for the project. Since there probably are legal constraints against commencing construction until the appropriate permits are in hand, it is assumed that a provision in the contract keys the start of contract construction time to the issuance of the necessary prerequisite permits.

## Construction Funding

Written documentation should be provided concerning construction funding. This documentation should include a clear identification of the construction funding, with a confirmation that the amount of funding covers the total of the contract plus an anticipated contingency for unexpected changes. In addition, the documentary evidence should clearly confirm that the construction funding from the funding source will be distributed in strict accordance with the terms and conditions of the owner/contractor agreement and that there are no unilateral provisions for withholding of these funds from the general contractor owing to actions by the owner or other parties. Where funding is to be provided through a public agency, both the source of authority and methods of disbursement should be confirmed to be in full compliance with the owner/contractor agreement. Careful attention should be given to issues with respect to retainages and holdbacks, specific actions to be taken and funding to be released at substantial completion, and other matters of interest related to funding in the context of the contract.

## Notice to Proceed

Since not all of the prerequisites to starting of construction in the field may be accomplished at the time the owner/contractor agreement is executed, most contracts provide a process to clear these prerequisites after the contract has been executed. Because some time may expire between execution of the contract and accomplishment of all the prerequisites, documentation, usually the issuance of a *notice to proceed,* is required so that this time is not accounted to the contract time.

The notice to proceed is issued by the owner to the general contractor in confirmation that the four other prerequisites to commencing construction have been accomplished, including the full execution and delivery of the contract; full acknowledgement of the drawings, specifications, and technical data; acquisition of building and other permits which may be required; and confirmation that funding is not only available but that it will be distributed according to the terms and conditions of the contract. The notice to proceed

established the date for the commencement of the contract time and is, therefore, an essential document.

## CONTROL SYSTEM PREREQUISITES TO COMMENCING CONSTRUCTION

In order for the control system to be effective, both major controls in integrated cost and schedule control must be in place and fully operative prior
to the commencement of construction. Each has its own special importance
for effective project management and control.

### The Budget

Usually viewed as the most important control system prerequisite is the budget. The budget provides not only a statement of expectations but also bench-
marks for measuring the cost performance of the project. The budget is
used for monitoring the cost control of a project and for measuring the
cost effectiveness of the purchasing processes, and is the basis for measuring
productivity and the production of the work force on the job. It is especially
valuable in the management information process when used to calculate cost
performance ratios of the project in work. All these indicators add to the
knowledge and information about job performance in a time frame when
corrective measures may be designed and implemented. The budget, therefore,
must be in place prior to the commencement of construction.

### The Plan and Schedule

The schedule, too, should be in place prior to the commencement of construction. For the most effective project control, the schedule and budget should
be fully integrated by the processes which have been presented. The schedule
is, in effect, a *time* budget and provides a statement of expectations with
respect to the time performance of the several activities involved in a project
and, consequently, of the entire project.

### Derived Performance Models

With the budget and schedule in place, the derived performance models
which have been previously discussed in detail may now be developed from
the two primary performance models. These include the cost curve, the production curve, the Schedule of Values curve, the income curve, and the cash
requirements curve. The preparation and utilization of this information signifi-

cantly enhances knowledge about the project and its planned performance and, consequently, the ability to control both the actual cost and schedule. From these derived performance models and raw data captured on actual performance, both time performance ratios and production performance ratios may be calculated. These, along with cost performance ratios, provide a quickly derived and convenient set of management information on overall job performance in a contemporary time frame.

## The Team

There are actually two teams which must be in place prior to the commencement of construction. The first, or *primary,* team consists of the owner, the designer, and the constructor, who together work toward the fulfillment of the owner's needs and the delivery of construction projects. The *contractor's* team is the one assembled by the general contractor to assist in the physical construction of the project (subcontractors), the preparation of special materials for the project (principal vendors), and a whole host of suppliers and other vendors who contribute various tools and equipment and supplies to the ongoing construction process.

Since the total team is involved in the entire game plan, it is in the best interest of the construction project to have the team assembled prior to the commencement of construction. If this is done, the team is able to review the overall requirements of the project and to validate the collective ability of the team to perform the project within the time and cost constraints which have been proposed through the schedule and budget.

The effective team effort in construction is significantly reduced when one or more members of the team are not be in place when construction commences. Getting the team in place is primarily a purchasing responsibility. However, the selection process associated with purchasing and the presentation of the team opportunities is a vital part of the process.

## OTHER PREREQUISITES TO COMMENCING CONSTRUCTION

The larger part of the prerequisites to construction have been discussed, including the selection of the team of subcontractors, principal vendors, and suppliers. It is assumed that all subcontracts will have been awarded and the subcontractor documents executed before work is commenced, likewise that all principal vendor items will have been purchased and confirmed by the appropriate purchase orders or other purchase documents; and that all supplier items will have been purchased. Each of these items are office related administrative prerequisites, frequently handled with little coordination or input from the field construction team.

## Field Supervision and the Field Team

The superintendent for the project, the clerk of the works, and other field assistants should be nominated as early as possible in the preparation for construction. To the fullest extent practicable, they should be provided an opportunity to review thoroughly all the contract drawings, specifications, the budget, and the schedule for the project. Input from the field team should be encouraged and made a part of the preparation of the budget and schedule and also the selection of the subcontractor and principal vendor team. Early involvement of field supervision significantly reduces the amount of time it will need to become familiar with the project once construction has been commenced. The field superintendent may also be helpful in some of the principal vendor or supplier purchase items as part of the project familiarization process. During its preconstruction involvement, field supervision may be charged with arranging for all temporary facilities and utilities which are required for the project, including such things as on-site mobilization, placement of field, office, and storage trailers, storage enclosures, temporary power, temporary sanitation, and location, hoisting facilities. The arrangement for security and temporary enclosures may also be directed by field supervision.

## Insurances

No physical work should be allowed to proceed in the field until all of the insurances required by the owner/contractor agreement, by all subcontracts, and for all purchase orders have been acquired, are in place, and firm evidence of effective dates of coverage have been received by the general contractor from the insurers. Especially careful attention should be given to those insurances which are required by the general contract. Fire and extended coverage, commonly known as builder's risk, should be in place. When these are to be provided by the owner, evidence of insurance placement should be provided and confirmed by certificates of insurance. When the coverage is to be provided by the general contractor, similar certificates of insurance should be in hand before the start of construction. In addition to the builder's risk insurance, property damage and liability insurances in the appropriate amounts, workers' compensation and unemployment insurances must be in place with confirmation of coverage as well as the effective dates.

All insurable risks should be carefully considered and placed under an adequate and complete insurance coverage program. Particular attention should be given to the deductible provisions in the insurance coverages. These should be within reasonable ranges and the cost effectiveness of increasing the amount of deductible insurance should be carefully examined. Higher deductible amounts are frequently selected under the false impression that these are less costly than lower deductible limits, thus the need for a close scrutiny of comparative rates. Certificates of insurance in confirmation of coverage should be carefully examined to confirm that the amounts of the coverage and other contractural requirements are met by the insurance in place.

In addition, the effective dates of the insurance should be carefully noted and an appropriate followup system established for those insurance policies which will expire prior to the planned completion of the project. Further, the notice provisions related to possible cancellation of insurance should be confirmed. At least a 30 day notice should be required before any insurance can be cancelled.

Where appropriate, certain parties may be added to the insurance as named insureds as in the case of the builder's risk insurance furnished by the owner. In this kind of coverage, the general contractor should be an added named insured. The same is true with respect to the builder's risk insurance furnished by the general contractor; in this case, the owner should be added as a named insured.

In addition to all of the above, careful consideration should be given to blanket, overall, or wraparound insurance coverages to provide complete protection where certain job specific insurances may have limited provisions. This is particularly important for property damage and liability insurances.

## AUTHORIZATION OF PRODUCTION

This chapter thus far has been a rather lengthy listing of items prerequisite to commencing construction. As a practical matter, what takes place as the prerequisites are completed is the change of the lead responsibility for the project from project management to field supervision. The way in which the details of the project are communicated from project management to field supervision will have a direct bearing on how effective the transfer of the lead responsibilities will be. Two specific things are recommended as a part of this procedure: first is a production authorization meeting in which all the details of the project are reviewed; second, as part of the production meeting, to physically convey copies of all contract and production related documents to field supervision for their use in the construction of the project. The production authorization should be a written document confirming for the general contractors organization that the project is ready for construction, although the authorization to proceed with the work may be communicated *verbally*. In either case, a production authorization meeting and some formal conveyance of authorization clearly establishes the lead change from the office to the field for the project.

## MAXIMIZING PRODUCTION

Once production has commenced in the field, the production of the project should proceed at the maximum practical rate through its final completion and closeout. Because at times a slower-paced rate of production may be required, this should be carefully built into the game plan for the project and all parties of interest should be properly informed of this strategy. Maximum rates of production may also be tempered by available workforce, materials, or equipment, or may otherwise be controlled by maximum efficiency when such occurs at less than the maximum rate of production available. Most important here are the cautions presented in Chapter 19 concerning the effect that rising or accelerated production has in increasing cash demands and reducing available cash.

## THE GAME PLAN

Although the prerequisites for commencing construction in the field represent in one way or another significant parts of the project game plan, the game plan becomes the basis for addressing the requirements for the completion of the project. It allows all parties of interest to be equally informed on what may be expected in terms of project performance and, specifically, what demands may be made on each party. The game plan also provides a basis for measuring the performance in a contemporary time frame. A complete game plan is easily modified and adjusted to special job related circumstances when such is required.

## SUMMARY

The construction industry is geared to *physically* building things, the real excitement of construction begins when the bulldozers start to clear the site and continues to mount until the project is substantially completed. Historically, completing the last 5% is more difficult administratively than completing the previous 95%. In a much more serious way, completing the administrative prerequisites at the start of construction is even more difficult than the administrative tasks involved in closing out a job.

The main danger is impatience: there is always a high interest on the part of the construction team in getting the work physically started in the field environment. Unfortunately, when pressure is applied to start the work prematurely without proper administrative preparation, the consequences of the inevitably poor resulting start of the project will most likely extend throughout its entire life. Good administrative procedures preliminary to starting construction facilitate the flow of construction once it has started and maximize production and profitability while controlling the cost and the schedule. Good preconstruction administrative processes also confirm that contracts are properly executed and the major provisions surrounding the contract addressed before construction starts. Drawings, specifications, and other technical data, necessary to define the scope of the project and to provide the information required to build the project, are clearly identified and acknowledged.

Construction cannot proceed in most jurisdictions without an appropriate building permit. Frequently other permits also must be in hand before production can proceed. Because of potential legal consequences which could slow or halt a project already under way, necessary permits and the like tend to receive particularly close preconstruction attention. Another essential prerequisite is firmly established construction funding. Only *assuming* that construction funding is in place and that it will be disbursed according to the terms and conditions of the contract is an unnecessary risk that can be avoided by good administrative oversight. There is also a need to formally document the start of construction through the notice to proceed.

A whole variety of control systems including those concerned with accounting for contract time, controlling the contract budget, controlling the contract schedule, handling accounts payable, handling accounts receivable, job costing, and a range of other items must be monitored routinely and efficiently. Once the project is committed to construction, the processes and procedures for each of these must be well established and in place prior to the start of construction. All of the team members must also be in place and ready to go to work as the contract work commences in the field. Securing the necessary insurances relating to all field operations and appropriate documentation of the insurance coverages through certificates of insurance is an additional preconstruction administrative task of significance.

The preparation of performance models facilitates communications between project administration and the field environment as the project is released for field construction. A significant change in lead responsibilities occurs as the project moves into field construction and field supervision assumes the full lead responsibility for the project. With the project in construction, the goal should be maximum production consistent with job conditions and contract requirements. Close attention to the preconstruction administrative tasks will greatly facilitate the flow of production as well as information and data relating to production once the project is underway.

# 22
# THE GAME PLAN AND MAKING IT WORK

In the previous chapters, a variety of materials has been presented concerning the technical methods for integrated cost and schedule control, for generating original performance models, for extracting derived performance models, and similar functions. Implementing the project strategy or game plan, including selecting the team and completing certain administrative details, has also been discussed. A major factor which not yet discussed but which may in fact control the ultimate outcome of the project in terms of time, cost, and profitability is establishing and maintaining a commitment to the "game plan." While the ideas and concepts to be presented are less technical and less precise, and while each construction organization will have its own approach to establishing both the game plan and the necessary commitment to it, the ideas expressed in this chapter will be, to say the least, helpful in these crucial areas. The most neatly articulated plan, schedule, and budget which have been implemented but lack the commitment of the parties responsible for carrying it out will never even approach their potential.

## GAME PLAN COMMITMENT AND OWNERSHIP

There are several major factors to be considered in successfully establishing and maintaining a commitment to the game plan: understanding of the participants, the extent to which all participants are consulted in the process of planning, their participation in the actual planning, and the resulting ownership of the planning process and the plan itself.

The level of understanding among the team members presents a primary challenge. The knowledge, training, and skills of the team members will vary widely. This can be seen as presenting major communications opportunities. These opportunities relate to collecting data and information, distributing the data and information, and providing the background and instruction necessary for understanding of the informational processes in integrated cost and schedule control.

These understandings are greatly facilitated when there is continual consultation during the preparation of the game plan. Those directly responsible for the preparation of the game plan should be in routine communications with other team participants even when there is feeling that such consultation is not necessary. While it *may* be true that consultation is not necessary with respect to the technical aspects of the plan, it is certainly both necessary and appropriate with respect to establishing the *level of communication* which will subsequently be essential in establishing commitment to the game plan. Consultation is, therefore, very useful among team members, and with the inevitable gain in insights from such give-and-take communication, adds a new dimension to the planning process.

## Participation In the Planning Process

Even more effective than consultation is actual participation in the planning process. This is instructional as well as informational: participation allows team members less skilled and less familiar with contemporary planning and scheduling techniques to be instructed in this methodology; and it allows direct involvement and input into the game plan while it is being formulated— so much more effective in producing commitment than does critiquing a completed game plan. The completed game plan may be somewhat intimidating to those less skilled in the methodologies involved. Participation of the less skilled also allows them to see their ideas, procedures, and methods for constructing a job converted to articulate and sophisticated job strategies. The result is partial ownership of the plan and a heightened likelihood of the consequent commitment to it.

## Avoiding Team Division

As a construction organization moves from informal planning and scheduling processes to more contemporary methods, a natural division will occur unless specific actions are taken to prevent it. Those who are skilled and involved in the newer methodologies will tend to push their procedures and methodologies especially when they have primary charge or responsibility for planning, scheduling, and budgeting. When the less skilled in these processes are not involved, they will tend to ignore the game plan and project strategy. Subsequently, when a project is placed in construction, there may be a reversion to old, comfortable, well known methods of production ignoring the carefully articulated game plan. The probable unfortunate result is that two game plans and two informational systems will spring up.

While there are many disadvantages of formal and informal systems operating within the same project environment, the most devastating of these from a production point of view is the vastly increased amount of effort required on the part of the team to effectively manage two systems and comfortably translate information out of one system into another. This extra, energy-draining effort reduces the total production capacity of the organization and further aggrevates the more or less natural antagonism which may already exist between the office and field work environments. These potential difficulties which would drain an organization's productive capability and people resource may be significantly reduced by appropriate training of all personnel involved in the systems and methods to be used to the point where there is a common language and a common working knowledge of the system. Involvement of the team in the planning and scheduling processes will further assist in establishing and maintaining the game plan and making it work.

## THE GAME PLAN

The previous chapters have discussed in detail two major elements of the game plan: the statement of expectations and the project strategy. The following discussion attempts to clarify the relationship between these two, and their absolutely essential place in the game plan.

### The Project Strategy

The project strategy is the plan for how the job will be put together. In its most complete sense, the project strategy is articulated as a time scale network derived from a thorough critical path analysis of the project which has considered not only the logic relationship among activities but their appropriate sequencing for maximum production. The project strategy normally addresses these logic relationships and by analytical processes converts them to a calendar dated schedule for the performance of the work.

### The Statement of Expectations

The statement of expectations adds to the calendar dated schedule to express expectations with respect to both time performance and cost performance. Together, then, the project strategy and the statement of expectations provide the basis for the game plan.

### The Game Plan

The game plan, however, is not complete until the full resources to be utilized have been established and the complete team selected for the performance of the project. When the schedule has been completed and is in place, when the budget has been completed and is in place, and when the resources for the performance of the work and the team members have been selected, the whole game plan is then in place. Especially note that the game plan requires that *all* the team members as well as resources for the performance of the work be clearly identified. When a project is physically started in the field environment without all of the subcontractors, principal vendors, and suppliers in place, the project is working under the handicap of a partial team and, consequently, an incomplete game plan.

It cannot be overemphasized that this condition is the most disruptive to the production of the project and to both cost and schedule control. It is very much like a football coach deferring the selection of the quarterback until the team has the ball on offense or a baseball manager not selecting the starting pitcher for an at-home game until the bottom of the first inning. Any such gap in the construction team as well, especially in the field, places enormous pressure on the remainder of the team and hinders job performance. This difficulty, to repeat, can be avoided by making the selection of the team a primary prerequisite to the completion of the game plan and to physically starting work in the field environment.

## Getting the Job Started

As we have seen, getting the job properly started will have a major effect on overall job performance. The job performance mentality is established within the early stages of a job and once established this mindset is very difficult to modify. Also as we have seen, there are two major factors involved in getting a job properly started: to make sure (1) that there is a complete game plan in place; and, of equal importance, (2) that the game plan in its entirety is fully communicated to all parties of interest.

Now the focus of attention is to maintain a high commitment to the game plan, based upon actual job performance resulting from each team member contributing to the overall success of the project in the context of the game plan.

## THE PROGRESS MEETING

Once the game plan has been established, few things can be as effective in maintaining a commitment to the game plan and in maintaining appropriate levels of production and overall control of the project then regularly held and properly conducted progress meetings. Each progress meeting should involve all parties of interest including the owner, the architect, the general contractor, all subcontractors and principal vendors. There are three major objectives to be accomplished in every progress meeting.

## Review Production and Schedule

The first of these is a review of the schedule and production to date and a comparison of these to the game plan. This review will reinforce the positive performance and will bring to light areas of difficulty in a time frame when corrective actions may be taken.

## Identification of Problems

The review of the schedule will naturally lead to the discovery of items which are impeding the flow of the work. This is the second major objective and perhaps the most important. While it is interesting and useful to know which parts of the project are proceeding within game plan expectations, it is much more important to know those items which are falling outside of acceptable performance since these are the items which will ultimately create difficulties in the project. Items of coordination and, in some cases, disagreement may be discussed and resolved through the regularly held progress meetings.

## Recording Action Required and Results

While there should be no limitations on the job related items to be discussed in the progress meeting, some skills will be required on the part of the person conducting the progress meeting for maximum effectiveness. As each item comes to light, it should be recorded. As each item is resolved and the course of action agreed upon, both the item and the person responsible for the action required should be noted along with the expected schedule for the completion of the action.

## USING THE PROJECT ACTION CHECKLIST

There are many ways of recording proceedings of the progress meeting. Typically, some written scenario is prepared of the discussions, which, while interesting from a historic point of view, may not be that useful in recording the particular action required by the responsible party and just when that action must be completed in light of project circumstances. Whether or not a separate scenario of the progress meeting is developed, the most effective way of recording the action required, the responsibility for action, and the due date is through the project action checklist (see Exhibit 22-1).

The recommended uses of the project action checklist in the progress meeting and as a follow-up are worth reviewing. Some person other than the person conducting the meeting should maintain the project action checklist.

## Exhibit 22-1. The Project Action Checklist

PROJECT ACTION CHECKLIST

OWNER:_____  PAGE:_____ of _____

PROJECT NAME:_____  JOB NO.:_____

DISTRIBUTION:_____  REPORT NO.:_____

_____  DATE ISSUED:_____

ATTENDED BY:_____

| ITEM NO. | DATE ISSUED | DESCRIPTION | ACTION REQ'D. BY | DATE | COMPLETED | REMARKS |
|---|---|---|---|---|---|---|
|  |  |  |  |  |  |  |
|  |  |  |  |  |  |  |

This will allow the meeting conductor to focus on the issues, to maintain the meeting on the subject at hand, and to make certain that the necessary action steps are recorded. If, for example, the project manager conducts the meeting, the job superintendent may maintain the project action checklist.

The second recommendation is that the project action checklist be maintained and handwritten during the course of the progress meeting. Some skill will be required in recording a description of each item to be entered. The purpose is not to recount the whole history of a particular action but rather to identify it as distinctly as possible for follow-up purposes. Usually a two or three word description will be sufficient for this purpose.

Some clerical details of managing the project action checklist within the progress meeting are important in keeping these procedures as simple as possible:

1. The item numbers should be assigned to an item when it comes to light. This number remains the same for that item throughout the job.
2. Only a brief description of the item—two or three words for identification purposes—is needed.
3. The team member who will be responsible for taking the appropriate actions must be identified and recorded when the action is agreed upon. This is a particularly important step since it captures the agreement and assigns responsibility for action to a specific party.
4. It is equally important to record *when* the action which has been agreed upon will be completed.

This use of the project action checklist will result in a complete and succinct record of the proceedings of the progress meetings, and establish responsibility for each item and an expected completion date. The remainder of the project action checklist is for recording when the required actions are completed. This information should be reported, captured and recorded as a routine part of the agenda of each progress meeting.

## Information Flow

The immediate distribution of information coming out of the progress meeting is absolutely essential. This is one of the reasons the handwritten recording of these proceedings in the progress meeting itself is recommended. When a copying capability is available at the site of the progress meeting, copies of the project action checklist may be immediately prepared and distributed to the attendees with a mail follow-up to the parties of interest who were not in attendance. When a copying capability is not available at the job site, it is absolutely essential that the project action checklist be copied and mailed to all parties of interest the same calendar day as the progress meeting. Two major items should be added to the project action checklist before it is distributed: the location, time, and date of the next meeting; and the first item on the agenda for the next meeting, which is a review of the project schedule.

Upon receipt of a project action checklist, all parties of interest should review the list in its entirety. Particular attention must be given to the action required, who is responsible for action, and when. By identifying one's own name in this column, each responsible party may generate an immediate list of required actions and note when these actions need to be completed in order to maintain the schedule. Every participant should be prepared at each progress meeting to provide information on the status of the action items which were previously presented.

## The Agenda for the Progress Meeting

The project action checklist from the previous progress meeting serves as a very convenient agenda for the subsequent progress meeting. Beginning with a review of the schedule, the agenda may immediately proceed to clearing those items from the agenda which have been completed, for reviewing status of those items which have not yet been completed, and for adding new items which require action. There will be a tendency to try to reduce or keep to a minimum the number of items included on the action checklist. But there is no reason for trying to limit the items since the very purpose of the action checklist is to give attention to *any* and *all* items affecting the flow of the work. The open agenda process allows complete freedom of expression on the status of the job and items affecting the work and, therefore, improves

**Exhibit 22-2. Partially Completed Project Action Checklist**

PROJECT ACTION CHECKLIST

OWNER: OFFICE DEVELOPERS                                    PAGE: 3  of  3

PROJECT NAME: DEVELOPERS OFFICE BUILDING                   JOB NO.: 1146

DISTRIBUTION: ALL VENDORS AND PARTIES OF                   REPORT NO.: 25

INTEREST                                                    DATE ISSUED: May 25, 1985

ATTENDED BY: SEE ATTACHED LIST

| ITEM NO. | DATE ISSUED | DESCRIPTION | ACTION REQ'D. BY | DATE | COMPLETED | REMARKS |
|---|---|---|---|---|---|---|
| 183 | 5/25 | Brick Discoloration | Twin | 6/1 | | |
| 184 | 5/25 | FH valve functional | Burg Bell. | 6/1 | 6/7 | |
| 185 | 5/25 | Final Inspect Plumbing | Burg. Bell. | 5/31 | 6/7 | |
| 186 | 5/25 | Air filters set | Clean Air | 5/26 | 6/7 | |
| 187 | 5/25 | Set condensers | Clean Air | 6/3 | | |
| 188 | 5/25 | Ceiling tile | Dining | 6/8 | | |
| 189 | 5/25 | Vinyl base-straight | Dining | 6/15 | | |
| 190 | 5/25 | Next meeting - June 1, 1985 - 8:30 A.m. | | | | |

**COMPLETION JUNE 15, 1985**

the proper supervision and flow of information with respect to the project.

In order to provide additional visibility to the projected schedule for the project, the projected completion date may be shown in bold writing across the face of the project action checklist, as illustrated in Exhibit 22-2.

There are both advantages and disadvantages in this procedure. When the project is on schedule, production is at the levels expected, and the job is otherwise going well, the bold statement of the expected completion date will *encourage* performance. When, however, there is considerable slippage in the schedule, constant revision of the schedule and the scheduled completion date may have a *negative* effect on the progress of the work.

## Administrative Detail

The project action checklist should not be considered solely in terms of production or job site issues. It should include the various administrative processes: reviewing the status of shop drawings, inspections, permits, deliveries of materials, processing of accounts receivable, processing of accounts payable, and any other similar item of general interest with respect to the project.

## Current Valid Information

Maintaining a commitment to the game plan is inseparably bound to the quality and timeliness of information provided to all members of the team. Information and follow-up is part of an essential attention to detail which is required if the expected level of production for the project is to be maintained. A special effort should be made to maintain a complete and open flow of information with respect to the project and to use the complete capability and experiences of the team members to maintain the commitment to the game plan. Job circumstances may dictate modifications to the game plan from time to time. These should be established in consultation with the team members and implemented in ways that will result in reestablishing the game plan, in maintaining the commitment to the revised game plan and making the game plan work.

## THE IMPORTANCE OF A GOOD START

The construction of a project is no different in this regard than any other enterprise. Getting the project off to a good start is one of those vital one-time opportunities to influence the total performance of the project. It involves adequate information properly distributed, a commitment by the team members, and maintenance of discipline and professionalism along with an awareness of and attention to detail by those managing the project.

Early compromises in the game plan itself should be resisted aggressively and vigorously to establish that significant deviations from the game plan, especially when such actions are unilateral on the part of any team member, simply will not be tolerated. When a time schedule problem may develop early on, it should be addressed by creating alternate solutions and implementing those solutions for the purpose of maintaining orderly job site production. This will assure that previous work is completed and ready to accept subsequent work as the follow-on work is called for. Setting both the tone and pace for the job in the early stages is crucial to the ability to maintain the game plan and to make it work.

## Attention to Detail

The construction process, especially in the field environment, requires close attention to detail. This means that the job site must have full details on such fundamental items as feet and inch dimensions and pounds of nails required. Construction does not proceed in the field based on generalities but must be accomplished by concrete and precise detail. Recent studies on large construction jobs indicate that productivity generally runs below 30%. Detailed investigations reveals that the significant or controlling factors of this very poor performance involve such items as a workforce available but no materials; materials available but no workforce; workforce and materials available but no specific construction details; workforce, materials, and construction details available but no equipment; workforce, materials, details and equipment available but no temporary power source; and so on up the chain of blunders.*

To repeat, attention to detail must be a major component of any serious attempt to control both cost and schedule.

## Follow-up Procedures

Attention to detail and revealing items which may affect the job through the progress meeting procedures is not in itself sufficient. Adequate follow-up procedures must be in place and working to make certain that those things requiring attention receive it within the time frame required. Good attention to detail and good follow-up procedures do not happen automatically; the specific processes must be in place and working to be effective. No detail is too small for both attention and follow-up if it has the potential of affecting the completion of the work. The follow-up and attention to detail usually requires sufficient clerical support to maintain the game plan to enhance production and productivity. Production may be considered as the rate of placement of product involving the selection of the major production methods. Productivity may be viewed as the refinements of a particular method of production. Obviously, production holds greater opportunity in the construction environment. Once the appropriate methods of construction have been selected, productivity may enhance or improve on those particular methods.

## The Enemies of Production Control

There are two real enemies of production that relate to cost and schedule control: cost overages and time delays. Cost overages may normally be limited by purchasing work packages on a lump sum basis. Those items which are applied on a *unit price* or *incremental* basis such as one's own labor have

---

* American Association of Cost Engineers, *1980 Transactions.* John D. Borcherding and Scott J. Sabastian, "Major Factors Influencing Craft Productivity in Nuclear Power Plant Construction."

the greatest potential for cost overages. The first problem with such items is that there is no quantity/cost closure. This means that unless there is constant supervision, monitoring, and evaluation of performance, there will be no control. The information processes must reveal negative trends sufficiently early to allow corrective measures. Time delays similarly affect production. Since many construction costs are time related, time delays have an immediate cost consequence.

Neither cost overages nor time delays should be considered inevitable with respect to construction projects. Each project should be planned carefully and then accurately monitored for the purpose of taking corrective action as any such problem first emerges. This requirement reemphasizes the need for proper informational systems and for unimpeded flow of information to minimize the impact of these two enemies, cost overages and time delays, of production in the construction environment.

## Informational Systems

The importance of current valid data and information as related to on job performance has been emphasized for a number of different environments in construction. Actual performance data captured in a contemporary time frame, manipulated and analyzed and presented as management information is not, however, the answer to cost and schedule control; nor is actual performance information directly useful by itself.

The key is to be found in the definition of the game plan itself. When there is no project strategy which expresses how a job is to be built in terms of the activities which must be completed in its construction, how each of these activities relates to the others, how long each will take, and the composite time effect of the project plan as represented by the project schedule, there can be no game plan.

Each job must be thoroughly planned and accurately scheduled. The plan and schedule must clearly reflect *how* the project is to be built and managed. When the project strategy is prepared in this way and when the actual performance data in terms of time is captured in exactly the same way, the project strategy becomes relevant in the field environment and the actual performance data may be accurately and validly compared to the planned performance data.

The same general comments may be made with respect to the statement of expectations relating to the combined effect of cost and schedule. The statement of expectations adds to the project strategy the dimension of resources to be used and the cost effect of those resources. Cost as a reflection of planned resources is as essential to measuring actual cost performance as the plan and schedule is to measuring time performance. Careful and valid budgeting provides information on expected cost performance and is used to create management information which compares actual cost performance to planned cost performance. This facility, the statement of expectations, allows a direct comparison of actual cost to planned cost for the purpose of producing management information which is both valid and current.

## Actual and Planned Comparisons as Part of the Game Plan Management

The comparisons of actual and planned cost and time are key issues in the overall performance informational processes. In each case, thorough planning is absolutely essential to provide the basis for comparing actual to planned performance. Planning should be carried out in the same way the job is expected to be built.

The game plan, however, is still not complete until it includes the names of specific resources. In addition to the project strategy and the statement of expectations, the people, organizations, and companies to be assigned and obligated to perform specific activities and groups of activities summarized as work packages must be nominated and committed to the work.

## Commitment to the Game Plan to Make It Work

The game plan is not the result of some theoretical or some psychological mystery concerning the game plan for a specific project. Rather, it is a result of thorough detailed and meticulous planning of the project and the enlistment of a team of willing participants who are collectively determined to make the game plan work.

The project strategy and the statement of expectations must be the primary informational source to be used in enlisting the production team. This is true as the owner enlists the designer and the constructor in addressing construction needs. It is equally true as the constructor or general contractor enlists the team of subcontractors, principal vendors and suppliers in addition to his own personnel in the performance of the project. The kind of commitments involved also convey ownership which in turn becomes a key to completing the game plan and to making it work.

## SUMMARY

A complete game plan for a construction project involves a development of a complete detailed project strategy consisting of a plan and schedule. It involves the development of a complete statement of expectations relating to budget, schedule, cash income and cash requirements and it involves enlisting a team willing to claim ownership of the game plan and which is committed to making it work. The team for any construction project is large and complex. As a result, there is a wide variety of interest and motivations. Maintaining a commitment to the game plan is directly proportional to the timeliness and validity of information provided to the team and to the ways in which the team is encouraged to respond to maintain the essential elements of the game plan.

The five prerequisites including the contract; the drawings, the specifications, technical data; building and other permits; funding; and notice to proceed have already been discussed in detail. Since each of these is controlling with regard to some aspect of project construction, it is important that each one be confirmed prior to the commencement of construction and that construction not commence until all five have been confirmed and are in place. These control system procedures which are prerequisite to construction are especially important to the orderly flow of information resulting from field construction and to the field construction itself.

The commitment of the team begins with involvement in the planning process. The team's perceptions must be negotiated so that different interests resulting from different perceptions may be integrated into the game plan. Without a doubt, the most important action is getting the job properly started. On the assumption that the planning, scheduling, and budgeting processes have been thoroughly addressed, the information for a proper start of construction is at hand. The interpersonal relationships and the system of communication become especially important at this stage—the beginning of physical construction.

Once the job is underway, routine progress meetings provide the best form of communication for all the members of the team. The administrative and clerical details associated with the progress meetings and the use of the project action checklist are designed to facilitate the communication processes and to enhance information flow. The progress meetings should also include necessary administrative detail which will affect on-site production. By any definition, the construction of a project to meet an owner's needs is a detailed and often tedious process. Close attention to these details will directly affect how well the project comes together, how well cost and schedule are controlled, and ultimately how profitable the job will be.

The major enemies of production control are cost overages and time delays. When these become apparent, they must be attacked vigorously; and with the use of good, valid, current information comparing actual performance to planned performance, remedial measures designed and corrective action taken.

# 23
# PRODUCTION MANAGEMENT

Throughout this reference, the point has been made repeatedly that integration of cost and schedule control are dependent on adequate planning, scheduling, and budgeting. The maintenance of production is equally important, also a key to integrated cost and schedule control. Maintenance of an adequate level of production undergirds the basic requirements for both cost and schedule control. Production in turn is maintained by good production management.

The importance of production management can scarcely be overestimated. It provides an orderly framework for applying resources in the field to the completion of the activities making up the project. Production planning requires a careful examination of the relationship between activities and a full recognition that not all activity relationships are controlled in the same way. Take, for example, the simple logic network shown in Exhibit 23-1.

Within the diagram, activity A must be completed before B can start; activity B must be completed before C can start. As has been pointed out previously, it is not possible to tell from the diagram whether the logic constraints are derived from physical limitations or whether the planner, for whatever reasons, may have *chosen* to do A before B before C. While there is no convenient way to show these differences within the logic network diagram, careful production planning is a prerequisite to production scheduling and will go a long way toward realistically relating planner's choices to what is expected in the field environment. Some flexibility, however, is required to accommodate variances between planned relationships and what actually happens in the field environment.

## PRODUCTION MANAGEMENT AND COST CONTROL

Production management has a direct impact on all issues related to cost control. The general contractor needs to effectively manage not only the work performed by the general contractor's resources but also the work performed by the subcontractors and principal vendors. The production plan in this context becomes a primary communications tool for advising all team

**Exhibit 23-1. Network Logic Constraints**

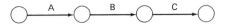

members when the resource of each will be required in the field environment and how this resource will relate to the resources of other members of the production team. In order for cost control to be effective, it must function for *all* the team members, not merely the general contractor. The ability to properly set out a production plan, to implement and manage the plan, and to communicate deviations from the plan on a timely basis are of vital interest to all team members. Every subcontractor, for example, has a right to expect his workforce to be able to work productively at the job site when the project plan requires the application of that workforce. To the extent that the general contractor is able to safeguard not only his own productivity but also the productivity of the subcontractors and principal vendors, the production plan works for all parties of interest and virtually assures a successful and happy project.

## PRODUCTION PLANNING AND THE SCHEDULE

Production planning is not only essential to schedule control but it is also the primary prerequisite to scheduling. Most contemporary planning and scheduling techniques such as the critical path method (preferred for construction) require that a sensible plan be articulated and diagrammed as a prerequisite to scheduling. Production planning includes a number of major considerations in relationship to the scheduling process.

The first of these is to establish the logic relationships among the activities to be included in the production plan. A clear definition and regard for the difference between relationships which are physically constrained and those which involve planner's choices need to be established. Especially where planner's choices are involved, the production team should provide input to the fullest extent possible into the relationships within the production plan. Early input from the team members having interest in the relationships among a particular group of activities will reduce, if not eliminate, the number of in-production changes to the production plan.

### Activity Duration

The second prerequisite to scheduling involved in production planning is to determine the duration of each activity. The number of duration units required to complete a particular activity is a function of the method of production, the size of the work force or resource to be applied, and the total production value of the item. The production method should be considered first. Once the method of production has been agreed upon by the team members, the calculations may be made to determine duration. It is worth noting at this point that the production activities are those activities which take place *in the field environment.*

It is the level of detail at which the project plan is to be implemented that becomes of vital importance. Planning a level of detail greater than the level of control intended to be exerted provides useless detail. Planning with less detail than is required creates difficulty in the areas of data capture and generation of management information. Once the appropriate level of detail has been built into the project plan, the field production of each activity is then controlled by labor or equipment or both. In either case, the duration calculations are the same:

> Duration = budgeted cost for the activity/
> cost per duration unit of resource to be used
> Example: D = $10,000 ÷ $1,000 per day = 10 days

These calculations work on the assumption of a straight-line application of resource from start to finish. Due consideration should be given to both the start-up and completion phases of each activity, particularly on any activity extending over a period of more than several days. Using the budgeted value of the work divided by the cost of the resources per duration unit assumes straight-line production. What may be necessary is to adjust the duration units to account for a slower rate of placement in the early stages, and a slower rate of placement in the wrap-up stages of each activity as compared with the level workforce requirements in the primary section.

While the sample calculation shows a duration of 10 days, practical considerations may indicate an additional day start-up and an additional day for final completion. The duration of the activity given these practical considerations would be:

> Duration = start-up + calculated duration + wrap-up
> = 1 + 10 + 1 = 12  days

## Empirical Estimates of Duration

It is not at all unusual to have durations established empirically by the supervisor of the work force to be utilized. Even when this method is used, it is possible to confirm the validity of the data by converting the dollar value budgeted into duration as illustrated. Expert data and its confirmation provide two separate bases for the estimating of duration. The planner may use both sets of information to determine the duration to be used for scheduling calculations.

Scheduling computations may proceed once a duration for all production activities has been established. The standard critical path methodology of calculating time boundaries, establishing the critical path itself, and calendaring the schedules may be used. On the assumption that the basis for the project plan is the roster of activities also contained in the budget, the foundation will have been laid for integrated cost and schedule control.

## FORMAL PRODUCTION PLANNING AND COMMUNICATIONS

One of the primary functions of production planning is communications. There is a large group of people interested in the production plan including the general contractor, all of the subcontractors, all of the principal vendors, all of the suppliers, the owner, the architect, public agencies, financial institutions and others. A properly articulated production plan may be used as a principal communications tool for project coordination.

Exhibit 23-2 shows an example of planned activity performance data.

This table also shows a number of informational groups which have been previously discussed. These include such groups as:

- Description of activity;
- Selected team members;
- Primary planned performance data;
- Derived performance data.

In addition, the components making up the job game plan are also shown and include:

- Job strategy;
- Job statement of expectations;
- Job game plan.

Note that the statement of expectations becomes the game plan only after the production team has been recruited.

A properly articulated production plan also greatly simplifies data capture procedures which become the basis for comparing actual performance to planned performance. The raw data collected in the field has the maximum validity when it is captured as close to the physical site of the work and as close to the performance of the work in time as is practical. Proper production planning can provide an adequate basis for data capture. The data capture

**Exhibit 23-2. Planned Activity Performance Data**

ACTIVITY DATA, INFORMATION, AND RESOURCES

| | PLANNED | | | BUDGET | | | SCHEDULE | | | VALUE/TIME MODELS | | | | |
|---|---|---|---|---|---|---|---|---|---|---|---|---|---|---|
| COST CODE | RE-SOURCE | DE-SCRIPTOR | SUB-CONTRACTOR PRINCIPAL VENDOR SUPPLIER | QUAN. | UNIT PRICE | BUDGET COST PAYM. | START DATE | DURA-TION | COM-PLETION DATE | COST GRAPHIC | PROD. VALUE GRAPHIC | SCHED. OF VALUES GRAPHIC | CASH INCOME GRAPHIC | CASH REQMT GRAPHIC |
| 4200 9 . . 3 | | Unit masonry Firewall | | 2952 SF | 3.130 | 8113.47 | $10/18/83$ | 4 | $10/24/83$ | 8,113.47 | 8,750.38 | 9,187.90 | 9,187.90 | 8,113.47 |

Description of Activity

Selected Team Members

Primary Planned Performance Data

Secondary (Derived) Performance Data

Job Strategy

Job Statement of Expectations

Job Game Plan

Values Distributed over Duration

SAMPLE OF PLANNED DATA

of actual performance provides real information; it generates valid management information through comparisons with planned performance.

The importance of maintaining production as a cost and schedule control procedure cannot be overestimated. When significant changes in the job plan occur, maintaining the planned level of production becomes even more important. The effect of disastrous delays in completing a project may be significantly reduced or minimized by recognizing the planner's choices within the production plan and thereby modifying the production plan to maintain an acceptable rate of product placement as reflected by its dollar value per duration unit. This allowance of flexibility in project management then becomes a tool of great import in maintaining production, to be analyzed in a following section.

Another major concept in production management is the *planning horizon*. The planning horizon determines the degree of visibility available at any point in time to the requirements of the project.

## THE LONG TERM HORIZON

In most construction projects, the long term planning horizon is the one which covers the project from its inception to its final completion. This represents the complete project plan, from which the complete project schedule is derived. From the plan and schedule along with the budget, the cost and production curves are generated.

In most construction projects in the field environment, the long term planning horizon presents more detail than is immediately useful to field supervision.

### The Mid Term Horizon

The long term planning horizon, while providing information on total project expectations, should be divided into several shorter planning period horizons. A convenient mid term horizon is the monthly horizon keyed to the production period and to the corresponding accounts payable, accounts receivable, and requisitioning procedures. This time period becomes very useful since it allows the preparation of certain forecasting information related to cash management and cash flow. It also allows all members of the team to do monthly planning of resource requirements related to the project.

### The Short Term Production Horizon

The mid term planning horizon of a month, as it turns out, is also longer than is practical for field supervision at the job site. What is required at the job site is a planning horizon which deals with the daily, weekly and, on occasion, the biweekly activities requiring attention. A planning horizon for the project superintendent which deals with long range issues that may change with time and production encumbers the informational process with no benefit.

The short term planning horizon should be selected collectively by the project team. Usually one or two weeks keyed to payroll periods is the most effective. Having visibility a week ahead is usually sufficient for field activities, for which resources are customarily assigned on a day by day basis. When adjustments in planning and production extend beyond those items included within the one or two weeks planning horizon, the master schedule list may be referenced to pick up these items so that proper reporting of activities can be maintained.

## The Value of Each Horizon

Another issue involved in choosing between the long term, mid term, and short term project plan for production management is relative validity. The project plan using a short term planning horizon tends to have a much higher relative validity then the mid term horizon, which in turn has a much higher validity than the long term horizon.

In an *interactive* environment, where the schedule is updated periodically, the relative validity of the short term planning horizon is much higher since each update is based on what has just been completed. Of course all production planning and production management is keyed to general long term goals. Mid term goals are most effective in dealing with monthly production and cash management, while short term goals are most useful in addressing the day to day production management and control of cost and schedule in the field environment.

## THE NEED FOR FLEXIBILITY AND INTERACTION IN PRODUCTION MANAGEMENT

Great emphasis has been placed on the need for production planning and on the need to use that plan to manage production. What will now be presented may seem very much a contradiction to what has already been presented. There is a constant need for flexibility and interaction in the production management once the project is in construction. Physical construction in the field environment is a dynamic process subject to a wide variety of variances. Some of these can be anticipated. Other variances, however, occur in ways that are beyond the control of the planners. Issues such as inclement weather, availability of materials, availability of work force, and disruption of transportation systems all can drastically affect the production plan. A key requirement is planning production to make it interactive and to provide the flexibility required for maximum production when unexpected contingencies occur. It is not at all unusual, as a result of variances in the field environment, for an alternate plan to evolve to compensate for unanticipated circumstances or to work around delays in the project when such occur.

Since the production plan and schedule are the primary communications tools providing information to all team members, several actions are required to maintain the necessary flexibility in the project plan. The first is an updating of the plan and schedule preferably through a logic network analysis to reestablish logic relationships which are now an appropriate basis for the plan. These then form the basis for a recalculation of schedule. When activities change in the schedule, the budget activities (but not the budgeted cost) should similarly change to account for variations from the original project plan in order to maintain integrated cost and schedule control.

## MAINTAINING PRODUCTION

The primary issue is to maintain production as measured by the *overall rate of placement of product* within the master plan even though the exact plan as originally presented is not being followed. It is in this primary regard that the production planning and control differs from cost and schedule control. Cost and schedule control deal with specific activities from a time and cost point of view while production deals principally with rate of placement of product without regard to the composition of product being placed. The schedule, the budget, and other performance models may be affected by adjustments in the plans for the project and must likewise be adjusted and kept current.

### The Project Action Checklist

One of the most effective ways of keeping a log of raw data needed for scheduling adjustments is the project action checklist. The project action checklist has been recommended as the primary basis for reporting the proceedings of weekly progress meetings. To the extent practical, the project action checklist provides a clear reflection of status on the roster of activities for the project—another step in the integration process, allowing the performance of each activity to be more closely tracked.

## AN IMPORTANT OVERVIEW

Some kind of planning and scheduling goes into every construction project. In the best context, a complete plan, schedule, budget, and all related performance models are prepared. These are the means for production management control and for communicating job expectations to all production team members. Where such information is not available or the appropriate vehicles are not in place to communicate this information to the team, production planning and management are left generally to field supervision. Reasonably competent supervision will use some planning method and some system to assign resources in the field environment. This kind of planning, however, with no basic framework, represents a significant additional burden for field supervision, both unnecessary and unwise.

Field supervision can and should expect a complete project plan and schedule to be available, including short term planning horizons for managing production in the field environment. The value of the *complete* planning for field supervision is dramatically illustrated when, in fact, changes to the plan are required. With an existing plan, the exact nature of the change and the effect of those changes can easily be measured. The alternate plan or strategy dictated by field circumstances can also quickly be established.

When *no* plan exists, the collection of meaningful data on actual performance is difficult if not impossible. The exertion of production control is equally difficult hence there is no clear goal or objective to be achieved. Management of production is left to what amounts to hand to mouth methods and the basis for effective communications within the production team is seriously compromised.

## PERFORMANCE DATA CAPTURE

On a typical project, the concepts developed thus far have taken the project through planning, scheduling, and budgeting. All the necessary performance models have been created. The team has been assembled and the project placed in construction. Since it is unlikely that all of these techniques combined will result in the project proceeding entirely within expectations, some appropriate method of measuring performance must be considered. There appear to be three major components to the performance measurements: (1) collecting data on actual performance of the work, (2) converting the raw data into useful management information, and (3) making appropriate comparisons between what is actually taking place in the construction environment and what has been planned.

### The Data Capture Concept

A term which has been popularized to describe the process of collecting information on performance in the data processing environment is *data capture*. This term has come into relatively common use with the advent of computer technology, but the principles involved in data capture are of long standing. Data capture may be defined as the systematized collection of raw data on some specific performance in some predetermined format. For any data capture procedure, whether manual or automated, to be effective, it must be able to collect the information on actual performance quickly and conveniently—with a minimum of clerical effort—in a way that (preferably) requires little or no recasting of the raw data collected before it is entered into the data processing or analytical environment to produce management information.

## Manual Data Capture

One of the major drawbacks in manual data capture is that it frequently provides raw data in a form not conveniently matching the requirements of the analytical processes consequently requiring considerable recasting before it can be used. Even where data may be neatly matched to one particular work environment, subsequent recasting of the data for other work environments may turn out to be as difficult and time consuming as improperly captured data.

## SHARED RAW DATA

The multiple work environments involved in construction require that data capture processes be matched as nearly as possible to the statement of expectations and to the requirements of the various disciplines which need the data collected.

## Anticipating All Data Requirements

The first major criterion for data capture for systems integration is anticipating the data needs of all the work environments. The requirements of field production will differ from those of project management, which in turn will likely differ from those of financial management and accounting, which again in turn will probably differ from those of administration. In each case, the data capture processes should provide the raw data required for each discipline and in a form convenient to the discipline.

This basic first concept is true without regard to which work environment may be designated as the first recipient of the raw data. The notion that one work environment may have proprietary rights to the data and resulting information can be very detrimental to systems integration. The whole process of useful data capture for *all* work environments therefore becomes of vital interest to all work environments. The design of data capture processes should therefore receive input from all the involved disciplines.

Considerable space in this reference has been given to developing and expanding the concepts of statements of expectations. The statement of expectations clearly sets forth (or should) the planned or expected job performance in considerable detail. The criterion for the level of detail in the statement of expectations is the greater detail required for either cost or schedule control. The usual result for a typical construction project is a statement of expectations which includes literally hundreds of items. Data capture procedures must therefore be developed with due recognition of the statement of expectations.

## Level of Detail and Systems Integration

After appropriate form, then, the second major criterion of data capture for systems integration becomes level of detail. The data capture processes must be designed to match exactly the level of detail, item by item, contained in the statement of expectations. Failure to address this second basic concept in the design of data capture processes can result in the collection of raw data which is extremely difficult to use under the best of circumstances and, under the worst circumstances, completely useless. This commonality of detail and activity within the data capture procedures is pervasive throughout all work disciplines. For example, if the schedule contains a certain list of activities for the concrete work, that identical detail should be found in the budget. It should also be found within the subcontractor's payment schedule where the included items are to be subcontracted. When this basic principle is meticulously followed, very little time and effort will be consumed in the useless recasting of raw data, and the data will be useful within each of the several work environments without further recasting.

## Single Data Capture

The third major criterion for data capture for systems integration is single data capture, discussed in Chapter 4. The term *single data capture* also comes out of the data processing environment. While single data capture may be perceived as more easily achieved in a data processing environment, the plain fact is that carefully contrived manual data capture systems can effectively accomplish the same end.

There are actually two separate components involved in the concept of single data capture. The first of these is capturing all of the data. The second is doing all the data capture at a single most convenient time. While the conjunction of these two components may require more detail to be captured at a particular time, capturing all of the data at a single time greatly simplifies the data capture process and keeps the post data capture clerical effort to an absolute minimum. Specific illustrations and methods including some proposed forms will be provided below to develop this concept.

## Source Documents and Data Capture

The fourth major concept concerning data capture involves the design of both the data capture and the analytical processes so that the data capture (source) documents commonly used (such as a time sheet for labor) become the (source) documents used in the analytical process. Again, the goal is to eliminate wherever possible the required recasting of raw data before it can be entered into the analytical process.

## PLANNED AND ACTUAL DATA CAPTURE

Some degree of complexity is added to the data capture process as a result of system integration. There is, however, no substitute for being able to develop simultaneously management information concerning both cost and schedule control from a single data capture process. Since the basic concepts which have been developed deal with both budgeting and scheduling by activity, each activity should have associated with it not only budget and schedule data but actual cost and actual schedule information as well.

The integration process and the data capture procedures associated with it provide the kinds of information on any activity shown in Exhibit 7-2 and expanded in Exhibit 23-3.

The actual cost and actual time are the data to be collected by the data capture procedures. The management information required for effective management is generated by comparing actual data to planned data in any of the categories. The schedule environment contains only two items for comparison: planned duration and actual duration.

The standard performance ratios are the time performance ratio (TPR), the cost performance ratio (CPR), and the production performance ratio (PPR). These may be quickly calculated at any summary level, beginning with an individual activity and proceeding through entire jobs, or even multiple job summaries for that matter.

The performance ratio opportunities are by no means exhausted by these three. Many others may be calculated simply and conveniently as they become of interest to management. If data is captured on quantities of material actually used, a quantity required ratio (QRR) can be calculated to measure yield of product or as a measure of quantity surveying accuracy. A unit price ratio (UPR) may also be determined after the actual unit price is calculated:

$$\text{actual unit price} = \text{actual cost} \div \text{actual quantity}$$

## 23-3. Planned and Actual Data Capture Descriptions (with Calculated Performance Ratios)

ACTIVITY DATA, INFORMATION AND RESOURCES

| ACTUAL | | | | ACTUAL COST | | | ACTUAL SCHEDULE | | | ACTUAL/TIME GRAPHICS | | | | |
|---|---|---|---|---|---|---|---|---|---|---|---|---|---|---|
| COST CODE | RE-SOURCE | DE-SCRIPTOR | SUB-CONTRACTOR PRINCIPAL VENDOR SUPPLIER | QUAN. | UNIT PRICE | COST | DATE STARTED | DURA-TION | DATE COMPL | COST GRAPHIC | PROD. VALUE GRAPHIC | SCHED. OF VALUES GRAPHIC | ACTUAL CASH INCOME | ACTUAL CASH REQ'D |

SAMPLE OF ACTUAL DATA

| | | | | | | | | | | | | | | |
|---|---|---|---|---|---|---|---|---|---|---|---|---|---|---|
| 4200 | 9 . . 3 | Unit Masonry Firewall | Ace Masonry Co. 555-1234 | 2.600 | 3.00 | $7,800.00 | $10/10/85$ | 6 | $10/17/85$ | $7,800.00 | $8,750.38 | $9,187.90 | $9,187.90 | $7,800.00 |

| Performance ratios* | | | | QRR | UPR | CPR | TRP | | | PPR | | CIR | CRR | |
|---|---|---|---|---|---|---|---|---|---|---|---|---|---|---|
| Calculated performance ratios (Performance ratio–actual ÷ planned) | | | | 1.003 | 0.958 | 0.961 | 1.500 | | | 1.000 | | 1.000 | 0.961 | |

*PLANNED DATA FROM EXHIBIT 23-1

| | | |
|---|---|---|
| QRR | Quantity required ratio | PPR Production performance ratio |
| UPR | Unit price ratio | CIR Cash income ratio |
| CPR | Cost performance ratio | CRR Cash requirements ratio |
| TPR | Time performance ratio | |

Cash income ratio (CIR) and cash requirements ratio (CRR) can be of great value as a part of an overall management information system. These are all made possible and relatively easy by the systems integration processes which have been presented.

Three procedures are involved in the data capture processes. The first of these is the collection of raw data *in a form* which exactly matches the statement of expectations; raw data of itself is seldom useful for management purposes. The second is the data processing procedures which are required to convert the raw data into useful information. These vary significantly depending on the work environment and the complexity of information to be generated. Frequently multiple data processing is required to accommodate the informational needs of the several work environments in construction.

## MANAGEMENT INFORMATION

The final step in processing data is generating management information. Management information is generated primarily by comparing the statement of expectations on an activity by activity basis, planned vs. actual; a work package by work package basis; or at any summary level throughout the project to the actual performance data. There are several major considerations involved in making this process work effectively. The first is that the project must be *implemented* in exactly the way that it has been *planned*. This means that the project strategy must be utilized for the regular and routine day by day production management in the field environment.

Any modifications or deviations from the plan must be recorded and made the subject of an adjustment to the planned schedule and budget. This is such a crucial issue because it is directly related to the data capture procedures. Not only should the job be implemented the way that it has been planned, but the data on actual performance should be captured exactly the way the work has been planned and the way the work has been assigned to the field environment.

The key issue is to plan the work the way the work is expected to be built. Then build by the plan. That is, implement the project strategy and capture the data by the plan. When this is done, the generation of management information involves the use of a series of performance ratios including the cost performance ratio, the time performance ratio, and the production performance ratio. The detailed method of calculating each of these performance ratios will be discussed in a later chapter. The fundamental concept of capturing data on actual performance is an essential procedure allowing simplified comparisons of actual cost to planned cost, actual time to planned time, and actual production to planned production on an activity by activity basis at a work package summary level, at a summary level equal to the standard

Construction Specifications Institute codes, or from an overall job point of view.

Data capture procedures based on capturing data matching the plan also simplifies interdiscipline informational requirements. For example, payroll data may be captured on the specific performance of activities handled through a labor distribution procedure, allowing a direct update of job costs and, on the payroll cycle basis, the use of the identical raw data for the complete processing of payroll. Similarly, data capture with respect to performance of activities within a subcontractor or principal vendor work package also allows the immediate updating of the job cost and the simultaneous updating of the accounts payable files; when it is time to process accounts payable, the data required will be already in place.

Another good example falls in the accounts receivable area. Each activity has an assigned scheduled value associated with its completion. As the activities are completed, or a percentage of completion reported on each of the activities, the activity's value as an accounts receivable amount is already established through a calculation based on the Schedule of Values. It is thereby possible to maintain current information on the value of product in place. When it is time to generate the accounts receivable, the files will have already been updated and the generation of the requisitions becomes a simple and direct process.

Similarly, the basic management information concept is based on planning how you intend the job is to be built, building according to that plan, and capturing data on performance according to the job plan versus the way it is being built. There are several other advantages to these processes. The principal one, which has already been discussed, is minimizing the recasting of raw data for either its use in any work environment, or, in particular, its use among different work environments. Again, the key to the integration of cost and schedule control is planning, scheduling, and budgeting by activities and implementing the work also by activities. This overall process becomes the foundation of simplified data capture which allows the immediate and direct utilization of raw data for the purpose of generating management information without impeding or complicating essential management services.

It is worth noting that field data capture consists of three time related pieces of raw data:

- Actual start of activity;
- Percent completed at reporting time;
- Actual completion of activity.

All other data on performance is keyed to field data. For example, when an activity is reported to be complete, other sources provide information on the effect of that completion. When the unit masonry firewall is completed, $8,113.47 of budget cost has been earned (from Exhibit 23-2). Exhibit 23-3 reveals that its production value is $8,750.38. Its scheduled value for payment purposes is $9,187.90. The payment to Ace Masonry currently due is $7,800.00 less the retainage. The current amount to be paid by the owner is $9,187.90 less the retainage. Because of systems integration, all of this data and much more becomes known simply from the report of one activity's completion in the field.

The process of generating *exception reports* will be discussed in detail in Chapter 29. It should be noted here that the performance ratios become the primary informational source for exception reporting. Some variance allowance should be established to define a performance exception since some variance between planned and actual performance may be routinely expected in construction. Suppose cost performance ratio variance of 5.0% is selected. Exception status occurs anytime the cost performance ratio falls outside of 0.95 to 1.05 inclusive for any activity or work package. When the cost performance ratio has been calculated, all activities falling within 0.95 to 1.05 inclusive are considered to be performing satisfactorily and require no immediate attention. Those falling outside 0.95 to 1.05 inclusive are performance exceptions and require further management attention.

## SUMMARY

Integrated cost and schedule control is built around three major premises. The first of these is that each job prior to its commencement in the field environment will be thoroughly planned and scheduled and that a valid cost budget will be prepared. This planning work allows the preparation of derived performance models to enable a more thorough examination of the actual performance of the project and comparison of its actual performance with its planned performance. Without regard to the exact methods to be used, this planning process must be thorough and it must relate closely to the actual way in which the job will be built.

The second major premise assumes that the job will be built in essentially the way that it has been planned and that the plan, schedule, and budget will be used as the primary production management tools in the field environment. The main concern here is that the actual performance of the work provides the basic framework for data capture so that the data collected on actual performance is an exact match to the way in which the job was planned and to the way in which it is being managed in the field environment.

This neat matching of planned and actual data is far more than a neat procedure or a technical nicety, leading to the third major premise of integrated cost and schedule control: generating management information quickly and conveniently by simple ratio calculations in the three primary performance areas of cost, schedule, and production. While a number of other performance ratios may be calculated, integrated cost and schedule control focuses on how well the costs are being controlled compared to budget, how long the activities are taking to complete compared to schedule, and the actual rate of placement of product compared to actual production.

All three of these performance measures are essential to provide complete management information on performance. For example, the *cost* performance ratio does not directly measure *time* performance. *Time* performance on each activity or on groups of activities does not directly relate to *rate of production*. The rate of production places in a time and cost frame how well the job is being managed and whether or not the performance expectations for the project are being met.

The processes described are thus designed so that actual performance data may be captured simply, conveniently, accurately, and on a timely basis. While these data capture procedures may be greatly facilitated and the associated clerical efforts somewhat reduced with computer capability, the methods described may be used successfully in a manual system. Another important feature of data capture procedures involves anticipating all data requirements and arranging for the sharing of raw data among all work environments. Single data capture is, of course, an important part of these procedures and significantly reduces further the clerical work associated with data capture.

Data capture procedures built upon thorough planning, scheduling, and budgeting have the capability of generating management information of a great variety and significance upon short demand.

# 24
# THE TURNAROUND DOCUMENT
# AND ITS USE

The turnaround document, which communicates all information and data which has been generated on an activity, work package, and job performance, has been introduced in connection with planning and scheduling, budgeting, resource assignment, and data capture. It is now appropriate to consider the turnaround document as the primary communications tool between the office environment and the job site.

The name *turnaround document* is a functional description of the processes for which it is used. All the information required by the field environment for production management and schedule control is communicated to the field in a specific and organized way. To expedite the flow of information from the field environment to the office on actual performance, additional space is provided on the turnaround document (the form used to communicate expected performance data to the field) for the recording of actual performance and returning (or "turning around") that information to the office environment.

## THE BASIS CONCEPT OF THE TURNAROUND DOCUMENT

The turnaround document deals with fundamental, practical, and everyday issues relating to improving the validity and accuracy of field data capture while simultaneously reducing the clerical work and restatement of descriptors required in the field environment. Many of the items which have been presented in detail may have seemed somewhat theoretical. Methods of production planning and the ways in which these relate to the project statement of expectations and game plan have been fully developed. The basic concept of planning by activity using the critical path method for planning and scheduling has been developed and demonstrated to be well suited for integrated cost and schedule control. The object of all of these processes and procedures has been to describe a full range of specific data and information concerning each activity, including:

1. Activity description.
2. Activity cost code.

3. Activity resource code.
4. Activity budget.
5. Activity early start date.
6. Activity duration.
7. Activity early completion date.
8. Budgeted value of activity.
9. Production value of activity.
10. The scheduled value of activity.
11. Purchased value of activity.
12. Name of the principal vendor, subcontractor, or resource responsible for performing the work.

It is obvious that a substantial portion of the information now available for every activity within the project needs to be communicated to the production team and, in particular, to the field supervisory personnel. Certain information which controls the generation of management information on performance is particularly important.

## Both Production Management and Data Capture

Both production management and data capture concepts have been discussed. The importance of capturing data as close as possible in time and physical proximity to the actual performance of the work is a major factor in the validity of data. With all of the information thus generated on project performance, there are a series of major questions which should be addressed:

1. How to facilitate the flow of information from the office to the field on planned performance.
2. How to facilitate the flow of raw data from field to office on actual performance.
3. How to "turnaround" information as quickly and accurately as possible.
4. How to simplify the whole process of data capture in order to provide performance information.

## Primary Communications, Office to Field

The basic concepts previously presented suggest that the development of any procedure should anticipate the ultimate use of the data to be captured by the system or process design. The kinds of information which need to be communicated from the office to the field environment include an appropriate description of what is to be done, who is to do it, and when it is to be done. Careful examination of these items clearly indicates that this is nothing more than certain elements of the planned data with respect to each activity. It would seem that a simple form could be designed to communicate this information from project management to field supervision.

**Planned Performance Data.**   Exhibit 24-1 is an example of the office to field information communication.

A number of important pieces of information are provided in a succinct summary form:

1. A general job reference.
2. The cost code.
3. The cost type, work class and level of indenture references.
4. The standard activity descriptor.
5. A comment field for additional information.
6. The planned early start.
7. The planned duration.
8. The planned early completion.
9. On the second line between activities, such other information as the subcontractor, principal vendor or resource assigned the responsibility for the performance work, including telephone number and purchase order reference.

While there is a wide variety of other information which could be included in the communication to the field, these provide the basic information required for the management of production in the field environment. They represent the planned performance with respect to production management and schedule control.

## Primary Communication, Field to Office

Since all of the information required to identify an activity is available, including the resource responsibility for its performance, an expansion of this form might allow quick and convenient data capture in the same way as the planned information has been presented. The kinds of information required for data capture may be presented in tabular form matching the planned data. An example of the data capture portion, which may be added to the planned performance data, is shown in Exhibit 24-2.

## Actual Performance Data

Since the complete identifiers have already been provided, it becomes unnecessary for data capture purposes to repeat this information. The amount of

**Exhibit 24-1. Planned Performance Information Office to Field Communication**

```
      JUNE  6, 1985                    GENERAL CONSTRUCTION CO., INC.              MONDAY 11:20 AM
                                  Actual Schedule sorted by Durations
PS075    JOB 9010 CONSTRUCTION MANAGEMENT SAMPLE        1 FLRS  75000 SQ FT                ESTIMATE  PAGE 04
         Daily Production Activity Check List & Time Report                               Report No  :  1
         Period Covered    0 -  3/ 1/85 to   88 -  7/ 1/85              Work Day: 71       Report Date: 06/07/85
=============================================================================================================

   Activity        Description           Comment  Early Start  Completion Dur. Actual Actual Percent   Res  Pay  Hold Close
Vendor Number/ Name                                                           Start Compl Complete Quant Adjust Pay  Out
                                                                             A d j u s t  R e a s o n / C o m m e n t s

  3320 (9    1) CAST-IN-PLACE FLOOR SLAB                06/14/85  07/05/85  15____:_____:_____:_____:_____:____:__

  3330 (9    1) SITE CAST-IN-PLACE CONCRETE             06/27/85  07/18/85  15____:_____:_____:_____:_____:____:__
  437 3365-1

  4200 (9    1) UNIT MASONRY              ALL MASONR *04/19/85  06/07/65  35_05/62_:_____:____95_:_____:_____:____:__
  273 2356-1 GOOD MASONRY, INC.          P.O.# 5185

  4200 (9    2) UNIT MASONRY              MASY FILL  04/19/85  06/07/85  21_04/29_:_05/30_:__100_:_____:_____:____:__
  273 2356-1 GOOD MASONRY, INC.          P.O.# 5187

  5120 (9    1) STRUCTURAL STEEL          ALL STRUCT 06/21/85  07/12/85  10____:_____:_____:_____:_____:____:__
  997 7211-1 STURDY STEEL CO., INC.      P.O.# 5193

  7150 (9    1) DAMPPROOFING                            05/19/85  05/20/85   1_05/19_:_05/20_:__100_:_____:_____:____:__

  7206 (9    1) CAULKING                  CNTRL JNTS *05/10/85  05/27/85 * 13_05/10_:_05/22_:__100_:_____:_____:____:__
  273 2356-1 WATERPROOFERS, INC.         P.O.# 5206

  8100 (1    1) METAL DOORS & FRAMES      INSTALL    *05/10/85  05/10/85   0_05/10_:_____:____45_:_____:_____:____:__
             EMP #    EMP NAME     HRS :EMP #   EMP NAME    HRS :EMP #   EMP NAME    HRS :EMP #   EMP NAME     HRS
MONDAY   :                             :                       :                       :
TUESDAY  :                             :                       :                       :
WEDNESDAY:                             :                       :                       :
THURSDAY :                             :                       :                       :
FRIDAY   :                             :                       :                       :
SATURDAY :                             :                       :                       :

  8100 (2    1) METAL DOORS & FRAMES      MATERIAL   *05/10 85  05/10/85 * 0_05/10_:_05/10_:__100_:_____:_____:____:__

  15400 (9   1) PLUMBING SYSTEMS          UNDER SLAB *03/29/85  04/12/85 * 4_04/02_:_04/13_:__100_:_____:_____:____:__
  587 0013-1 GREAT CONNECTIONS, LTD.     P.O.# 5212

  16111 (9   1) CONDUITS                  UNDER SLAB *04/01/85  04/04/85 * 1_04/01_:_04/04_:__100_:_____:_____:____:__
  384 1053-1 POWER PLUS, INC.            P.O.# 5216

  16400 (9   1) ELECTRICAL SERV.& DISTRIBUTION         06/01/85  06/08/85   5____:_____:_____:_____:_____:____:__
  384 1053-1 POWER PLUS, INC.            P.O.# 5216

  81110 (4   1) Builders Risk Insurance   BY OWNER   06/01/85  06/01/85   0____:_____:_____:_____:_____:____:__

                                                                              CONFIDENTIAL

WEATHER:_____:_REMARKS:_____
TEMP___:_____:_____
FACTORS CAUSING DELAYS:_____:_____
                                           :_BY:_____DATE:_____
```

## Exhibit 24-2. Data Capture Field to Office Communication

```
         JUNE  6, 1985              GENERAL CONSTRUCTION CO.,' INC.                    MONDAY 11:20 AM
                                  Actual Schedule sorted by Durations
PS07S       JOB 9010  CONSTRUCTION MANAGEMENT SAMPLE        1 FLRS   75000 SQ FT              ESTIMATE  PAGE 04
         Daily Production Activity Check List & Time Report                           Report No  :  1
         Period Covered    0 - 3/ 1/85 to    68 -  7/ 1/85                Work Day: 71   Report Date: 06/07/85
=========================================================================================================

    Activity          Description            Comment  Early Start  Completion  Dur. Actual Actual Percent  Req  Pay  Hold Close
Vendor Number/ Name                                                                  Start  Compl  Complete  Quant Adjust  Pay  Out
                                                                                     Adjust   Reason/Comments

  3320 (9   1) CAST-IN-PLACE FLOOR SLAB                  06/14/85   07/05/85   15____!_____!____!_____!_____!___!____

  3333 (9   1) SITE CAST-IN-PLACE CONCRETE               06/27/85   07/18/85   15____!_____!____!_____!_____!___!____
  437 3365-1

  4200 (9   1) UNIT MASONRY                 ALL MASONR *04/19/85   06/07/65   35_05/02_!_____!____!__25_!_____!___!____
  273 2356-1 GOOD MASONRY, INC.             P.O.# 5185

  4200 (9   2) UNIT MASONRY                 MASY FILL   04/19/85   06/07/85   31_04/29_!_05/30_!_100_!_____!_____!___!____
  273 2356-1 GOOD MASONRY, INC.             P.O.# 5187

  5120 (9   1) STRUCTURAL STEEL             ALL STRUCT  06/21/85   07/12/85   10____!_____!____!_____!_____!___!____
  997 7211-1 STURDY STEEL CO., INC.         P.O.# 5193

  7150 (9   1) DAMPPROOFING                              05/19/85   05/30/85    1_05/19_!_05/20_!_100_!_____!_____!___!____

  7206 (9   1) CAULKING                     CNTRL JNTS *05/10/85   05/27/85 *  13_05/10_!_05/27_!_100_!_____!_____!___!____
  273 2356-1 WATERPROOFERS, INC.            P.O.# 5206

  8100 (1   1) METAL DOORS & FRAMES         INSTALL    *05/10/85   05/10/85    0_05/10_!_____!____!__45_!_____!___!____
             EMP #    EMP NAME         HRS !EMP #   EMP NAME       HRS !EMP #    EMP NAME     HRS !EMP #   EMP NAME      HRS
  MONDAY    :                              !                           !                         !
  TUESDAY   :                              !                           !                         !
  WEDNESDAY :                              !                           !                         !
  THURSDAY  :                              !                           !                         !
  FRIDAY    :                              !                           !                         !
  SATURDAY  :                              !                           !                         !

  8100 (2   1) METAL DOORS & FRAMES         MATERIAL   *05/10 85   05/10/85 *  0_05/10_!_05/10_!_100_!_____!_____!___!____

  15400 (9  1) PLUMBING SYSTEMS             UNDER SLAB *03/29/85   04/12/85 *  4_04/02_!_04/13_!_100_!_____!_____!___!____
  587 0013-1 GREAT CONNECTIONS, LTD.        P.O.# 5212

  16111 (9  1) CONDUITS                     UNDER SLAB *04/01/85   04/04/85 *  1_04/01_!_04/04_!_100_!_____!_____!___!____
  384 1053-1 POWER PLUS, INC.               P.O.# 5216

  16400 (9  1) ELECTRICAL SERV.& DISTRIBUTION            06/01/85   06/08/85    5____!_____!____!_____!_____!___!____
  384 1053-1 POWER PLUS, INC.               P.O.# 5216

  81110 (4  1) Builders Risk Insurance      BY OWNER    06/01/85   06/01/85    0____!_____!____!_____!_____!___!____
```

CONFIDENTIAL

```
WEATHER:_____!_REMARKS:_____
TEMP___:_____!_____
FACTORS CAUSING DELAYS:_____!_____
                                   !_BY:_____DATE:_____
```

field clerical work required for data capture and to record performance has been reduced to simple handwritten data entries to record:

1. Actual start.
2. Actual completion.
3. Percentage completion.
4. Required quantity.
5. Pay adjustment.
6. Hold.
7. Closeout.

As each activity is started, the actual start date is recorded. This corresponds to, and compares with the planned start. When desired, information on percentage completed may also be captured and entered, with the percentage completed noted as time specific data valid only for the date and time of data capture. When an activity is completed, the actual completion date may also be entered. This corresponds to and compares with the planned completion date. The foregoing items complete the basic information which is required to update the schedule. Since both cost and schedule are fully integrated and since all of the related information with respect to budget, purchased, production value, scheduled value is known for each activity, any management information required may be quickly and simply generated.

## Other Informational Processes

While information is being collected, a number of additional data capture spaces may be provided depending on the informational requirements of the end user. For example, required quantities may be captured to support a quantity unit price data base. Additional control tools may be provided for field supervision by adding a capability of holding payment and a capability of closing out an activity.

## COMPUTER GENERATED TURNAROUND DOCUMENT

Exhibit 24-3 shows a computer generated turnaround document with actual performance entered. What is illustrated is a convenient form of presenting planned information to the field environment and, for the field environment, to capture performance data and to return the information to the office. This data may also be used in generating management information and in supporting all downstream systems including accounts receivable, accounts payable, payroll, job costing and cash management. The key process is to present to the field planned performance data and for the field to "turnaround"

## Exhibit 24-3. Computer Generated Turnaround Document (TAD)

JUNE 6, 1985 GENERAL CONSTRUCTION CO., INC. MONDAY 11:20 AM

Actual Schedule sorted by Durations

PS075 JOB 9010 CONSTRUCTION MANAGEMENT SAMPLE 1 FLRS 75000 SQ FT ESTIMATE PAGE 04

Daily Production Activity Check List & Time Report Report No : 1

Period Covered 0 - 3/ 1/85 to 88 - 7/ 1/85 Work Day: 71 Report Date: 06/07/85

| Activity / Description<br>Vendor Number/ Name | Comment | Early Start | Completion | Dur. | Actual Start | Actual Compl | Percent Complete | Rem Quant | Pay Adjust | Hold Pay | Close Out |
|---|---|---|---|---|---|---|---|---|---|---|---|
| | | | | | Adjust Reason/Comments | | | | | | |
| 3320 (9 1) CAST-IN-PLACE FLOOR SLAB | | 06/14/85 | 07/05/85 | 15 | | | | | | | |
| 3330 (9 1) SITE CAST-IN-PLACE CONCRETE<br>437 3365-1 | | 06/27/85 | 07/18/85 | 15 | | | | | | | |
| 4200 (9 1) UNIT MASONRY<br>273 2356-1 GOOD MASONRY, INC. P.O.# 5185 | ALL MASONR | *04/19/85 | 06/07/85 | 35 | 05/02 | | 95 | | | | |
| 4200 (9 2) UNIT MASONRY<br>273 2356-1 GOOD MASONRY, INC. P.O.# 5187 | MASY FILL | 04/19/85 | 06/07/85 | 21 | 04/29 | 05/30 | 100 | | | | |
| 5120 (9 1) STRUCTURAL STEEL<br>997 7211-1 STURDY STEEL CO., INC. P.O.# 5193 | ALL STRUCT | 06/21/85 | 07/12/85 | 10 | | | | | | | |
| 7150 (9 1) DAMPPROOFING | | 05/19/85 | 05/20/85 | 1 | 05/19 | 05/20 | 100 | | | | |
| 7206 (9 1) CAULKING<br>273 2356-1 WATERPROOFERS, INC. P.O.# 5206 | CNTRL JNTS | *05/10/85 | 05/27/85 * | 13 | 05/10 | 05/27 | 100 | | | | |
| 8100 (1 1) METAL DOORS & FRAMES | INSTALL | *05/10/85 | 05/10/85 | 0 | 05/10 | | 45 | | | | |

| | EMP # | EMP NAME | HRS | EMP # | EMP NAME | HRS | EMP # | EMP NAME | HRS | EMP # | EMP NAME | HRS |
|---|---|---|---|---|---|---|---|---|---|---|---|---|
| MONDAY | | | | | | | | | | | | |
| TUESDAY | | | | | | | | | | | | |
| WEDNESDAY | | | | | | | | | | | | |
| THURSDAY | | | | | | | | | | | | |
| FRIDAY | | | | | | | | | | | | |
| SATURDAY | | | | | | | | | | | | |

| Activity / Description<br>Vendor Number/ Name | Comment | Early Start | Completion | Dur. | Actual Start | Actual Compl | Percent Complete | Rem Quant | Pay Adjust | Hold Pay | Close Out |
|---|---|---|---|---|---|---|---|---|---|---|---|
| 8100 (2 1) METAL DOORS & FRAMES | MATERIAL | *05/10 85 | 05/10/85 * | 0 | 05/10 | 05/10 | 100 | | | | |
| 15400 (9 1) PLUMBING SYSTEMS<br>587 0013-1 GREAT CONNECTIONS, LTD. P.O.# 5212 | UNDER SLAB | *03/29/85 | 04/12/85 * | 4 | 04/07 | 04/13 | 100 | | | | |
| 16111 (9 1) CONDUITS<br>384 1053-1 POWER PLUS, INC. P.O.# 5216 | UNDER SLAB | *04/01/85 | 04/04/85 * | 1 | 04/01 | 04/04 | 100 | | | | |
| 16400 (9 1) ELECTRICAL SERV.& DISTRIBUTION<br>384 1053-1 POWER PLUS, INC. P.O.# 5216 | | 06/01/85 | 06/08/85 | 5 | | | | | | | |
| 81110 (4 1) Builders Risk Insurance | BY OWNER | 06/01/85 | 06/01/85 | 0 | | | | | | | |

CONFIDENTIAL

WEATHER: | REMARKS:

TEMP :

FACTORS CAUSING DELAYS:

| BY: DATE:

actual performance data on the same document. The design of the turnaround document presents planned job information. The field environment inserts on the same form the actual performance information for updating the schedule and for the other downstream processes.

Note that field data capture concerns *only* field performance data such as start date, percent completed, or completion date. The consequence of this performance varies widely depending on the processing area in which it is used. In the budget, it tells the budgeted amount earned, the production value earned, and the amounts payable to be incurred and consequently the job cost. But it also provides accounts receivable and cash management information, all as a result of specific field data on activity start, completion, or percentage complete.

The turnaround document eliminates the common cost coding errors since document's information is presented fully. The turnaround document has several uses in addition to capturing raw data. It may be designed in a form to serve as the data entry vehicle when data processing systems are used. It should be pointed out, however, that turnaround documents for data capture serve equally well in either manual or computer systems.

### Labor Reporting

Several additional features may be added to the turnaround document to facilitate the informational flow. For example, when an activity is to be performed by one's own labor, cost type 1, the informational spaces may be expanded to allow direct reporting of payroll, illustrated in Exhibit 24-4.

This form expanded to account for the reporting of labor is consistent with the single data capture process of collecting raw data in the field and also consistent with the single data capture which allows the raw data, once captured, to be utilized in a variety of informational systems including payroll and job costing.

## THE TURNAROUND DOCUMENT AS A PRODUCTION MANAGEMENT AND CONTROL TOOL

One of the common pitfalls which will be experienced in moving toward full systems integration of cost and schedule control is the temptation of field personnel to maintain two separate production management systems. The first system will be the one with which they are already familiar. The second system will be the formal one discussed in this reference for integrated cost and schedule control. The use of dual systems is, of course, a waste of time and effort and should be discouraged. Some additional instruction and development to field personnel in the use of turnaround documents can signifi-

## Exhibit 24-4. Turnaround Document Labor Reporting

```
        JUNE  6, 1985                    GENERAL CONSTRUCTION CO., INC.              MONDAY 11:20 AM
                                   Actual Schedule sorted by Durations
PS075        JOB 9010  CONSTRUCTION MANAGEMENT SAMPLE        1 FLRS   75000 SQ FT            ESTIMATE  PAGE 04
      Daily Production Activity Check List & Time Report                                    Report No  :  1
      Period Covered   0 - 3/ 1/85 to   80 - 7/ 1/85                Work Day: 71            Report Date: 06/07/85
===================================================================================================================

      Activity         Description              Comment  Early Start  Completion  Dur. Actual Actual Percent  Res  Pay  Hold Close
Vendor Number/ Name                                                              Start  Compl Complete Quant Adjust Pay  Out
                                                                                 A d j u s t   R e a s o n / C o m m e n t s

  3320 (9    1) CAST-IN-PLACE FLOOR SLAB             06/14/85   07/05/85   15_____:_____:_____:_____:_____:_____:____:

  3330 (9    1) SITE CAST-IN-PLACE CONCRETE          06/27/85   07/18/85   15_____:_____:_____:_____:_____:_____:____:
  437 3365-1

  4200 (9    1) UNIT MASONRY              ALL MASONR *04/19/85  06/07/85   35.05/02_:_____:___85_:_____:_____:_____:____:
  273 2356-1 GOOD MASONRY, INC.           P.O. # 5185

  4200 (9    2) UNIT MASONRY              MASY FILL  *04/19/85  06/07/85   31.04/29_:_05/30_:__100_:_____:_____:_____:____:
  273 2356-1 GOOD MASONRY, INC.           P.O. # 5187

  5120 (9    1) STRUCTURAL STEEL          ALL STRUCT  06/21/85  07/12/85   10_____:_____:_____:_____:_____:_____:____:
  997 7211-1 STURDY STEEL CO., INC.       P.O. # 5193

  7150 (9    1) DAMPPROOFING                          05/19/85   05/30/85   1_05/19_:_05/20_:__100_:_____:_____:_____:____:

  7200 (9    1) CAULKING                  CNTRL JNTS *05/10/85  05/23/85 * 13_05/10_:_05/23_:__100_:_____:_____:_____:____:
  273 2356-1 WATERPROOFERS, INC.          P.O. # 5206

  8100 (1    1) METAL DOORS & FRAMES      INSTALL    *05/10/85  05/10/85   0_05/10_:_____:___45_:_____:_____:_____:____:
  ▲  EMP #     EMP NAME      HRS :EMP #    EMP NAME     HRS :EMP #    EMP NAME    HRS :EMP #    EMP NAME     HRS
MONDAY   :                           :                        :                        :
TUESDAY  └────Cost Type 1—Labor Time Report
WEDNESDAY:                           :                        :                        :
THURSDAY : 301  Smith Wm. 8.0 :315  Brown E. 6.5 :                 :
FRIDAY   :                           :                        :                        :
SATURDAY :                           :                        :                        :

  8100 (2    1) METAL DOORS & FRAMES      MATERIAL   *05/10 85  05/16/85 * 6_05/10_:_05/10_:__100_:_____:_____:_____:____:

  15400 (9   1) PLUMBING SYSTEMS          UNDER SLAB *03/29/85  04/12/85 * 4_04/02_:_04/15_:__100_:_____:_____:_____:____:
  587 0013-1 GREAT CONNECTIONS, LTD.      P.O. # 5212

  16111 (9   1) CONDUITS                  UNDER SLAB *04/01/85  04/04/85 * 1_04/01_:_04/04_:__100_:_____:_____:_____:____:
  384 1053-1 POWER PLUS, INC.             P.O. # 5216

  16400 (9   1) ELECTRICAL SERV.& DISTRIBUTION       06/01/85   06/08/85   5_____:_____:_____:_____:_____:_____:____:
  384 1053-1 POWER PLUS, INC.             P.O. # 5216

  81110 (4   1) Builders Risk Insurance   BY OWNER    06/01/85  06/01/85   0_____:_____:_____:_____:_____:_____:____:

                                                                                          CONFIDENTIAL

WEATHER:_____:  REMARKS:_____
TEMP   :___                             :
FACTORS CAUSING DELAYS:_____
                                        :  BY:_____  DATE:_____
```

cantly reduce (if not completely eliminate) the dual systems approach which may otherwise be experienced. Field supervision will require training to use integrated systems effectively as a production management tool.

## Field Use of the TAD

Here are some helpful suggestions. In the field when a subcontractor is to be assigned an activity of work, the following sequential steps should be taken:

1. The activity to be performed should be located and clearly identified on the turnaround document as that subcontractor's responsibility.
2. When the activity has been identified, it should be confirmed that it is within the payment schedule for the subcontractor responsible for performing the work.
3. Once the activity identification in both the turnaround document and the subcontractor's payment schedule has been confirmed, the subcontractor's workforce should be assigned to the performance of that activity.
4. The commencement of performance should be recorded on the turnaround document with specific instructions to the subcontractor's supervision to report immediately upon completion of the activity.
5. When the completion of the activity is reported, that completion must be confirmed by the job superintendent.
6. Once the completion has been confirmed, the actual completion date should also be recorded on the turnaround document. Where interim steps require a percentage of completion report, this can be recorded in the space provided. Customarily, the percentage of completion may be used at the end of a requisition or billing period when a significant item may span across the billing period. The percentage completed may be included in the requisition from the general contractor to the owner and in the billing from the subcontractor to the general contractor.

Maximum validity on performance data, however, may be obtained when each activity is considered to be an indivisible unit of work. This eliminates the field measurements and other frequent pitfalls associated with determining percentages of completion. A prerequisite to working on the basis of completed activities is the definition of "completed." For the purpose of these discussions, "completed" means the following:

1. The activity requires no more time.
2. The activity requires no more resources.
3. The activity will not appear on any punch list.

With these criteria of "completed" in place, the answer to the question "Is an activity completed?" becomes a simple "Yes" or "No." Any response other than an unequivocal "Yes" means that the activity is not complete. When the payment schedule attached to each subcontract establishes completed activities as the basis for payment, significant economic control is established relating to the performance of the activities included in any particular subcontract.

## SUMMARY

The turnaround document, properly designed to present to the field planned performance data and to capture from the field actual performance data, can significantly enhance the flow of information between the office and the field. This is two way communications; all information available from the office environment on planned performance should be provided to the field. Conversely, all information on actual performance should be presented to the office by way of the turnaround documents.

A relatively common occurrence is the reassignment of an activity to a subcontractor, principal vendor, or other resource after having been assigned to another subcontractor, principal vendor, or resource. This kind of change in contract responsibility initiates a whole series of administrative processes which need to be completed on a timely basis in order to maintain administrative control of the project. Such things as issuing credit purchase orders to the original subcontractor, issuing supplementary purchase orders to the new subcontractor or vendor, and modifying subcontract control to reflect the changes must be attended to on a timely basis.

In its best sense, the turnaround document can be a very effective interactive communications tool between office and field. Both simplicity and convenience should encourage the generation of current valid information on actual job performance.

The actual performance data reported through the turnaround document is the data which drives all downstream processing and information systems. The processing systems dependent on the TAD data include accounts payable, accounts receivable and most job related accounting functions as well; the informational systems driven by TAD data include job costing, all under-expensed reporting, all management information performance ratios, and all cash management information systems.

Within the context of integrated cost and schedule control, the turnaround document and its effective use is singularly the most important procedure once the project has been properly planned, scheduled, budgeted, and placed in construction.

# 25

# PERFORMANCE MEASUREMENTS

Two separate and distinct procedures involved in performance measurements must be effectively utilized to produce useful management information: valid and timely data capture and timely performance calculations.

## ACTUAL PERFORMANCE DATA

Assuming that a turnaround document or other appropriate data capture vehicle is available and being properly utilized, the first major consideration is the collection of raw data on actual performance. Actual performance data provides information in three primary arenas: cost control, production control, and schedule control. Each of these in its own way significantly controls profitability in the job.

The processes and procedures which have been discussed and developed in some detail provide an adequate source of raw data for performance measurement. Usually some computational or analytical routine is required to convert the raw data into performance measurements useful to project management and field supervision. Performance data is usually presented in a simple, factual, and concise way. For example, how much did an activity cost? How long did it take? When did it start? When was it completed? What was the rate of placement per duration unit? Other questions also may be of interest to a particular systems user, but the foregoing appear to be the primary items for consideration.

### The Activity as the Basic Unit of Work

In integrated cost and schedule control procedures, the activity, work package, Construction Specifications Institute (C.S.I.) Division, or other summary levels may be used to present information about performance. The *activity*, however, remains the lowest basic unit of work and is the common descriptor by which various performance data is presented. Where one's own resource (such as labor, materials, or equipment) is utilized, cost, time, and production are closely related functions. When a resource is purchased for a lump sum amount, the cost is fixed and is therefore independent of performance since the cost is guaranteed by the subcontract.

Not to be overlooked, however, are the related issues such as general requirements, insurances, and taxes, which are functions of time and which can be directly affected by time extensions or low productivity resulting from the performance of a subcontractor or principal vendor. In these cases, while the cost is guaranteed with respect to the value of the specific activity, *indirect* cost may be significantly affected by poor time and production performance.

## Job Cost Reports

Historically, most performance measurements have been presented to management, project management, and field supervision in the form of job cost reports. Job cost reports as a basis for performance measurement will be discussed in considerably more detail subsequently. Suffice it to say here, however, that in addition to the problems of providing timely job cost information through most accounting systems, job costing does not usually provide specific comparisons between actual performance and planned performance.

This raises the second major issue of performance measurement, which involves comparing actual performance to planned performance. The basic question to be asked concerning any activity, work package or cost coding classification (or for the entire project for that matter) is: How is the project performing compared to how it was expected or planned to be performed? What is required is some convenient and expeditious way of converting raw data into measured performance information which presents to management a comparison of actual to planned performance. At the very foundation of integrated cost and schedule control is the *matching of data on an activity by activity basis.* This data matching allows the use of simple ratio computations to generate highly valid, useful, and contemporary management information.

## BASIC HISTORIC PERFORMANCE MEASUREMENTS

Before we proceed with analysis and ratio computations, let's examine in some detail historic methods of presenting performance measurements.

### Profit and Loss Statements: the Most Basic Performance Measurement

The profit and loss statement is a pure accounting presentation involving, usually, the operation of the entire company and may include a single job or many jobs without regard to their diversity, complexity, or duration. The profit and loss statement provides what may be referred to as the "final bottom line." Information is presented with regard to profit or loss resulting from business activities over some period of time usually ranging from three months, or quarterly, to six months, or semi-annually, to, most frequently, a full fiscal year. From a production management point of view, such data is almost always so far out of phase with actual field construction and with the ability to influence outcome that it is virtually useless for contemporary cost or schedule control.

This is *not* to say that the information is entirely useless. It is, on the contrary, essential information for management in terms of overall business activities, and the whole business operation. It is less than useful in providing immediate or contemporary feedback on actual job performance to detect adverse trends while they are correctable. It has been apparent for many years in the construction industry that more timely management information measuring project performance has been required.

## Job Cost: the Next Most Basic Performance Measurement

Job cost has evolved over the years as a vehicle to provide information more closely related to actual performance than profit or loss statements. In most construction organizations, job costing is a primary accounting function. As a primary accounting function, the basic raw data for job cost is derived from two major accounting functional areas: accounts payable and payroll. Since payroll usually involves a weekly cycle of processing and consequently of performance measurement information, and since accounts payable usually involve a monthly cycle, it is not possible to get job cost information as current as required for project management. Costs are posted to the job cost records not at the time they are incurred in the field environment but rather at the time the accounts payable invoices are received and when the payroll is posted.

In work environments which are not interactive in nature, both accounts payable and payroll may be posted to job cost on a monthly cycle. Since there is normally a two to four week processing delay from the close of the month until job cost information is prepared, the performance information presented by most job costing routines may be between four and eight weeks old at the time it becomes available to management. In most work environments, this time lag places job cost information far outside the time frame for detecting adverse trends while they are correctable. It is obvious that some better vehicle for providing performance information based on performance measurements must be derived.

## THE VALUE OF INTEGRATED COST AND SCHEDULE CONTROL

The value of integrated cost and schedule control has been presented in a variety of different ways. As it relates to the timely availability of job cost performance information, the value of integrated cost and schedule control takes on new meaning. On the assumption that the project has been planned, scheduled, and budgeted in anticipation of integrated cost and schedule control, and since this includes full integration of all other informational systems using the roster of activities, an extraordinary amount of information concerning each activity is immediately available based on the performance data provided by the turnaround document (TAD). This information includes the budgeted amount for each activity, the amount to be paid to the subcontractor or principal vendor for the performance of the activity, the production value of the activity, and the scheduled value of the activity.

Contemporary performance data from the turnaround document allows the immediate determination of any of the different values for any specific activity, subset of activities, or group of activities within a job costing summary category or in the entire project. Note that any of these values is determined simply by looking up the activity and reading the particular value of interest; no calculations are needed to determine the specific values for completed activities. The primary calculation required is multiplying an activity value by a percentage when a percent completed is reported for an activity. A listing of activity values for an integrated system includes:

1. Estimated value (bid value at cost level).
2. Budgeted value (fair value at cost level).
3. Production value (fair value at contract level).
4. Scheduled value (accounts receivable value, A/R).
5. Purchased value (accounts payable value, A/P).

In addition to these five values, a wide range of other data is known about each activity. Exhibit 23-1 presents in tabular form a summary of planned activity performance data. Actual performance data capture descriptions are presented in Exhibit 23-2. Certain planned and actual data has been inserted into these two exhibits to demonstrate the convenience and accessibility of performance information associated with integrated cost and schedule control.

A detailed examination of the data presented in tabular form reveals the kinds of high quality information immediately available upon receipt of turnaround document data that indicates the activity is started, partially completed, or fully completed.

The essential actual performance data for each activity to be reported on the TAD consists of three items only:

- Actual start.
- Percent complete at report time.
- Actual completion.

These items are discussed in detail in the previous chapter. Actual start simply records the commencement of the actual duration of the activity. Percent complete and actual completion are the keys to other performance information.

Since the completed activity provides a more concise basis for performance measurement, assume the turnaround document reports that activity 4200 9_3 from Exhibit 23-2 has been completed. The values assigned in Exhibits 23-2 and 23-3 to this activity are shown in Exhibit 25-1.

In this case, the budgeted amount of $8,113.47 has been earned because the activity is completed. The activity's production value is $8,750.38 and represents the amount earned in contract terms. The tabular data reveals an agreement to pay the Ace Masonry subcontractor $7,800 for this activity.

## Immediate Knowledge of Future Accounts Payable and Other Values

The requisition from the subcontractor for this activity, which is now complete, may not be received for several weeks; but when it is received, it will be in the amount of $7,800 less retainage. It is unnecessary to wait for the subcontractor's invoice to know exactly the amount to be billed based on the subcontractor's activity payment schedule and the (completed) performance data now shown on the turnaround document. This, by the way, also provides the independent verification of accounts payable.

This information becomes instantly available for a management job cost report without a wait until the end of the month or billing period to be informed by the subcontractor that he expects to be paid $7,800 less retainage for the item. This is the current accounts payable (purchased) value of the activity when complete.

### Exhibit 25-1. Tabular Values for a Specific Activity

|  | 4200 9_3 | Unit masonry | Firewall |  |
| --- | --- | --- | --- | --- |
| PLANNED | ESTIMATED (Not shown) | BUDGETED $8,113.47 | PRODUCTION $8,750.38 | SCHEDULED $9,187.90 |
| ACTUAL | PURCHASED $7,800 |  |  |  |

Similarly, it is also immediately known that the production value of the item (cost plus the spread) is $8,750.38. It is apparent from the tabulation that the scheduled value of this activity is $9,187.90, reflecting some weighting of the item. Its current accounts receivable value is $9,187.90 less retainage.

It is an extraordinary discovery to recognize that simply by knowing this activity is complete (from the TAD), all of these values are confirmed and immediately usable for all management information in all work environments.

## Contemporary Job Cost Information

Using the foregoing procedure, it is possible to immediately determine for every activity the amount to be subsequently expensed (by accounting) and its current accounts payable value as well as the amount earned in terms of the budget upon its completion or even while it is being completed if percentage completion data is captured. Although some accounting procedures may insist on expensing job cost items only from actual payroll and accounts payable invoices, it is clearly possible to generate "unexpensed" or "under-expensed" job costs which can provide valid and current management job cost data (both planned and actual) as current as the latest turnaround document. Thus job cost data through the current day can be made available consistently if the data is captured with a daily turnaround document.

It is neither necessary nor desirable to compromise the formal job costing accounting procedures which are driven by accounts payable and payroll to generate this kind of information. It does, however, represent a classic use of single data capture where the activity data to be captured in the field environment through the turnaround document is the actual start, percent complete, and actual completion of an activity. This raw data is used to determine job cost incurred to date. As the formal job cost accounting procedures receive and confirm job cost data from payroll and accounts payable, these items become a part of the accounting job cost report.

Due to the normal time lag from cost incurred in the field to posting payroll data and accounts payable invoices to job cost, there will always be some portion of cost incurred which will not appear on the current job cost report. This portion then becomes the basis for the under-expensed report which, when added to the formal job cost report, provides instant, complete and accurate information on job cost to date.

This process contains a high degree of validity since the actual cost is determined either by the payroll cost reported for the activity or from the payment schedule for the subcontractor or vendor responsible for performing the work. Except for unexpected costs, the purchase value less retainage is the amount to be paid to the subcontractor and should be exactly the same as the invoice for that activity when it is received. These amounts are included in the payment schedule for that activity within the subcontract. As time progresses and the actual invoices are received and actual payroll is recorded, there should be, of course, confirmation that the expected billing and/or payroll is in keeping with the payment schedule or the payroll reported for the activity.

## The Under-expensed Report

The under-expensed report prepared by the processes described provides new, contemporary job cost information which, then, has high validity. Even more interesting and exciting from a management information point of view is the relative timing of the informational process. Subject to the limitations of clerical effort available and/or data processing capability, it is possible, as noted above, to generate valid job cost information to the current day. The method of calculating job cost to the current day is:

Total job cost to date = expensed job cost + under-expensed job cost

where *expensed* is formal accounting job cost reported through the end of the most recent reporting period, and *under-expensed* is the payment or purchase value, or payroll for activities started, completed, or percent complete since the last job cost report.

The enhanced job cost data results from a combination of accounting job cost data plus the under-expensed report information. The job costing and under-expensed processes involve detailed analyses but represent significant improvement in the quality and timeliness of data available. In certain ways, however, even these lack the ability to provide management information within a sufficiently short time period to detect adverse trends while they are correctable.

## CONTEMPORARY MANAGEMENT INFORMATION:
## PERFORMANCE RATIOS

What is lacking, then, in the performance measurement techniques just discussed is the capability of providing quick and concise information on comparative performances for project management purposes. That is, in this context the question again must be raised: "How *are* we doing compared to how we *expected* to be doing?"

## Comparative Performance Information

The answer to the question of comparative performance may be found if a simple way could be developed to compare actual performance to planned performance. Within the context of a project which has been planned, scheduled, and budgeted, anticipating integrated cost and schedule control, the capability is immediately available for providing this kind of contemporary information. Some significant research into appropriate methods for providing contemporary and valid performance measurements and performance information for integrated cost and schedule control has led to the development of the *performance ratio*. Performance comparisons using the processes and procedures which have been previously presented may be quickly and conveniently made. There are three primary performance ratios: cost performance ratio, time performance ratio, and production performance ratio which provide the most interesting and useful information.

The conceptual framework upon which the performance ratios are based is the simple notion that

"If we plan the expected performance of an activity from a cost, time, and production point of view, and if we measure actual performance on the same cost, time, and production parameters, we can generate management information via simple performance ratios in these categories for any activity within the project plan."

Since the maximum value is generated by comparing actual performance to planned performance, all performance ratios may be calculated by the following formula:

Performance ratio = actual performance/planned performance

The basis for this formula is that the planned or expected performance represents the standard by which actual performance should be measured.

This simple ratio calculation provides quick, accurate, and valid information, beginning with the individual activity and proceeding through any summary level for the project. The performance ratio thus generated, when expressed as a percentage, provides immediate information which is easy to assimilate on how actual performance compares to planned performance. With the conceptual framework in place, it is now possible to discuss the specific performance ratios which are most commonly of interest in the construction environment.

## The Cost Performance Ratio (CPR)

The cost performance ratio is the easiest of the performance ratios to calculate. Cost performance ratios may be calculated for individual activities, or for completed activities within the work package or at any other summary level. The formula for calculating the cost performance ratio is:

Cost performance ratio (CPR) = actual cost/planned cost

The planned cost is the amount budgeted for an activity. The actual cost of activities being performed by subcontractors and principal vendors is the amount taken directly from the payment (or purchase) schedule appended to the purchase document. In the case of the tabular data provided in Exhibits 23-2 and 23-3, the cost performance ratio is calculated as follows:

CPR = actual/planned = $7,800.00/$8,113.47 = 0.961

Note how quickly and conveniently the information is derived and the other data which may be immediately drawn from it. Since the cost performance ratio is less than one, it means that this activity is below budget. When the ratio of 0.961 is expressed as 96.1%, it immediately becomes apparent that actual costs are running below budgeted cost; that is, this item is below budget by about 4%.

**Composition of Actual and Planned Cost Data.**   A word of caution is essential concerning the dollar values of actual cost and planned cost. In calculating the cost performance ratio, the erroneous assumption may be made that all incurred costs to date are the actual cost to date and that all planned (budgeted) costs to date are the planned cost to date. Using these kinds of raw data may not result in valid cost information. That is, it is absolutely necessary for valid cost information at any point in time that the *activities* making up both actual and planned cost be the same. The list of activities and/or their percentage completion which make up the actual cost in the numerator must be identical to the list of activities and/or their percentage completion in the denominator.

Confirming that the list of activities is identical in both cases may be facilitated with tabular data. In making the calculations, one should make sure with side-by-side comparisons that the activities included in both actual and planned cost are identical. Where percentages of completion are used for individual activities, the same percentage of completion must be utilized in both cases.

**Purchases and Cost Control.** Another interpretation of the CPR of 0.961 is that approximately 96.1% of the planned dollars were spent to earn 100% of the planned cost. The cost performance ratios are of particular interest throughout the project. It is worth noting, however, that when most of the activities to be bought and performed are subcontracted on a lump sum basis, the cost performance ratio is established at the time the payment schedule for the subcontract or for the purchase order is established.

There is a potential risk, however, in assuming that the purchase order amount, or cost, is the same as actual cost. The first danger lies in the fact that the purchase cost does not become actual cost until the work is performed. The second danger is that claims may develop or added costs may be incurred in the process of completing a particular activity. When the assumption *is* made that the purchase cost is the same as actual cost, the tendency will be to understate the actual cost or to present a more favorable picture of the cost performance ratio than may be warranted when claims and unanticipated expenditures are incurred.

Assuming the purchased value and actual cost are the same has another significant disadvantage in integrated cost and schedule control. This assumption focuses on the part of job cost most likely to be under control, while reducing visibility to claims and to items *not* purchased hence less likely to be under control and more likely to create both cost and schedule problems.

## The Time Performance Ratio (TPR)

The time performance ratio is calculated in a manner similar to the cost performance ratio. The three different planned and actual data sets which may be used to calculate time performance ratio include:

1. Lapsed project time.
2. Summation of durations.
3. Production values.

The composition of the items for actual time performance and planned time performance must be identical. The calculation of the time performance ratio is as follows:

$$\text{Time performance ratio (TPR)} = \text{actual time/planned time}$$

There are several ways in which time can be counted for both actual and planned time within the context of a particular project. Exhibit 25-2 shows planned and actual time scale schedules which demonstrate two different methods of counting time.

**Exhibit 25-2. Planned and Actual Schedules for Calculating Time Performance Ratios (TPR)**

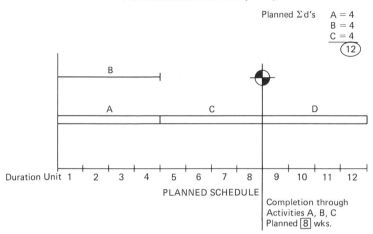

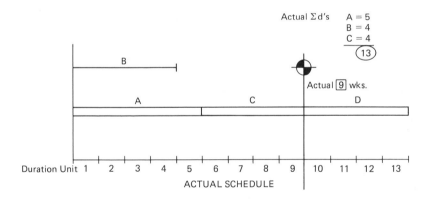

The first of these is dividing the actual time lapse from the actual schedule within the project by the planned lapse within the project as taken from the planned schedule. For example, if the lapsed time of activities completed within the overall project schedule is 9 weeks and the planned time for the same activity completion is 8 weeks, the time performance ratio is calculated as follows:

$$TPR = 9/8 \text{ or } 1.125$$

When the ratio, 1.125, is converted to a percentage, it becomes immediately evident that the project is running 12.5% behind schedule.

**Summations of Durations.** Another method of calculating the time performance ratio is to take the summation of durations for the completed activities compared to the planned summations of durations (or reported to a percentage completion); the composition of activities in the numerator and denominator must be identical. This data source is also illustrated in Exhibit 25-2.

For the planned performance, the planned duration of all the activities reported to be completed are totaled. Similarly, the actual time to complete each activity is, likewise, totaled. These two total figures are used to calculate the time performance ratio.

In the example, the planned summation of durations is 12. The actual summation of durations is 13. The time performance ratio calculated by this method is:

$$TPR = 13/12 = 1.083$$

According to the summation method, the project is about 8% behind schedule.

The ease and convenience of the time performance ratio calculated by summation of duration has the disadvantage of not taking into account (or showing) the effect of total project duration derived from the logic network diagram with recognition to concurrency of activities and float. Both methods, however, have their particular applications and the user should choose the one that provides the most meaningful information by way of the time performance ratios.

Another way of calculating the time performance ratio uses, to some degree, production information. Typically, the production curve for a project will take the shape of a standard "S" curve. The "S" curve indicates that a project starts with a relatively low rate of production which continues to accelerate through the inflection point and decelerate through the completion of the job. To calculate the time performance ratio, the comparison is the time planned to achieve a certain percentage of completion of the project as represented by the product in place vs. the actual time required to reach the same location on the "S" curve. To make the calculation, the planned and actual production curves (or equivalent tabular data) are needed. Exhibit 25-3 provides the information required.

Note that both contract sum and time are shown as percentages. The calculations are the same whether these or other measures are used.

To achieve a 60% completion, the diagram shows that 50% of the time has been planned to achieve this level of completion. Examining the actual

**Exhibit 25-3.  Planned and Actual Production Curves Used to Calculate Time Performance Ratio (TPR)**

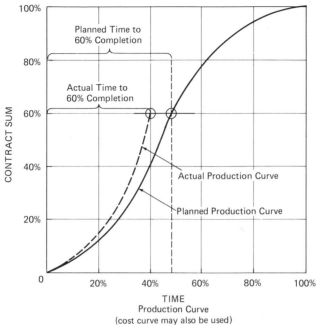

Production Curve
(cost curve may also be used)

production determines that only 40% of the time has been actually used to achieve a 60% production level. The time performance ratio given this data is calculated as follows:

Time performance ratio (TPR) at 60% completion = 40%/50% = 0.80

Converting the performance ratio to a percentage (80%) immediately reveals that the project is ahead of schedule by 20.0% at the 60% level of production. Actual time can be used in place of percentage of time, but a percentage provides quick and valid information on performance whether (the time performance ratio is) determined by the actual project lapsed time, the summation of duration, or by the time to a particular percentage of completion. Summation of duration is the easiest to calculate and while usable on a comparative basis is less likely to be valid in an absolute sense.

It is apparent that each computational method has its own advantages and disadvantages. The production values method is the easiest to calculate but is the least informative about actual job site progress. The summation of duration method addresses the duration of each activity but not the composite effect in the context of overall project. Lapsed project time ratio provides the most complete information but is the most difficult to calculate since the actual production schedule must be first updated to provide valid source data. Each environment will, however, use the method most suited to its requirements.

## The Production Performance Ratio (PPR)

The production performance ratio is calculated on the basis of the actual dollar value of product in place, divided by the planned dollar value of product in place at a given point in time. The most logical reference points for the production performance ratio to be calculated are at the billing or requisitioning periods though PPRs may be calculated based on any time. Production performance ratios are calculated by the following formula:

Production performance ratio (PPR)
    = total value of actual production/total value of planned production

In this case, the only matching criterion is that the actual data and the planned data cover the identical time period or interval, for example, a particular monthly billing period, or the job from inception through the current date. Unlike the cost performance ratio and the time performance ratio, the production performance ratio is not concerned with an exact matching of activities. The primary concern with the production performance ratio is to maintain a level of production in actual performance approximating the level of production which has been planned. A principal production management criterion is substituting other productive work at the job site when delays may be incurred on the activities which are planned for production. Exhibit 25-4 provides the graphic data for calculating the production performance ratio at 60% of planned time.

The calculation:

$$PPR = 70.0\%/80.0\% = 0.875$$

The job is running 12.5% behind planned production. Of course, dollar values of production can be substituted for percentages; the computations are the same.

Another criterion for calculating the production performance ratio involves the rate of application over a specific time interval. The computation is similar except that the actual rate of product placed over a particular time period is compared to the planned rate of product placement for that same time period. The formula for these calculations is:

Production performance ratio (PPR)
= actual rate of placement of product for a time period/planned
rate of placement of product for the identical time period

These values may be expressed as a dollar value per month, such as $100,000 of product in place per month. In these calculations, the matching of the time interval is particularly important. Shorter term intervals are best for comparisons since in the typical production curve for a project (the "S" curve) the rate of application is constantly changing, starting slowly, accelerating through the point of inflection, and then decelerating through the end of the project.

## MANAGEMENT INFORMATION AND INTERPRETATION OF PERFORMANCE RATIOS

There are a number of aspects to consider in the use of performance ratios as primary management information. Their high importance in creating per-

**Exhibit 25-4. Planned and Actual Production Curves Used to Calculate Production Performance Ratio (PPR)**

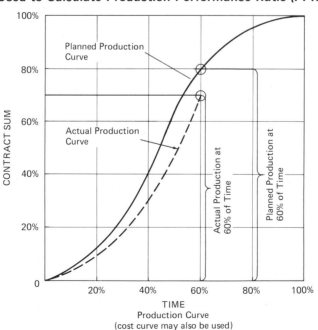

formance models and statements of expectations have now been developed in considerable detail.

For consistency, actual performance is always compared to planned performance. That is, the raw data for actual performance will always be the numerator of the ratio, and the planned data the denominator of the ratio. This convention in calculating performance ratios allows the ratios to provide a direct and immediate response to the question, "How are we doing compared to how we expected to do?"

It should also now be apparent how crucial the accurate interpretation of the statement of expectations is to whole process of integrated cost and schedule control, and, equally apparent, how crucial the effective use of the turnaround document as the primary data capture procedure for performance is in the whole process.

## The Significance of the Turnaround Document

The use of the turnaround document can now be further examined. The activity as the common descriptor throughout all work environments has been firmly established. Each work environment maintains its own set of values with respect to the statement of expectations for each activity. For example, administration maintains the budget; project management maintains the schedule, production administration maintains the production value, and accounting maintains the scheduled value. For each of these separate work environments to determine the exact status of each activity, all that is required is the field information as to when the activity is started, its percentage of completion, and when it is completed. In addition, the turnaround document may provide supplementary information about the activities requiring attention. Providing raw data to each work environment is even further simplified when projects are set up and managed using the concept of completed activities as the primary cost and schedule control vehicle.

It is possible, once the processes and procedures which have been presented are fully understood, for every work environment to be provided information on job performance as current as the information from the field to the office via the turnaround document. Each work environment will have the immediate capability of calculating the appropriate performance ratios it needs to generate high quality and contemporary management information on performance.

A final word concerning the interpretation of the cost performance ratios is appropriate. In each case, as we have seen, a performance ratio of 1.000 indicates nominal performance or performance at the expected level. In the case of the cost performance ratio, a cost performance ratio of less than one means that the work is being performed below budget, which is good.

A cost performance ratio greater than one indicates that the work is being performed over budget, not good.

In the case of the time performance ratio, less than one indicates that the job is being completed in less time or ahead of schedule, which, again, is good. A time performance ratio greater than one indicates that more time is being taken and, as a consequence, the job is behind schedule, that is, performing poorly.

When both the CPR and the TPR are examined and interpreted, a generalization emerges which says:

"Less (cost, time) is better."

Note however from the point of view of the production performance ratio (PPR) that:

"More (product in place within a specific time) is better."

The less-than-one-is-good, greater-than-one-is-bad criteria reverse for the production performance ratio. A production performance ratio less than one indicates that less product is being placed than planned, which is poor performance. A production performance ratio greater than one indicates that more product is being placed than planned, which is good.

## Informational Processes and Summary Levels

The processes and procedures which have been fully developed and presented have the potential of revealing significant changes in the kind of information available, the quality of that information, and its timeliness. Immediate performance information is no longer a matter of cumbersome and involved analytical processes requiring major recasting of the data that it once represented. Instead, simple performance ratios serve to create sufficient high quality and contemporary information for effective project control.

## The Benefit of High Quality Performance Information

The purpose of the kinds of performance information provided by performance ratios needs to be reexamined and reemphasized. The kind of information which now becomes available creates visibility to potential problems, enabling them to be identified, examined, and corrected, particularly while they remain correctable. By any standard, performance ratios as a basis for management information are universal in nature. Project management information is normally summarized at the work package, trade grouping, or project level.

The great value of these summary performance measures is found in the ways the summary data is created. Validity of the information processes is achieved and maintained by measuring performance at the level of the activity, the lowest common unit of work. The activity is the easiest unit to identify, the easiest for which to isolate performance, and the easiest on which to evaluate performance. It fortunately also offers the best way to control the validity of the information provided.

The method of constructing management information through summarization beginning at the level of the individual activity is discussed in detail in Chapter 5. Exhibit 5-1 presents a matrix of summary levels which clearly indicate how management information is generated as a result of measuring performance at the individual activity level.

The nicety of the matrix of summary levels is that management has access to information and data in both directions within the matrix. Management information summarized by individual jobs may be examined. An investigation to isolate poor performance using management by exception procedures may be narrowed to a single activity performed by a specific subcontractor on a specific job. Also provided within the matrix of summary levels are separate performance measures of subcontractors and principal vendors and, when wanted or required, on other contractors' field personnel as well.

Each organization will find particular parts of the summary methodology of particular interest. To the extent that the basic concepts and procedures involved in integrated cost and schedule control are not violated, there is great flexibility to the user related to individual needs and interest.

## SUMMARY

The primary purpose of this chapter on performance measurements is to bring into focus as an organized system of performance measurement many of the concepts, procedures, and processes which have been previously developed. The assumption is that, since any worthwhile project is entitled to appropriate planning, scheduling, and budgeting, why not expend this effort within the context of an organized system which has other benefits associated with it as well, for example, simplicity of design, simplicity of data capture, and simplicity of converting performance measurements into management information.

The purpose of the historical overview is to create a perspective on how far management information science in construction had advanced, from the traditional utilization of the profit and loss statement through the contemporary use of exotic and finely detailed job cost reporting systems. Another purpose, however, is to seek better ways to provide management information through the question "How are we doing compared to how we expected to be doing?" This is, of course, a comparative question, and it is the comparative approach which measures actual performance against planned performance—the key to contemporary management performance information.

In reviewing integrated cost and schedule control systems, it becomes apparent that the key to systems integration is not so much in more exotic routines as it is in anticipating the needs for information and for the data required to produce that information, and in building processes and procedures around these needs. From this concept emerges the value of the activity as the lowest common unit of work and as the universal communication and data transmitting vehicle which makes authentic systems integration possible.

Contemporary management information is possible once the framework of systems integration is in place and being utilized. Performance ratios are especially useful since these have the power to immediately present performance information as a ratio comparing actual performance to planned performance, and as a (percentage) increment above or below expected performance. While a wide variety of performance ratios may be calculated, the three which, as the most significant, are developed in detail are the cost performance ratio, the time performance ratio, and the production performance ratio.

Performance ratios in management information systems abetted with the turnaround document—which needs only to capture data on activity start, percent completed, or its completion—automatically key in a wide range of data on performance, both planned and actual. A performance ratio of 1.000 is nominal or expected for both cost and time, of which "less is better;" ratios less than one indicate good performance and greater than one indicate poor performance. In the case of production performance ratios, for which "more is better," above 1.000 indicates good performance; below 1.000, poor performance.

# IV
# Project Monitoring and Control

Integrated Cost and Schedule Control deals with two major and separate issues. First is integration of informational systems throughout all work environments and disciplines; the second is control of the ultimate outcome. In other words, the ultimate purpose of integrated systems is not only reporting but also the utilization of this output to control the cost and schedule of projects.

In this context, Chapter 26 covers schedule updating and control. Given particular attention is maintaining schedules in a current status so that activities yet to be performed fully recognize historic performance. That is, a major function of the informational process in integrated cost and schedule control systems is maintaining the game plan in a current status.

An informational issue constantly of concern to management involves the same basic question concerning both cost and schedule: "Where will we be at completion?" The answer is based upon historic performance. Chapter 27, then, deals with specific methods of projecting both cost and schedule to completion.

For more general business management, production projections are of high concern. Chapter 28 deals with methods of updating production projections. The implications of production projections for sales and marketing are also presented.

A multiproject construction environment contains so much detail requiring attention that some special methods are required to direct the attention of the limited people resource to problems requiring action. Contemporary techniques to facilitate project management are covered in Chapter 29. Exception reporting procedures and methods are discussed in some detail, including the ways in which these may be utilized in the construction environment. There are two purposes in exception reporting processes: (1) to establish a level of confidence which routinely confirms that most of the job related activities are being carried out within planned expectations, and (2) to give management visibility to those activities which are not. In essence, project management by exception reveals where management should be expending time and effort to effectively control the cost and schedule of a project.

Certain specific issues which relate to cost control are particularly important in the construction environment, and are discussed in detail in Chapter 30. One of the major issues is that the ability to control cost decreases as the project moves from concept into construction. Once a project enters into the field construction phase, only *negative* opportunities for cost control remain, that is, cost escalations resulting from failure to control cost.

Finally, Chapter 31 cover aspects of the schedule and other informational processes which need to be presented but which have not found a neat context in other areas.

In the final analysis, the material covered in Part IV, Project Monitoring and Control, gives meaning and substance to all the other information presented in this reference.

# 26

# SCHEDULE UPDATE AND CONTROL

The previous chapters have dealt with the basic issues, techniques, and processes which are involved in getting a project properly set up for construction. These have also dealt in considerable detail with the major issues involved in implementing a project strategy for maximum production. On the assumption that all of these procedures and processes have been followed, it is now appropriate to turn attention to the monitoring of the progress of the work and to focus on integrated cost and schedule control.

Before proceeding, it is necessary to reemphasize the basic thrust of integrated cost and schedule control and to review the basic differences between job cost accounting and cost control. Job cost accounting deals primarily with the historic environment, reporting on what has taken place in a time frame which leaves no opportunity to influence job cost results. The review of schedules in a time frame similar to the job cost cycle will result in information on the schedule and on the production performance of the job, but usually does not allow the outcome to be influenced.

In this context, cost control and schedule control are significantly different from review of historic job cost and historic schedule. The thrust of integrated cost and schedule control may be viewed as:

Using primary and contemporary information systems to routinely compare current performance with planned or expected performance for the purpose of influencing the outcome of negative cost and schedule trends while these are yet correctable.

At the very heart of integrated cost and schedule control, and its primary justification, is the ability to influence the outcome of adverse cost and schedule trends. What has been presented thus far is the informational framework which establishes expected performance in ways that allow the generation of quick, sufficient, and accurate data on performance to focus on adverse trends and to allow the opportunity to design corrective measures and implement them in time to influence a project's ultimate outcome.

It is now appropriate to consider the informational processes and resultant actions involved in integrated cost and schedule control.

## MAJOR FACTORS OF SCHEDULE CONTROL

There are several major factors which control how long it takes to complete a particular activity, a group of activities comprising an interrelated work package, and the entire project.

### Precedent Activities

The first of these factors is the completion of prerequisite activities. Activities which are precedent by design constraints or planner's choice must be completed before an activity may be commenced. From a practical point of view and in terms of field production control, no distinction should be made between physical constraints and planner's choices. It should be assumed that the game plan requires equal and diligent attention to both. The completion of prerequisite activities or, more technically correct, precedent activities, is the first of the major issues involved in field production control. The informational processes must routinely confirm that prerequisite activities are being completed as expected.

### Material Requirements

The second major prerequisite is having the necessary materials at the actual site of the work for placement by the workforce. Bricklayers cannot build walls with brick or block that doesn't exist; carpenters cannot construct structural framing without the necessary lumber and connecting devices; ironworkers cannot erect structural frame which has not yet been delivered. The informational and administrative processes must confirm that materials are at the work site. In fact, the check-off list should include materials, tools, and equipment.

### Workforce Requirements

The third major consideration in production control is workforce availability. The consideration with respect to workforce includes the proper crew mix (craftsmen, journeymen, apprentices, or helpers and laborers). It also includes the physical limitations of the work. One physical work location may not accommodate the normal full crew size while another work area may have the capability of accommodating several full size crews.

The foregoing three considerations are all part of the planning detail which should go into the job and are customarily implemented at the field level with field supervision rather than in master project planning. Detailed planning does, however, need to be properly and formally addressed in order to maximize both production and productivity.

## Supervision and Construction Details

In addition to the foregoing items, recent studies indicate two other significant factors involved in loss of both productivity and production in the field environment. It is entirely possible to have all prerequisite activities completed; all materials, tools, and equipment waiting at the job site; and the necessary work force available and still fall short of optimum productivity and production. The absence of specific construction information or detail and inadequate supervision cause poor production even when all of the prerequisites to productivity and high production have been carefully addressed.

## CRITICAL PATH METHOD WEAKNESSES AND SCHEDULE CONTROL

A major false perception, which developed early with the critical path method, leading to its early disfavor for construction planning and scheduling, must be addressed. The false perception is that detailed planning and scheduling of the kind available through the critical path method will automatically result in proper schedule control. This is, of course, not true. The early discovery of this fact led to some disenchantment with the critical path method. The problem, however, is not the method itself. The problem is the expectation that the project plan and schedule can simply be wound up, turned loose, and left to run their own course, resulting in proper schedule control.

## Implementation, Monitoring and Control

What is required in support of the extraordinary analytical and informational capability of the critical path method is proper implementation. A system of monitoring and controlling performance is also required. Updated schedules must be reimplemented whenever significant variances occur related to job conditions. Proper planning and scheduling are crucial to a well controlled project; it is equally true that proper implementation, monitoring, rescheduling, and reimplementation are also required. This process is cyclical and should be continued throughout the construction of a project. Exhibit 26-1 shows the repetitive cycle which should occur at selected intervals.

**Exhibit 26-1 Flow Diagram of Planning and Scheduling,
Production, Monitoring and Control Cycles of Work**

PLANNING and SCHEDULING CYCLE of WORK

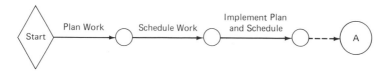

PRODUCTION CYCLE of WORK

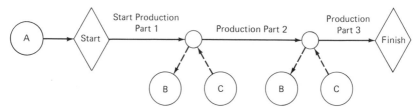

MONITORING and CONTROL CYCLE of WORK

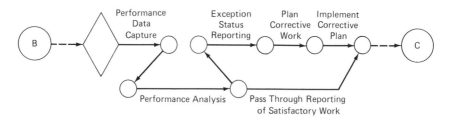

The flow diagram assumes the presence of some management by exception capabilities to address poor performance while passing satisfactory performance. It also assumes routine data capture procedures and performance analysis, which may take the form of time performance ratios (TPR).

## THE REALITIES OF FIELD CONSTRUCTION

As noted, not all of the logic relationships shown in the project plan are the result of physical constraints. As a consequence, the actual conditions in the field environment may require adjustments from time to time in either the sequencing or the phasing of the work to be accomplished. This of itself should cause no great discomfort either in the field environment, among the planning and scheduling resource, or with project management.

## Field Responsibility for Schedule Control

The major focuses in field construction should continue to be high production, productivity, and schedule control. Cost and schedule informationally bear a vary close relationship. Controlling the schedule, for example, will go a long way toward controlling the cost of time related activities. Directly controlling cost itself in construction, though necessary, will not automatically be helpful in controlling the schedule. What is required is the control of cost and schedule together, but with schedule being the primary focus in the field environment, where there is little opportunity to directly affect cost outside the schedule.

## Schedule Flexibility

Maintaining effective cost and schedule control involves two principal requirements. The first of these is to have a valid plan and schedule for the project. The second is a dynamic response capability which allows for field variances without upsetting or disrupting the production control processes and the downstream elements involved in integrated cost and schedule control.

## The Schedule as a Time Budget

The most effective way to view the original schedule is to consider it as a time budget for individual activities, for groupings of activities by work packages or trade groups, and for the entire project. The original schedule should be maintained with the same degree of integrity as an historical document as is the cost budget. The original schedule will be used for generating time performance ratios and is a significant component in production performance ratios as well.

## THE PLANNED AND ACTUAL SCHEDULES

As the job progresses, a schedule recording actual performance is generated. The actual schedule is similar to job costs in some respects. It is different from job costing in that the actual schedule becomes the basis for planning the remaining work carrying the project through to completion; a later chapter will deal with the projections to completion for job cost.

Since there will be differences between the planned schedule and the actual schedule, the data capture processes on actual performance become especially important. The data capture vehicle which has been discussed in considerable detail is, of course, the turnaround document, of vital importance in capturing data on actual performance of activities. The primary record is on actual start and actual completion. Where appropriate, a percentage completion may also be reported. With the variances between the planned and actual start dates and the planned and actual completion dates on each activity noted, the historic performance data needs to be inserted into the actual logic network diagram for the project in place of the planned performance data, and the actual schedule recalculated.

This schedule recalculation is especially interesting from several points of view. First, it allows a new projection of the schedule to completion along with each schedule update. Where significant variances occur in production and performance, these become clearly evident as a result of the schedule update. Since field construction is always a dynamic environment, the schedule should be updated as frequently as practical. With a computer capability in place, schedule updates may be done as frequently as weekly. When the work is done manually, the schedule should be updated not less frequently than monthly.

## Schedule Updates and the Turnaround Document

A major benefit of frequent schedule updates is the dynamic and interactive nature of the turnaround document for each planning horizon. Take, for example, a project in work for which the planning horizon is a calendar week. The turnaround document provides information on actual performance for the calendar week. This data is then inserted into the actual schedule. The data allows the schedule to be recalculated with the actual performance data (assuming there are weekly updates from inception to completion). The updated schedule then becomes the basis for generating the turnaround document for the next planning horizon. When this process is meticulously followed, the turnaround document presented to field supervision for production management and control over the next planning horizon is based on what actually has taken place through the end of the previous period, not on some (planned) schedule which may have been prepared months earlier. The turnaround document with its other data capture capabilities also allows field supervision to provide essential information for schedule maintenance as well as schedule updates.

## PROCESSES FOR SCHEDULE UPDATE

It has been pointed out that the planned job schedule is maintained as a time budget. The actual schedule is the schedule that is updated for each planning horizon. Schedule updating includes the data capture processes involved with the utilization of the turnaround document. The data on the turnaround document reflects the current state for each activity, which has been either started, worked on, completed, or started in some earlier planning horizon but not completed.

Initially, the assumption is made that the durations and logic relationships of activities yet to be addressed will remain unchanged. The schedule is then recalculated using the actual data for the items in the most recent planning horizon. The result is a complete new schedule from the end of the most recent planning horizon through the end of the project.

This schedule should be evaluated for validity. When it is necessary to adjust other durations, this should result from the examination of the recalculated schedule and, to the extent necessary, the schedule should be updated and recalculated based on future anticipated changes. Where significant changes or modifications have occurred in the plan and schedule for the project, whether from some major delay or from some significant acceleration, it may be necessary to reconstruct the *logic* of the schedule from the end of the last planning horizon to completion.

While there may be some reluctance to undertake a major schedule revision, such should be undertaken at any time the validity of the historic schedule becomes suspect or major changes occur which make the historic schedule no longer applicable. It is important that the historic schedule be updated for significant changes and that the activities being addressed in the actual schedule bear some reasonable resemblance to the historic schedule. Otherwise, time performance and production performance ratios loose some of their meaning and value.

## SCHEDULE MONITORING

Schedule monitoring occurs as a result of the data capture on actual performance through the turnaround document. Schedule monitoring provides a clear indication of schedule performance and will be reflected in the time performance ratio calculations. Schedule monitoring alone, however, is not sufficient. The schedule must be updated at frequent intervals to maintain validity and a realistic relationship to what is taking place in the field. Each schedule update carries with it, to some degree, a reimplementation of the schedule, especially when significant changes are occurring as a result of actual job performance.

Part of the schedule monitoring process is to detect adverse trends while they are yet correctable. Adverse trends can occur in either the field production or administrative activities. In the field production activities, schedule monitoring should provide a clear indication and confirmation that adequate workforce, materials, tools, and equipment are at the job site for actual performance. Any activitiy falling outside the expected time performance must be examined in detail and the reasons for delays clearly identified. With the reasons for the delays identified, corrective action must be taken immediately.

The reasons for the delay may be many. The most likely problem may be an inadequate workforce. Closely associated with workforce may be inadequate supervision. Immediately behind in importance would be having adequate materials, equipment and technical information. Whatever the cause(s) for delay, it must be identified and corrected without delay.

## Administrative Activities and Schedule Control

Frequently overlooked when the primary focus is on field construction are those administrative activities which must be carried out to assure that materials and equipment are available at the job site when required for placement. Separate monitoring of administrative activities (which implement the processes required to purchase materials and deliver them to the job site in advance of their scheduled placement) must be an integral part of schedule control. This is best done through a series of milestones which need to be reported and confirmed on a regular basis: purchase of materials with a sufficient lead time, preparation of shop drawings and other required details, submission and approval of the details, fabrication of materials with adequate time allowances, and shipping materials to the job site to be on hand when required.

## THE IMPLEMENTATION OF CORRECTIVE ACTION

Corrective action must be based on current valid information on performance. No single process or procedure can provide the overall control cycle; it, rather, results from the effective use of a combination of techniques including schedule monitoring and updating, the weekly progress meeting, the project action checklist resulting from the weekly progress meeting, and the continued cooperation and good communications from all of the team members for the project.

One of the unfortunate features of schedule monitoring updating and schedule control is that it of necessity focuses on the negative aspects of project performance. While it is pleasant to focus on those 95 items out of every 100 which are performing as expected, the focus should be on the remaining five, the likely source of schedule control difficulties.

The potential for causing schedule problems is not related to the value of the activity. A relatively small item can cause schedule and related cost problems many times its own cost. It is therefore necessary to have the informational systems working to determine which 95 items are performing as expected so that project management can identify the five which are not performing satisfactorily. Once a clear focus on the problem is available, there is probable opportunity to correct the problem while it is correctable. (Management by exception reporting procedures will be dealt with in a later chapter.)

## SUMMARY

Schedule monitoring and control, as it turns out, is a matter of careful planning and scheduling, of schedule implementation, and monitoring of the schedule. Schedule monitoring and control assumes a properly planned and scheduled project. It also assumes a production team ready, willing, and able to address the project for maximum productivity and production.

The critical path method with its extraordinary analytical capabilities is an effective tool for schedule monitoring and control. Its weaknesses need to be recognized along with its most useful capabilities in the measure of schedule performance. The planned schedule serves as the benchmark of time performance in much the same way that the budget provides the basis for job cost performance measurements.* The actual schedule is maintained independently of the planned schedule and updated frequently based on actual performance data. With each schedule update, a new projection of time to completion is made. When major changes in schedule logic occur or major modifications occur in the project plan, significant revisions to the schedule may be required for both validity and relevance.

The realities of field construction are such that there will be changes in the plan and schedule for the project. An appropriate capability for responding to these changes is important.

A key component in performance data capture is the turnaround document. Data from the turnaround document serves as the basis for schedule updates and major revisions when these are required. Constant monitoring of the schedule in the field environment is important. Of equal importance is monitoring administrative activities since these may have a direct effect on the ability to control the schedule.

There are three cycles within the planning and scheduling processes which are repetitive in nature and are essential to integrated cost and schedule

---

* In practice, two schedules are established and maintained. The first is obviously the *planned schedule*, which, as the job progresses, may be referred to as the *historic schedule*. As the job proceeds, the *actual schedule*, recording actual time performance, emerges and may take several forms, the simplest of which is nothing more than the planned or historic schedule "marked up" with actual performance data. A completely separate actual schedule may also be maintained since there may be significant consequenial effects on work yet to be done as a result of work already completed. Comparative performance data can be obtained simply by comparing the two schedules, planned (or historic) and actual.

control. The first of these is the planning and scheduling cycle of work which establishes the basic job strategy. Next is the production cycle of work which involves an implementation of the plan and schedule. The third, and perhaps the most crucial in schedule control, is the monitoring and control cycle of work. This cycle captures data on actual performance and through performance ratios compares actual performance to planned performance. When the performance of an activity is found to be satisfactory, no further attention is required. When an exception of status occurs, that is, an activity is performing outside acceptable limits, corrective strategies must be designed and implemented to maintain or regain full control of the schedule.

# 27
# PROJECTIONS TO COMPLETION

Throughout this reference, the procedures for setting up a project plan, schedule, and budget as primary performance models have been discussed in detail. Ways to extract secondary performance models from the budget and schedule for a job have been developed in detail. Implementing the schedule, maintaining production, and other issues have also been presented and discussed. What has not yet been considered involves an area of primary interest in relation to every construction project.

## JOB EXPECTATIONS

Most projects begin with a high degree of optimism concerning the project and its expected performance. The initial assumption is that the project will be completed in less than the scheduled time and for less than the budgeted cost. As the project progresses in the field environment, this initial optimism gives way to the realities of construction and the less optimistic view of the project's final outcome emerges. As the cost performance, time performance, and production performance ratios are meticulously calculated on a month by month basis, questions concerning the final outcome of the project and the need for some capability to project time and cost to completion become increasingly apparent.

### Some Basic Questions

Three continuing questions need to be asked and answered for every project in construction:

1. When will the project be completed?
2. What will the cost of the project be at completion?
3. What overhead and fee may reasonably be expected from the project at completion?

The ability of a construction organization to anticipate these questions, to collect appropriate data concerning the questions, and to respond with valid information to the questions becomes a clear measure of organizational maturity.

## Integrated Systems Data and Information

Integrated systems of cost and schedule control provide the raw data and informational base for answering the foregoing questions. Integrated systems built around activities is the basic informational source about the work required to complete the project. At some point in time, the focus of attention should shift from activities which have been completed to those which yet need to be completed. When the project has been planned, scheduled, and budgeted at the detail level at which control is to be exerted, the information is already available on the activities to be completed to complete the project itself.

Once construction has started, what is *not* available is information on the performance of the activities in the process of their completion. What is required in order to project time, cost, and profitability at completion is some basis for validating the relationship between the planned schedule and budget and the actual schedule and cost. This process involves comparisons between planned and actual to create new performance models which may validly be used to modify the planned schedule and planned budget for those activities which have not been completed to determine the effect of their individual completions on the completion of the project. In essence, actual performance may be used for creating new performance models to completion. This concept becomes the foundation of valid projections of time, cost, and profitability which may be anticipated at the completion of the project.

## SCHEDULE PROJECTIONS TO COMPLETION

In some respects, projecting schedule to completion is somewhat easier than projecting of cost to completion if for no other reason than the projecting of schedules is the more familiar process. The data capture process using the turnaround document provides the systems for data capture and for recording actual job performance. The schedule is recalculated based on actual performance.

## Significant Factors Affecting Schedule to Completion

There are several significant factors which may affect the duration of activities and subsequently the schedule. First, if there is a consistent pattern of activities taking longer than planned, the projection of time to completion must take this actual performance data into account. The time performance ratio is the best indicator of how well the actual durations of activities match the planned durations of activities. When appropriate, the planned duration of activities yet to be completed should be modified by the historic actual performance. This involves modifying the durations of each activity where this is appropriate to reflect a more valid schedule. The alternate in terms of time control is to reevaluate the resources and rate of application required to maintain the original schedule. Whichever choice is made, project management and field supervision together must make the decision and must be committed to encouraging the appropriate levels of production for the completion of the work. In the meantime, the actual schedule must be projected to completion and this should be done at regular intervals.

A second factor which may affect the time to completion is significant changes adding to or eliminating certain work in the original project. Such changes must be built into the schedule in the form of the activities affecting the change, with resource data used to determine the duration of each of the new activities. This information must, of course, be built into the logic network diagram before the recalculations may be made. An important factor in these processes is validating the resource requirements for the performance of activities and the availability of these resources.

In some cases, significant modifications in logic, which are reflected in the relationship of activities, will affect schedule to completion even when there is no or little change in the scope of the project. Circumstances in the field may require the sequence of work to be significantly changed when these are not physically constrained. Such changes in logic are partially picked up in the updating of schedule using the turnaround document data. Changes, however, may be of such magnitude that this approach does not result in a valid schedule. When this occurs, the activities which have been completed should be so noted and, in effect, a new logic network constructed for those activities yet to be completed.

## The Production Performance Ratio as a Time to Completion Indicator

When these procedures are followed, the projection of time to completion is a very dynamic process involving the constant monitoring and updating of schedule as frequently as weekly if required for validity. The projections of time completion also need to be considered in the light of production performance ratios. The production performance ratios provide information on the overall movement of the project from start to finish. Presented previously is the concept of a single inflection point, where the project accelerates from inception through the inflection point and decelerates into final completion. When a break in the production curve occurs in the form of a reduced production performance ratio prior to the point of inflection, the project must then be accelerated a second time to make up the lost production. When the lost production is not recovered through a reacceleration of the rate of placement, the completion of the job on schedule and on time will be unlikely.

The production performance ratio must, therefore, be used as an additional modifier of the projected time to completion. Production performance ratios significantly below one are clear indicators that the rate of placement of product is well below what is required to maintain schedule. The time performance ratio and the production performance ratio must be taken in concert in order to validly project time to completion.

## SCHEDULE PROJECTION COMPUTATIONS

Three steps are required to complete the schedule projection computations. To illustrate these methods, the network from Exhibit 9-2 is utilized in Exhibit 27-1. Exhibit 27-1 gives network notations showing the status of the schedule prior to projection or calculation of schedule to completion.

Certain actual schedule performance data has been added to the network. For activity A, 5 days were planned. The actual days required were 8 and these are indicated in the network. Similarly, for activity B, the actual time required was 13 compared to planned requirements of 12. The scheduling computations are carried forward for each of the events, concluding with the most recently completed activities. This calculation indicates a time earliest at event 4 of 8 and a time earliest at event 6 of 13. The time reconciliation will, therefore, be made through duration unit 13.

**Exhibit 27-1. Network Notations Showing Status Prior to Projection to Completion**

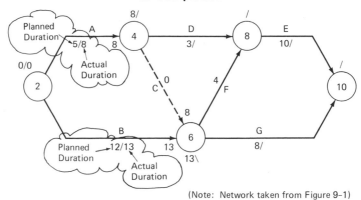

(Note: Network taken from Figure 9-1)

## Exhibit 27-2 Time Performance Ratio (TPR) Calculation and Revised Durations from Network (Exhibit 27-1)

$$\text{Time performance ratio (TPR)} = \frac{\text{Actual summation of completed duration}}{\text{planned summation of durations}}$$

$$= \frac{8 + 13}{5 + 12}$$

$$= \frac{21}{17}$$

$$\text{TPR} = 1.235$$

| ACTIVITY | REVISED DURATION FOR REMAINING ACTIVITIES | | | NOTES |
|---|---|---|---|---|
| | ORIGINAL DURATION | TPR | REVISED DURATION | (1) REVISED DURATION |
| D | 3 | 1.235 | 4 | (1) Values rounded |
| E | 10 | 1.235 | 13 | up to |
| F | 4 | 1.235 | 5 | nearest |
| G | 8 | 1.235 | 10 | whole |
| | | | | number |

The next step in the schedule projections is to calculate the time performance ratio for activities already completed and to use this to calculate revised durations for the remaining activities. These calculations are shown in Exhibit 27-2.

The method of calculating the time performance ratio is based on the summation of durations. The result of these calculations for the sample network is shown in Exhibit 27-2 as 1.235. In order to revise the durations for the remaining activities, an assumption is made that these durations will essentially reflect the historic performance to date. The basic premise is that the actual performance will continue unabated. Exhibit 27-2 shows in tabular form the activity designator's original duration, the time performance ratio and a revised duration rounded up to full duration units.

To recalculate the schedule through completion and to determine whether or not historic performance has affected the critical path, the revised data is added to the network, then the new scheduling computations are completed. These notations and calculations are shown in Exhibit 27-3.

For clarity, several additional notations are made. The actual durations are indicated by a subscript of A. The revised durations, based on the time performance ratio and indicated by R, are used to recalculate the schedule forward pass from events 4 and 6 and the backward pass from the objective to the origin. The result of these revised scheduling calculations indicate a

**Exhibit 27-3. Network Notations Showing Revised Durations and Revised Scheduling Computations**

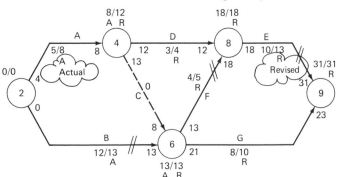

projected completion of the job in 31 duration units, as compared with the 26 duration unit scheduling computations shown in Exhibit 9-2.

## Delays in Construction

It is worthwhile to examine the effect of these delays in construction on production and on the production performance ratio. The production performance ratio compares production at a specific point in time or as a comparative rate over a specific time interval. To determine the effect of schedule delays on production, the data from Exhibit 27-3 is used to plot the revised time scale network which is shown in Exhibit 27-4.

Since no other information is available, it is assumed that the cost of each activity is not changed. What is significantly changed through duration unit 13 is the rate of application. The resource data is added to the time scale network and the daily rate of placement is calculated. This data is retabulated for cost at both the production unit and cumulative levels. The production values (cost plus the markup) are determined by multiplying the cost by 1.155. This multiplier comes from Exhibit 10-3 as the planned markup in the sample project being used. The result of these calculations indicates the production value of product through duration unit 13 is $10,748 actual.

In order to determine the planned value, reference is made to Exhibit 10-3. It is noted, however, that a summation is available through duration unit 10 and duration unit 15. What is needed, however, is a summation at duration unit 13. The simplest method of making this calculation is to use a straight line interpolation of the production value for production period 3.

## Exhibit 27-4. Revised Production Tabular Data

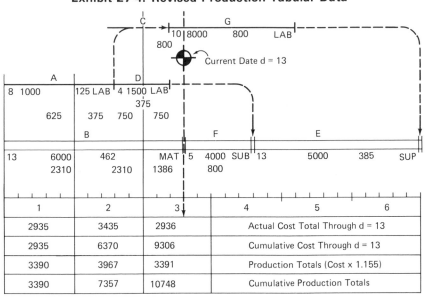

| 1 | 2 | 3 | 4 | 5 | 6 | |
|---|---|---|---|---|---|---|
| 2935 | 3435 | 2936 | Actual Cost Total Through d = 13 | | | |
| 2935 | 6370 | 9306 | Cumulative Cost Through d = 13 | | | |
| 3390 | 3967 | 3391 | Production Totals (Cost x 1.155) | | | |
| 3390 | 7357 | 10748 | Cumulative Production Totals | | | |

Since the end of the 13th duration unit represents three-fifths of the planned production, the calculation is:

$$\$8,085 \times .6 = \$4,851$$

which when added to the cumulative through duration unit 10 of $8,663 equals planned production through the end of duration unit 13 of $13,514. The production performance ratio using the interpolated data is calculated as follows:

$$PPR = actual/planned = \$10,748/\$13,514 = 0.795$$

This indicates a production lag of approximately 20.5%

A review of the time scale network in Exhibit 27-4 indicates significant changes in the rate of production for production period #3 indicating that a more specific examination of planned production for duration units 11, 12, and 13 might prove of interest. Using Figure 10-2 as the source document, duration units 11, 12 and 13 account for a total of $3,000 of additional cost which when multiplied by the mark-up factor of 1.155 indicates production in this time period of $3,465. When this amount is added to the summation through duration unit #10 of $8,663, a total planned production of $12,128

**Exhibit 27-5. Production Performance Comparisons**

1. Production actual through d = 13 (from Exhibit 27-3) = $10,748

2. Production planned through d = 13:

    a. Through d = 10 (from Exhibit 10-2)      = $ 8,663

    b. d = 11, d = 12, d = 13 (from Exhibit 10-2):
        ($1,000 + $1,000 + $1,000) × 1.155   = $ 3,465

    c. Total through d = 13             = $12,128

3. Comparison and conclusion:

    a. Production performance ratio (PRR)    $= \dfrac{\text{actual production}}{\text{planned production}}$

                                         $= \dfrac{\$10,748}{\$12,128}$

                                         = 0.886

    b. 1 − 0.886                          = 0.114

                                         = production lag
                                            of 11.4%

is indicated. Exhibit 27-5 provides production performance comparisons using these data.

These calculations indicate a production performance ratio of 0.886, or a production lag of 11.4%.

## Estimated Time to Completion

The calculations to determine the time performance ratio using the summations of durations method are shown in Exhibit 27-6.

Time projected to completion using scheduling calculations shown in Exhibit 27-3 of 31 days may now be used in calculating the time performance ratio. Projected time of 31 is divided by the planned time of 26. The result is a projected time performance ratio of 1.192, or a 19.2% time delay in the project.

## Explanation of Different Performance Ratios

In comparison with the 19.2% delay just calculated. When a straight line interpolation is used for duration units 11, 12 and 13, indicates a time delay

### Exhibit 27-6. Time Projected to Completion and Time Performance Ratio

1. Time estimated to completion:
   Projected time (from Exhibit 27-3) $\quad= 31$

2. Planned time to completion:
   From network calculations (Exhibit 9-2) $= 26$

3. Projected time ratio (PTR)—comparison $\quad= \dfrac{\text{projected time}}{\text{planned time}}$
   $$= \frac{31}{26}$$
   $$= 1.192$$

4. Conclusion: $1.192 - 1.000 \qquad = 0.192$, or a projected time delay of 19.2%

of 20.4%. These two calculations are expected to be relatively close because both work on the premise of straight line time effect concerning the delays. The lower production performance ratio of 11.4% lag in production results from two factors. First is that it clearly reflects a lower rate of production in the first three duration units of production period 3. Second, this production performance ratio is calculated early in the total project time, when the rate of applications tend to be lower.

In any case, it is clear through these calculations and through the performance ratios that the project is significantly behind schedule and that delays ranging from 12% to 20% may be expected if the historic performance continues unabated through the end of the project. Of course, the primary opportunity which this information presents is to identify the problems, determine whether these are correctable and, if so, to take immediate corrective action.

## COST PROJECTIONS TO COMPLETION

Projection of cost to completion is significantly more complex than projecting schedule to completion. Projection of cost to completion uses historic data as a performance model to project the cost to completion. There are, however, a number of different factors which must be taken into account in projecting cost to completion.

## Cost Projections and Cost

An examination of cost types 1 through 9 (App. C) clearly indicates that the variables affecting cost to completion are different with each cost type. For example, cost type 9, sub lump sum, is used when the cost of the performance of a portion of the work is established on a fixed price basis. While there may be a tendency to assume, therefore, that the costs are guaranteed, the possibility of items being discovered which have not been included in the lump sum price might affect the cost of the work adversely. The projection of cost to completion for cost type 9 items must recognize this contingency and be able to deal with it effectively. Cost type 5, sub labor, and cost type 6, sub material, are usually purchased on a lump sum basis. The same conditions, therefore, apply as in the case of sub lump sum.

In case of cost type 7, sub equipment, the basis for purchasing the services becomes a critical issue in determining the procedures to be used for projecting cost to completion. When subcontractor equipment is purchased on a lump sum basis, the same procedures as sub lump sum apply. When, however, the equipment is rented, the issue of duration becomes a significant factor and it is necessary, therefore, to project the schedule to completion in order to properly evaluate the effect of this item on the cost projected to completion.

## Labor Cost Issues

One's own labor is, perhaps, the most volatile of cost items and the most difficult to project to completion. The ability to control costs is directly related to not only the quality of supervision but the degree of organization in providing tools, materials, and equipment, along with supervision and necessary details for the production of the work.

The material one buys from suppliers is similar to labor but is less volatile since it can be quantified with some degree of accuracy. For example, the total number of cubic yards of concrete may be determined for a project within a reasonable tolerance. The variable then remains concerning how efficiently the material is being used and whether or not the purchased quantities are yielding the appropriate quantities of product in place. Yield measurements should be a part of monitoring project performance and should be taken into account in the projecting of cost type 2, material, completion.

One's own equipment is similar to cost type 1, labor, in that it is applied to a job as needed. The ability to control such items and to project requirements to completion contain similar difficulties.

Cost type 4, own cost others, and cost type 8, subcontractor cost others, tend to be relatively small and relatively fixed and usually present no great concern in projected cost to completion. A problematic factor of significance, however, intrudes when cost type 4 and cost type 8 include insurances and taxes, which are an arithmetic function of other less defined and less controllable cost items such as labor, material, and equipment. When arithmetic functions of other elements, the projection of the cost of other elements to completion allows the recognition of in-production changes affecting such items such as insurances and taxes so that they are appropriately ajusted in the same way they were initially calculated in the etimating process.

## GENERAL GROUPING OF ITEMS FOR PROJECTION OF COST TO COMPLETION

There appear to be several different kinds of items which require different treatment in projecting of cost to completion. These are as follow:

1. Items which are a function of duration and unit cost:

$$\text{Duration} \times \text{duration unit cost} = \text{projected cost}$$

2. The sub lump sum or other lump sum purchases for establishing known values for each activity.
3. The most unpredictable items: claims, missed items, items not covered in the budget, and the like.

Costs for lump sum purchases are fixed at the time the purchases are made. Cost fixation must be validated, however, upon the completion of each activity.

### Claims and Cost Projections

Claims, unfortunately, present the greatest difficulty in projecting cost to completion. They also comprise the most likely problem areas with respect to cost control. In the case of such items as claims, well established and well managed contract administrative procedures offer, perhaps, the only valid approach to providing data for projecting cost to completion. Where good contract administrative procedures are not in place and hence not operative, these items can prove very difficult to handle.

A probability analysis is the most valid means to deal with unresolved claims between the owner and the general contractor and with unresolved claims between the subcontractors and the general contractor when projecting cost to completion. Probability analysis may not be valid for individual items, but for the purpose of projecting cost to completion, probability analysis will provide a sufficiently accurate modifier of projected cost and projected contract sum.

## THE MASTER X LIST FOR CLAIMS ANALYSIS

The master X list, discussed in detail in Chapter 18, provides control information for owner/contractor claims. Some numerical series is used for owner/contractor claims. The administrative detail associated with claims should be kept current in an effort to pursue them to a conclusion as quickly as possible. The purpose is to minimize the need for a probability analysis on open claims by effectively converting all claims to change orders on a timely basis. The source of these claims may be any work environment where information comes to light affecting the cost of the work, the terms and conditions of contract, the time required to complete the project, or the conditions of work in the field.

Certain items, which may become the subject of an owner/contractor claim, will originate in claims from the subcontractor and principal vendors to the general contractor. The most direct sources of such information include the billings, requisitions, and invoices from the subcontractors and principal vendors. The requisitioning procedures for subcontractors and principal vendors should require that all claims be asserted not later than the end of the billing period in which they originate. This procedure allows addition of such data to the master X list, facilitating the claims' administrative pursuit.

In some cases, claims originating from a subcontractor or principal vendor will not be categorized as owner/contractor claims but will involve a separate series of X numbers which relate to charges and backcharges to principal vendors and subcontractors. A relatively common occurrence involves subcontractor A performing work which is the responsibility of subcontractor B, because B is unable, unwilling, or uninterested in performing the work. Administratively, this item should be assigned an X number in the backcharge series. A supplementary purchase order is issued to subcontractor A for performing the work and a supplementary credit purchase order issued to subcontractor B, reducing subcontractor B's contract sum by the amount to be paid subcontractor A plus a percentage handling fee. The percentage handling fee should, of course, be embodied in the subcontract and/or purchase order documents so that it does not separately become a source of dispute or disagreement.

## Tracking Claims from Subcontractors

Claims from subcontractors may be presented to the general contractor in a variety of ways. Most common is to include in the current billing claims for certain items occurring in the production period associated with billing. The accounts payable file for the subcontractor then becomes one of the primary sources of information on subcontractor claims.

A basic procedure and assumption, however, will keep the procedures for tracking subcontractor claims administratively as simple as possible. This approach involves a concept, presented earlier, which states, in essence, that the contractor is in a buy/sell position in his relationships with the owner and the subcontractors. The crucial followup assumption is that if the owner has a valid claim against the contractor, that claim is transferable to a subcontractor or principal vendor if it involves product or services purchased in compliance with the owner/contractor agreement. Conversely, if the subcontractor has a claim against the general contractor, that claim could be asserted against the owner if the general contractor sold to the owner what he bought from the subcontractor.

The net effect of this assumption is that the master X list can (and should) be used as the universal vehicle for keeping track of all claims including both owner/contractor and the contractor/subcontractor. All of the claims presented to the general contractor by the subcontractor through both payment and other procedures should be cross-referenced without exception to X numbers in the master X list. This administrative nicety and closure will minimize the probability of claims being lost or overlooked.

## Probability Analysis with Respect to Cost

On the assumption that the contract administrative procedures just discussed have been carefully followed, a probability calculation affecting contract cost projected to completion may be made. For each claim made by a subcontractor or principal vendor, a separate probability factor is determined. Probability factors range from 0.00 to 1.00; the higher the probability factor assigned, the more likely that the subcontractor or principal vendor will be successful in perfecting a claim. The probability of zero means that the claim is most likely not valid. Within each subcontractor work package, each claim is assigned the separate probability factor. This factor is multiplied by the amount of the claim to provide a dollar value for the claim, to be used in the work package summary. After each claim is extended, the calculated dollar probability of the effect on the work packages is totaled. This is illustrated by a sample contract summary, Exhibit 27-7:

## Exhibit 27-7. Contract Summary

| ABC PLUMBERS | | | | | |
|---|---|---|---|---|---|
| Base subcontract amount | | | $100,000 | | |
| Supplementary purchase orders to date | | | 7,500 | | |
| Adjusted contract sum | | | 107,500 | | $107,500 |
| Claims | | | | | |

| X NO. | ITEM | AMOUNT | PROBABILITY | VALUE | |
|---|---|---|---|---|---|
| 10 | P | $ 5,000 | 0.5 | $ 2,500 | |
| 13 | Q | 15,000 | 0.3 | 4,500 | |
| 14 | R | 4,000 | 0.7 | 2,800 | |
| 16 | V | 750 | 1.0 | 750 | |
| 19 | W | 12,500 | 0.4 | 5,000 | |
| Probability value of claims | | $37,250 | 0.42 | $15,550 | 15,550 |
| Work package cost projected to completion | | | | | $123,050 |

This process is continued for each work package within the contract. The result of the calculations is a projected effect of all claims on the cost of the work. This dollar amount will be added to the projected cost to completion as an additional modifying factor. The usual expectation is that the probability analysis of subcontractor and principal vendor claims will tend to increase the projected cost of the work, thereby decreasing the projected or calculated overhead and fee at the conclusion of the project.

### Probability Criteria

Some cautions are necessary about the use of probability in the projection of cost. The first concerns how the claim originated. If it is the result of a change in the work made by the owner and acknowledged by all parties but short of maturing administratively into a change order, perhaps a 0.90 probability is valid. If an authentic dispute exists concerning the inclusion of an item in a work package clearly required by the contract documents, perhaps a 0.50 probability is appropriate. Fallacious claims rate perhaps 0.10. While there is research and statistical support for this process, it is nonetheless an empirical method. Therefore, since nothing is to be gained by an over-optimistic evaluation of claims, a more conservative cost projection favors the use of higher probabilities. Then when the resolution of claims produces a more favorable result, the overall job position improves.

## Owner/Contractor Claims

Owner/contractor claims are handled in a similar manner since only one set of probability calculations is required. The master X list will contain a complete list of all claims and pending claims between the owner and the contractor. When these have been evaluated and submitted to the owner but no action has been taken, the proposal for claim amount is assigned a probability factor in the range of zero to one.

When the claims have not yet been assigned a dollar value, an approximate dollar value for the probability analysis must be determined. Under normal circumstances, an empirical evaluation is sufficient for this purpose. Similar to the processes for subcontractors and principal vendors, the value of the claim is multiplied by the probability factor. When all the claims are so extended, a total is calculated, representing the probable value of the claims when they are finally resolved. The probability calculations for owner/contractor claims to be used in projecting contract sum to completion are the same as those presented for contractor/subcontractor claims.

This probability analysis is *not* valid for individual items. As a general rule, any claim will be either accepted or rejected *in total.* In a summary of all claims, however, the probability analysis is valid. When the calculations for the individual claims are completed and all claims are totalled, this last amount represents a probable adjustment to the contract sum resulting from unresolved claims. Under most circumstances, these claims would be added to the contract sum, giving the projected contract sum at completion. The calculations for the projected contract sum at completion are illustrated in Exhibit 27-8.

The projection of contract sum to completion is somewhat easier and less cumbersome then the projection of cost to completion, especially in fixed sum contracts. The simple formula of basic contract sum plus all change orders plus the probability value of all open claims equals the projected contract amount to completion.

### Exhibit 27.8. Projecting Contract Sum to Completion

| Owner/Contractor | |
| --- | --- |
| Base contract amount | $2,500,000 |
| Change orders 1-27 total | 72,300 |
| Adjusted contract amount through change order 27 | 2,572,300 |
| Pobability value of claims from master X list | 55,000 |
| Projected contract amount at completion | $2,627,300 |
| Analysis | |
| 15 claims totaling $137,500 × average probability of 0.40 = $55,000 | |

## PROJECTING COST TO COMPLETION

The projecting of cost to completion is somewhat more complex than projecting contract sum to completion. Two factors underlie this added complexity. First, different kinds of activities and work packages will be affected differently by the cost performance ratio (CPR), the time performance ratio (TPR) and certain other performance ratios as well. The independent (fixed) variables control which of the performance ratios must be utilized in the calculations to project cost to completion. Second, work packages and other groupings of activities should be independently calculated and projected to a completion. These results should then be summarized to provide the completed cost projection to completion.

A significant variable which must be addressed to project cost to completion are subcontractor, principal vendor supplier claims arising out of the performance of the work. While there may be some validity to the premise that subcontractor and vendor claims tend to be offset by open owner/contractor claims and, therefore, can be ignored, better and more solid methods are available to address the issues of claims, especially in projecting cost to completion.

### Significant Variables in Projecting Cost to Completion

Two kinds of variables must be considered in projecting cost to completion. These are determined by how resources are committed to the performance of the work and how the job cost will be accounted for. Exhibit 27-9 presents the basic formula for calculating cost projected to completion and the fixed and dependent variables applying to each variable.

### Exhibit 27-9. Fixed and Dependent Variables in Cost Projections

| WORK PACKAGE GROUP | PRINCIPAL VARIABLE | COST PROJECTION = | | | | | |
|:---:|:---|:---:|:---:|:---:|:---:|:---|:---|
| | | QUANTITY × | UNIT PRICE × | TIME | = | COST | × FACTOR |
| I | Quantity | Variable | Fixed | 1 | | Variable | QPR |
| II | Unit price | Fixed | Variable | 1 | | Variable | CPR |
| III | Time | 1 | Fixed | Variable | | Variable | TPR |
| IV | None | Fixed | Fixed | 1 | | Fixed | 1 |
| V | Claims | Fixed | Fixed | 1 | | Fixed | PF |
| VI | Unit Price | Fixed | Variable | 1 | | Fixed (NTE) | UPR |

A basic formula consisting of

$$\text{Quantity} \times \text{unit price} \times \text{time} = \text{cost}$$

provides only part of the requirements for calculating cost projections. Once the cost is determined, it must be modified by an additional performance ratio or factor for validity. Exhibit 27-9 lists six work package groups which have different sets of variables and control aspects relating to projected cost. Also listed with each group type is the principal dependent variable which ultimately controls the cost projection.

Each of the six groups will be discussed in detail and the computational methods for projecting cost to completion presented by a series of examples. Involved in these calculations will be several additional performance ratios not previously discussed including the quantity performance ratio (QPR), the unit price ratio (UPR) and the probability factor (PF); their calculations will also be presented.

**Group I: Variable Quantities.**  Exhibit 27-10 presents group I calculations for variable quantities.

Cost projection calculations for variable quantities requires that a quantity performance ratio be calculated. This calculation is the same as all previously

**Exhibit 27-10. Calculations for Variable Quantities (Group I)**

$$\text{Quantity performance ratio (QPR)} = \frac{\text{actual quantities}}{\text{planned quantities}}$$

| ASSUME: | PARAMETER | SOURCE OF INFORMATION |
|---|---|---|
| Planned quantities | = 100 units | (from budget) |
| Percent complete | = 50.0% | (from TAD) |
| Actual quantities | = 60.0% | (from TAD) |

$$\text{QPR @ 50.0\%} = \frac{100 \times .60}{100 \times .50} = \frac{60}{50} = 1.200$$

Cost projected to completion (CPC):

ASSUME:
Unit price     = $1,000/unit
Planned cost = quan. × unit price = 100 × $1000 = $100,000

CPC = quan. × unit price × time × QPR
     = 100  × 1,000   ×  1  × 1.2 = $120,000

$$\text{CPR} = \frac{\$120,000}{\$100,000} = 1.200$$

presented performance ratios. It compares actual quantities to planned quantities by dividing the former by the latter. To present meaningful calculations, certain assumed parameters have been selected. The normal sources of the information concerning these assumed parameters in integrated cost and schedule control is also presented. The percent completed information and actual quantities, where quantities are a significant variable, must come from the turnaround document. Budgeted cost is quantity times unit price; both quantity (units) and unit price come directly from the budget at the activity level.

The formula required to calculate the cost projected to completion (CPC) is presented. Since time is not a variable in this calculation, the additional information needed is the quantity performance ratio. The result of these calculations is a cost projected to completion of $120,000 compared to a budgeted cost of $100,000. The cost performance ratio for this activity is 1.200. An example of the kinds of activities and work packages to which variable quantities may apply is the purchase of concrete on a fixed unit price basis but with quantities are determined by field requirements.

**Group II: Variable Unit Price.** Exhibit 27-11 presents the calculations required to determine the cost projected to completion (CPC) for activities and work packages when the unit price is the principal variable. An example of this kind of variable is one's own labor where the quantities are fixed, labor is applied on an open ended basis, and the unit price is subsequently calculated.

The data available on variable unit price activities and work packages allows the calculation of the cost performance ratio (CPR) directly. Note that the source of planned cost is the budget. Percent complete comes from the turnaround document and the actual cost from the job cost records or subcontract control. The method of calculating the variable unit price is shown in Exhibit 27-11. The planned cost in the illustration is $50,000. Calculations indicate a cost projected to completion of $51,950. These items are also tabulated in Exhibit 27-16 showing a projected cost of completion summary.

**Group III: Variable Time.** This group of activities and work packages relates to the resources applied to the job where the quantity is considered a single unit, the unit price is fixed but is a function of time, and the principal variable is time. In order to project the cost to completion, it is necessary to have certain planned information including planned time and planned unit price; it is also necessary to obtain from the turnaround document a percentage complete and a number of duration units complete. Exhibit 27-12 illustrates the computation.

## Exhibit 27-11. Calculations for Variable Unit Prices (Group II)

Cost performance ratio (CPR) $= \dfrac{\text{actual cost}}{\text{planned cost}}$

ASSUME:

| | PARAMETER | SOURCE OF INFORMATION |
|---|---|---|
| Planned cost | = $50,000 | (from budget) |
| Percent complete = | 38.5% | (from TAD) |
| Actual cost | = $20,000 | (from JCR or SC) |

$$\text{CPR @ 38.5\%} = \frac{\$20,000}{\$50,000 \times .385} = \frac{\$20,000}{\$19,250} = 1.039$$

$$\text{Actual unit price} = \frac{\text{cost}}{\text{quan.}} = \frac{20.000}{500 \times .385} = 103.90$$

Cost projected to completion (CPC):

ASSUME:

Planned quantity $= 500$ units

$$\text{Planned unit price} = \frac{\text{budget}}{\text{quan.}} = \frac{\$50,000}{500} = 100$$

$$
\begin{aligned}
\text{CPC} &= \text{quan.} \times \text{unit price} \times \text{time} \times \text{CPR} \\
&= 500 \quad\times \$100 \quad\times\ 1\ \times 1.039 \\
&= \$51,950
\end{aligned}
$$

## Exhibit 27-12. Calculations for Variable Time (Group III)

Time performance ratio (TPR) $= \dfrac{\text{actual time}}{\text{planned time}}$

ASSUME:

| | PARAMETER | SOURCE OF INFORMATION |
|---|---|---|
| Planned time | = 75 duration units | (from original schedule) |
| Percent complete | = 57.3% | (from TAD) |
| Actual time | = 35 duration units | (from TAD) |
| Planned cost | = 75 × $750 = $56,250 | (from budget) |

$$\text{TPR @ 57.3\%} = \frac{35}{75 \times .573} = \frac{35}{43} = 0.814$$

Cost projected to completion (CPC):

ASSUME:

Unit price $= \$750/\text{duration unit}$

Planned cost $= \text{quan.} \times \text{unit price} = 75 \times \$750 = \$56,250$

$$
\begin{aligned}
\text{CPC} &= \text{quan.} \times \text{unit price} \times \text{time} \times \text{TPR} \\
&\quad\ 1 \quad\times\ \ \$750\ \ \times\ 75\ \times 0.814
\end{aligned}
$$

$$\text{CPR} = \frac{\$45,788}{\$56,250} = 0.814 = \$45,788$$

The time performance ratio is calculated by comparing actual time to planned time. The result is a time performance ratio of 0.814, which indicates that this activity or work package is approximately 18.6% ahead of schedule at this point in time. The cost projected to completion is calculated and indicates a projected cost at completion of $45,788, which compares to a planned cost of $56,250. The net result is a cost performance ratio of 0.814. It is worth noting that this is the same as the time performance ratio.

This data is transferred to Exhibit 27-16 as part of the projection of cost to completion summary.

**Group IV: Fixed Price or Lump Sum Work Packages.** The work of most subcontractors and principal vendors will fortunately fall into this fourth group of calculations, dealing with fixed price or lump sum application of resources. The cost performance ratio requires a minimum of performance information including the planned quantities, unit price, and cost, which comes from the budget; and a percent complete, which comes from the turnaround document. The actual cost data for the percent complete report comes from the job cost report or subcontract control.

There are actually two versions of the calculations for fixed or lump sum work packages. The first of these omits adding effective claims to the cost projected to completion. These are the calculations indicated in group IV.

**Exhibit 27-13. Calculations for Fixed Price or Lump Sum (Group IV)**

Without claims:

$$\text{Cost performance ratio (CPR)} = \frac{\text{actual cost}}{\text{planned cost}}$$

| ASSUME: | | PARAMETER | SOURCE OF INFORMATION |
|---|---|---|---|
| Planned cost | $= Q \times UP = 30 \times \$250 = \$75,000$ | | (from budget) |
| Percent complete $=$ | 91.5% | | (from TAD) |
| Actual cost | $= \$64,625$ | | (from JCR or SC) |

$$\text{CRP @ 91.5\%} = \frac{\$64,625}{\$75,000 \times .915} = 0.942$$

Cost projected to completion (CPC):

ASSUME:

Planned quantities $=$ 300 units

Planned unit price $= \$250/\text{unit}$

$\text{CPC} = \text{quan.} \times \text{unit price} \times \text{time} \times \text{CRP}$
$\quad\quad = 300 \times \$250 \times 1 \times 0.942 = \$70,650$

Group V will subsequently illustrate how claims are factored in to the cost projected to completion.

The cost performance ratio for the data illustrated is 0.942. When this is inserted into the formula for the cost projected to completion, the result is a projected cost of $70,650 compared to a planned cost of $75,000. These data are also transferred to Exhibit 27-16 to be included in the summaries.

**Group V: Claims Factors.** Exhibit 27-14 provides an illustration on the ways in which claims may be factored into the cost projected to completion for any work package.

**Exhibit 27-14. Calculations for Claim Factors (Group V)**

ASSUME:                          PARAMETER      SOURCE OF INFORMATION
  Same data as above
  Unresolved claims     = $10,000      (work package claims)
  Combined probability =     .40        (composite probability)

$$CPR @ 91.5\% = \frac{\$64,625}{\$75,000 \times .915} + \frac{(\$10,000 \times .40)}{\$75,000} = 0.942 + 0.053 = 0.995$$

$$CPC = quan. \times unit\ price \times time \times CPR$$
$$= 300 \times \$250 \times 1 \times 0.995 = \$74,625$$

The data used in these calculations is the same as presented in group IV for the basic work package. Two additional items are required to calculate the effect of claims on the cost projected to completion. The first of these is the total value of unresolved claims. The second is the combined probability factors. The method of making these calculations is demonstrated earlier in this chapter. A cost performance ratio at 91.5% for the basic work package is obtained using the identical calculation for group IV items. To this, however, is added a cost performance ratio factor relating to the claim, expressed as a percentage of the total planned cost and shown in Exhibit 27-14. The cost performance ratio for the basic work package and the equivalent for the claims are added together, resulting in a cost performance ratio of 0.995. This new ratio is inserted into the formula for the cost projected to completion and results in a projected cost of $74,625, which when compared to planned cost confirms the cost performance ratio of 0.995. This data is included in the summary, Exhibit 27-16.

Claims probability analysis is especially useful within the context of individual work packages since it provides an appropriate arithmetic base for the calculations as just demonstrated though not valid for individual claims.

**Group VI: Fixed Quantities, not to Exceed Price, Variable Unit Price.** The purpose of group VI calculations is to illustrate how cost may

**Exhibit 27-15. Calculations for Fixed Quantities
not to Exceed Price and Variable Unit Price
(Group VI)**

Unit price performance ratio (UPR) $= \dfrac{\text{actual unit price}}{\text{planned unit price}}$

| ASSUME: | PARAMETER | SOURCE OF INFORMATION |
|---|---|---|
| Use group 2 data | | |
| Planned quantity | = 500 units | (from budget) |
| Planned unit price | = $100 NTE | (from budget) |
| Planned cost | = $50,000 NTE | (from budget) |
| Percent complete | = 38.5% | |
| Actual cost | = $18,000 | |

$$\text{UPR} = \frac{\text{actual unit price}}{\text{planned unit price}} = \frac{\text{actual cost}}{\text{actual quan.}} \div \frac{\text{planned cost}}{\text{planned quan.}}$$

$$= \frac{\$18,000}{500 \times .385} \div \frac{\$50,000}{500}$$

$$= \frac{93.51}{100} = 0.935 \quad \text{but NTE 1.00}$$

(Note: The not to exceed cost fixes the upper limit of the unit price ratio at 1.00.)

Cost projected to completion (CPC):

$$\begin{aligned} \text{CPC} &= \text{quan.} \times \text{unit price} \times \text{time} \times \text{UPRNTE 1.00} \\ &= 500 \quad \times \$100 \quad \times 1 \quad \times 0.935 \\ &= \$46,750 \end{aligned}$$

$$\text{CPR} = \frac{\$46,750}{\$50,000} = 0.935$$

be projected to completion where the quantities are fixed, the price is a not to exceed price which means that the variable is the unit price. These calculations are shown in Exhibit 27-15.

The unit price ratio is effective in making the calculations required; these calculations are shown. With input of the data provided from the sources indicated, the result of the calculations is a unit price ratio of 0.935. This means at 38.5% of completion the actual unit prices are running 6.5% below the planned or budgeted unit prices. A unique feature, however, of the the not to exceed price is that the unit price also has a limit which in turn restricts the unit price ratio, not to exceed 1.00. The cost projected to completion calculation is made using the processes previously described and illustrated with the result of a cost projected to completion of $46,750 compared to a planned cost of $50,000, giving a cost performance ratio of 0.935.

## Projection of Cost of Completion Summary

The result of the calculations from each of the six work package groups is used to present a projection of cost to completion summary for a job. It is assumed that each sample cost projection to completion, groups I–VI, represents a work package. The resulting summary of cost projected to completion is shown in Exhibit 27-16.

In a typical construction job, there would be many more work packages than shown in this exhibit. Most of the items or work packages included in the summary would fall in group IV with fixed quantity, fixed unit price, and fixed price for the work package. Tabulation, however, in Exhibit 27-16 does provide some indication of the kinds of information resulting from the summary of cost projected to completion. Planned or budgeted cost is $406,250, representing a cost performance ratio of 1.000. The purchased items total $397,550. When the purchased total is compared to budget, the indication is that the job has been bought about 2.3% below budget. To reconfirm what has been previously pointed out, there are several significant problems with using purchased as a basis for projecting cost to completion. The principal disadvantage of this method is that it focuses on those items which are most likely to be under control, namely, purchases, while ignoring the claims issues, which are most likely to create problems. Purchases also do not take into account the effect of time on the application of resources.

The previous work package calculations of cost projected to completion result in a projected cost of $409,763 which when divided by the budgeted cost indicates a cost performance ratio of 1.009.

In order to determine the cost projected to completion, it is necessary to have previously completed both the cost performance ratio and time performance ratio calculations for each group of activities. Where the variable

### Exhibit 27-16. Projection of Cost to Completion Summary

Assume each of the sample cost projections to completion for groups 1 through 6 represents a work package. The resulting summary of cost to completion is shown.

| VARIABLE | WORK PACKAGE/ GROUP | BUDGET | PURCHASED | COST OF COMPLETION | CPR |
|---|---|---|---|---|---|
| Quantity | 1 | $100,000 | $100,000 | $120,000 | 1.200 |
| Unit price | 2 | 50,000 | 50,000 | 51,950 | 1.039 |
| Time | 3 | 56,250 | 56,250 | 45,788 | 0.814 |
| Fixed price | 4 | 75,000 | 70,650 | 70,650 | 0.942 |
| Claims | 5 | 75,000 | 70,650 | 74,625 | 0.995 |
| Not to exceed | 6 | 50,000 | 50,000 NTE | 46,750 | 0.935 |
| | Totals: | $406,250 | $397,550 | $409,763 | 1.009 |
| | | 1.000 | 0.979 | 1.009 | |

groupings of activities and work packages is a primary function of time, the time performance ratios may be used for these calculations. Other performance ratios may also be useful when they are calculated by using the simple actual divided by planned at a particular percentage of completion.

## Other Methods of Calculating Cost Projected to Completion

While there are other, more detailed methods of projecting cost to completion, the ones which have been presented are preferred for several reasons. The first concerns the logical grouping of activities into work packages for the purpose of making cost projections. This isolates performance by individual work packages and allows specific calculation of performance ratios for the subcontractor or principal vendor responsible for that work. It also provides a convenient summary level which not only matches job costing and subcontract control procedures but provides for common variances in making the calculations. Since normal summaries of job performance including budgets, purchased, expensed, under expensed, and variances are normally presented by activities at the work package level, the methods of calculations for cost projected to completion remain consistent with these summaries.

Another method which may be used in cost projected to completion calculations follows;

Cost projected to completion
= (cost to date from job cost budget for uncompleted labor
× labor CPR
× TPR)
+ (cost of materials not yet in place
× materials CPR)
+ (subcontracts to completion
× CPR)
+ (equipment cost
× CPR
× TPR)
+ (subcontract claims
× probability factor)

The results of cost projected to completion using this method will be more crude and less useful than the other method suggested.

## Grades of Available Information

Important in projecting cost to completion is to use the best information available as the basis for making the cost projections. Here are some examples:

If an activity is known to be completed and all of the job cost associated with this activity has been expensed, the job cost represents the final cost and hence is the best data available for projecting to completion. The next best information occurs where an activity is partially completed and exact performance in terms of its value at that percentage completion is known; with this information, cost may validly be projected to completion. The third best data occurs where an activity has been purchased but no performance has occurred against the activity. In this case, there is no historical basis for calculating performance ratios or for projecting the cost to completion beyond the purchased figure. With the limitations of purchased amounts for projection of cost to completion noted, the purchased figure represents the best information available. Finally, when all else fails, the budget itself may be used provided it is adjusted to reflect known variables such as the cost performance ratio, the time performance ratio or the production performance ratio, which will affect the budgeted values.

## CONTRACT SUM, OVERHEAD, AND FEE PROJECTED TO COMPLETION

Methods of projecting contract sum to completion have already been presented. The basic component of contract sum projected to completion is the basic contract amount plus all change orders issued to date. To this, however, must be added the sum of all claims times the probability factor; keep in mind that the more conservative evaluation of the probability factor will result in a more conservative projection of the contract sum to date. Projected overhead and fee at completion is the simple arithmetic result of deducting the projected contract cost to completion from the projected contract sum.

Three general sets of calculations are involved in providing a total projection to completion of a job still in progress. The first of these is projection of the contract cost to completion, the second is projection of the contract sum to completion, and the third is the estimate of overhead and fee at completion.

These three sets of calculations will result in reasonably accurate projections of cost to completion, contract sum to completion, and projected overhead and fee at completion. These calculations should be routinely made at monthly intervals, and included in a regular monthly administrative evaluation of the project. This evaluation should include, among other things, the calculation of cost performance, time performance, and production performance ratios; the preparation of monthly requisitions; and the evaluation of all owner/contractor claims and all the contractor/subcontractor/principal vendor claims. Such a routine administrative review and evaluation of the project will significantly improve the quality of information available on project performance. It will also highly enhance the organizational control of the project.

## Purchasing and Subcontracting and Cost Projections

In most general contractor environments, a significant portion of the work produced and delivered to the owner is subcontracted or purchased by the general contractor from subcontractors and principal vendors. The subcontracting and purchasing processes present major opportunities for cost control but also represent the major areas for cost control problems. These issues center on how well matched the subcontractor and vendor work packages are to the work included in the general contract between the owner and the contractor. When these are well matched, the gap items are kept at a minimum. To the extent that they nonetheless occur, they are well within manageable parameters.

When these are not properly addressed and the definitions of scope vary significantly, major problems can develop. A basic concept develops from the recognition of these factors. When work is purchased from principal vendors or subcontracted, the general contractor must maintain the integrity of a buy/sell position. That is, the general contractor must buy from subcontractors and principal vendors exactly what has been previously sold to the owner. The relevant descriptions of the work included in the owner/contractor agreement must be repeated verbatim in the subcontracts and purchase orders if the buy/sell gaps are to be minimized. Particularly to be avoided are edited or rewritten descriptions of the scope of the work and listings of exceptions without an exceptions closure statement such as "except cutting and patching, there being no other exceptions." Such redundancy of descriptions of work has the potential of resulting in ambiguity of subcontractor and vendor requirements and becomes potentially the basis for a claim. Each organization should establish clear and specific guidelines with respect to basic owner/ contractor contract documents as the sole reference to be used in subcontracts and purchase orders to describe the scope of the work involved.

## The Purchase/Buy Account

A constant tendency in projecting cost, contract sum, and, as a consequence, profitability to completion is to be overly optimistic, particularly in the early stages of a job. The purchasing and buying processes may result in the improvement or reduction of actual cost below budgeted cost on certain items. Until the job is completely bought out and well along toward completion (perhaps beyond 50%), it is not useful to include the effect of purchasing in projections to completion except where purchased generally exceeds the budget.

A separate budget account called a purchase/buy account should be set up and carried as a regular budget item to handle cost reductions resulting from purchasing. Under this procedure, an item which may be budgeted at $100,000 and is purchased at $90,000 would show a net reduction in the cost of the work of $10,000. But this reduction is not shown; instead, $10,000 is *added* to the purchase/buy account to maintain the total budget and cost at the budgeted level. In the case of an item budgeted at $50,000 and purchased at $55,000, the same process would deduct $5,000 from the purchase/buy account. A management exception, however, should always be required before any purchase overruns are authorized. The total budgeted amounts would still remain constant. Purchase/buy accounts by these processes are utilized to create a reserve as well as a separate accounting for the net effect of the purchasing processes.

As the job moves toward maturity and the purchasing processes are completed, and assuming developing claims are clearly indentifiable, the purchase/buy account may be added as a contribution to overhead and fee when it is positive. However, in any situation in which the purchase/buy account is negative, this amount must be immediately deducted from the budgeted overhead and fee since the only source for funding a negative purchase/buy account is overhead and fee. The projected overhead and fee at completion obtained by the calculations given earlier cannot now be considered valid until it is modified with the addition of the purchase/buy account.

## Estimates of Overhead and Fee

Exhibit 27-17 is a copy of a computer generated profitability estimate based on a cost projected to completion analysis of a project and projected contract sum to completion.

Note the two separate groupings of calculations for overhead and fee. In each grouping, the percentage of overhead and fee is estimated. The first grouping does not take into account the purchase/buy account except when it is negative. The second group takes into account the purchase/buy account, negative or positive.*

---

* The first group of overhead and fee estimates presents a more conservative view especially appropriate early in the job before all buying is done and before job performance is firmly established. It minimizes the tendency to be overly optimistic about a job's early performance.

## Exhibit 27-17. Estimates of Overhead and Fee

```
                              GENERAL CONTRACTING COMPANY, INC.

        October 27,1985                         Thursday 8:05 PM

JC072NB      JOB     JOB PERFORMANCE ANALYSIS   3FLRS  58740 SQ FT  STEEL FRAME,MASY VENEER   ACTIVE   PAGE 001

JOB 7002 CONSTRUCTION PLAZA PHASE I Purchase/Buy ACCOUNT RANGE From 84000 To 84999
DATE OF PROJECTION  BUDGET      PROJECTED COST   VARIANCE
   10/31/85    $1,515,406     $1,558,816      $43,410-

PROJECTED O/H & FEE ANALYSIS  (EXCLUDING PURCHASE / BUY ACCOUNTS)

BUDGET O/H & FEE                              $199,682
PROJECTED VARIANCE                             $43,410-
                                             ----------
PROJECTED O/H & FEE                           $156,272

PROJECTED O/H & FEE AS MARKUP OF PRODUCTION COST
   (PROJECTED O/H & FEE) /
   (PROJECTED PRODUCTION COST)                  11.14 %

PROJECTED O/H & FEE AS % OF CONTRACT SUM
   (PROJECTED O/H & FEE) /
   (CONTRACT SUM)                               10.31 %

COST PERFORMANCE RATIO (CPR)
   (PROJECTED PRODUCTION COST) /
   (PRODUCTION COST)                            1.03

PROJECTED O/H & FEE ANALYSIS  (INCLUDING PURCHASE / BUY ACCOUNTS)

PROJECTED O/H & FEE                           $156,272
PURCHASE / BUY ACCOUNTS                        $7,183-
                                             ----------
MAXIMUM PROJECTED O/H & FEE                   $149,089

PROJECTED O/H & FEE AS MARKUP OF PRODUCTION COST
   (MAXIMUM PROJECTED O/H & FEE) /
   (REVISED PROJECTED PRODUCTION COST)          10.57 %

PROJECTED O/H & FEE AS % OF CONTRACT SUM
   (MAXIMUM PROJECTED O/H & FEE) /
   (CONTRACT SUM)                               9.83 %

REVISED CPR
   (PROJECTED PRODUCTION COST) /
   (REVISED PRODUCTION COST)                    0.99
```

## Differences between Management and Accounting Percentages of Overhead and Fee

Historic differences have always existed between a management statement of overhead and fee and an accounting statement of overhead and fee. The separate overhead and fee percentage calculations shown in Exhibit 27-17 are used to provide both kinds of data.

The difference occurs as a result of the way in which management and accounting account for overhead and fee. In most cases, management will consider overhead and fee as a *markup* over production cost. Accounting will consider the identical data stating overhead and fee as a *percentage distribution* of the contract sum. To illustrate the point, assume a project has the following distribution:

| | |
|---|---|
| Production cost | $ 800,000 |
| Overhead and fee | 200,000 |
| Contract sum | $1,000,000 |

When management calculates overhead and fee as a markup, the result will be $200,000 divided by $800,000, or 25%. However, when accounting uses the same data, it will be considered as a percentage distribution and calculated by taking a percentage distribution determined by $200,000 divided by $1 million, or 20%. Obviously, the raw data and the dollar markup in the job are the same in both cases. Since these different approaches to calculating overhead and fee commonly cause misunderstandings, both are presented in the analysis.

As the project moves toward completion, the second group calculations add in the purchase/buy account as a contribution to overhead and fee.

## SUMMARY

Projections of cost, contract sum, and profitability to completion may be made with a high degree of validity using the performance data generated by integrated cost and schedule control processes. Routine administrative procedures should be kept current in order to improve the quality and timeliness of the data. Claims between the owner and general contractor must be recognized in all projections. Claims from all principal vendors and subcontractors must also be taken into account in making the projections. Reductions in cost projections resulting exclusively from "buy" and reflected in the purchase/buy account should be deferred until all buying is completed and the project is well along toward completion. When the purchase/buy accounts are negative, however, they must be immediately introduced into the calculations as a reduction in overhead and fee.

Projections of both cost and schedule to completion can be very useful management tools in integrated cost and schedule control processes. Projections of schedule to completion tend to be less difficult simply because of a greater industry familiarity with the processes involved. Tedious though they may be, schedule projections to completion are vital issues in schedule control. Frequent updates of schedules for validity is a crucial aspect of schedule control.

Cost projections to completion are somewhat more complicated simply because they involve more detail and must consider more factors. No cost projection to completion and no projection of contract sum to completion can be considered valid unless taken into account are all claims which have come to light in the performance of the project. Probability analyses which are empirically based and easy to use add a dimension of validity to the cost projection processes. The notion that subcontractor and vendor claims tend to offset owner claims should not be accepted as the basis for management information. Similar probability analyses should be used for both owner/ contractor claims and contractor subcontractor/vendor claims. Of course, the most conservative projection of cost and profitability results from adding in all contractor/subcontractor/vendor claims to cost while ignoring any owner/contractor claims which might increase the contract sum.

There are a number of variables involved in the calculations required to project cost to completion. Each of these variables has its own impact on the ultimate cost of the project. Purchasing and subcontracting are important issues in making cost projections. The use of the purchase/buy account as a standard cost control procedure significantly enhances the validity of the cost, contract sum, and profitability to completion projections and at the same time reduces the difficulty of making these projections.

# 28
# UPDATE OF PRODUCTION PROJECTIONS

The issues discussed thus far have focused on single projects from a cost and schedule control point of view. The statement of expectations involving budget, schedule, and other performance models has been developed on a single job basis. However, in an ongoing business enterprise, a number of projects are usually in work simultaneously and at various stages of completion. Each stage of completion of an individual project has certain cost and schedule control characteristics and has a particular impact on the cash requirements, cash flow, and opportunities for profitability of the business enterprise. From an overall business management perspective, the composite effect of a number of projects in progress is both interesting and essential information for management.

Production projections deal primarily with income. In Chapter 13, concepts of extended production projections are discussed in detail. The six month production projection shown in Exhibit 13-4 actually presents a whole range of information beyond merely projecting production.

For example, the primary purpose of the multi-project production projection is to present information on the expected production in the months ahead. Simply by adding cost information (cost = production value less overhead and fee), a whole new set of information useful in cash management and sales and marketing becomes available.

## MANAGEMENT INFORMATION
## FROM PRODUCTION PROJECTIONS

In analyzing the receipts (production) and disbursements (cost) on each job and summarizing these for all jobs in work, useful information is provided on projected production, projected receipts, projected disbursements, net income on a monthly basis resulting from production, reductions of net income attributable to overhead and, with the addition of a starting balance, projected net cash position for each month. In addition, the production projection can be expanded to include projects expected to move into construction within the production projection planning horizon, giving new and additional meaning to the production projection with some forecasting capability.

This reference has a preference for the six months projection, because it seems to suit best the dynamic nature of construction while minimizing the decrease in the validity of projections with time. But production projections may be made over any reasonable time interval that bears a sensible relationship to typical project duration, for example, a fiscal year, a calendar year, even a three month period—whatever suits the requirements of both project and management.

## The Six Month Projection Format

Since all of the processes and procedures discussed thus far deal intensely with creating statements of expectations with respect to cost, production, profitability, cash income and cash requirements, all of the data required for the production projections is immediately available. The format for presenting the raw data and information generated through production analysis varies greatly. The preferred format is in Exhibit 28-1.

The particular nicety of the spread sheet display for production projections is that each project is displayed as a single line entry containing both receipts and disbursements throughout the production projection period. Positive or net income or negative net income is obtained by subtracting the larger of receipts and disbursements from the smaller and displaying the net figure under either receipts or disbursements to indicate an increase or reduction in net income.

The projection begins with the most recent historic month (or selected period) to create a historically validated benchmark of performance. The second interval is the current month (or period) for which the projection should have the maximum validity based upon the historic month (or period). The number of months or periods covered in the actual projection may vary, but four months of projection beyond the current month are preferred since the validity of the project decreases with time from the current month.

## Differences between Net Income and Cash Income

To some extent, the term *production* projection is somewhat misleading. What the projection presents is a view of the net income (not *cash* income) resulting from production. The significant difference is that the production figures include the dollar amounts to be retained under most construction contracts by the owner pending completion of the project. With this recognition, such projections are none the less valid since retainages are held over

## Exhibit 28-1. Form for Six Month Production Projection

GENERAL CONSTRUCTION CO., INC.,

Job # _____    Job Name: Recap all jobs

Summary of Receipts, Disbursements, Cash Flow
Projection and Cash Analysis.

Page __1__ of __4__

Current Month __Oct.__ 19 __85__

| | PREVIOUS MONTH | | CURRENT MONTH | | | | | | | | | | |
| ITEM DISCRIPTION (in $1,000s) | RECEIPTS (1) | DISBMTS (2) | RECPTS (3) | DISBMTS (4) | RECEIPTS (5) | DISBMTS (6) | RECEIPTS (7) | DISBMTS (8) | RECEIPTS (9) | DISBMTS (10) | RECEIPTS (11) | DISBMTS (12) | (13) |
|---|---|---|---|---|---|---|---|---|---|---|---|---|---|
| 1 Job A (1,000 10%) | -0- | -0- | 90.9 | 151.1 | 270.0 | 277.8 | 315.0 | 235.2 | 117.0 | 94.0 | 63.0 | 54.8 | |
| 2 Job B (5,000 10%) | -0- | 159.5 | 750.0 | 755.5 | 1350.0 | 1589.0 | 1675.0 | 1176.0 | 595.0 | 470.0 | 315.0 | 279.0 | |
| 3 Job C (3,000 8%) | -0- | -0- | -0- | -0- | -0- | 97.5 | 270.0 | 461.7 | 810.0 | 849.0 | 945.0 | 718.7 | |
| 4 Job D (2,500 8%) | -0- | -0- | -0- | -0- | -0- | -0- | -0- | 82.8 | 225.0 | 392.0 | 675.0 | 720.7 | |
| 5 Job E (8,000 5%) | -0- | -0- | -0- | -0- | -0- | -0- | -0- | -0- | -0- | 267.3 | 720.0 | 1260.3 | |

the primary construction period and are released as each project passes through the stages of substantial and final completion.

Since retainages being withheld and retainages being released tend to balance each other, valid projections of production volume are provided along with reasonable (though not detailed) approximations of net cash without complicated retainage calculations. The validity of this technique can be confirmed monthly as revealed by the production curve summarized for each project over time. In an ongoing business enterprise, the release of retainages for jobs being completed will balance out against the retainages being held for jobs in the early stages of production. An important caution is that in an accelerating production environment a distortion in the projection will occur which will not reveal a reduction or depletion of cash resulting from acceleration of production.

However, the separate cash analysis provides visibility to this problem and allows an appropriate cash management strategy to deal with the cash shortfall. The converse is true in decelerating production where jobs being completed will be releasing the retainages and give the impression of a larger cash flow then is realistic on a continuing basis.

With the recognition of these two distortions associated with accelerating or decelerating production, the tools for providing production information are in place.

## PRODUCTION PROJECTION UPDATES

Like any other informational process in a dynamic environment, the production projections must be updated on a regular basis. Production projections involve the production value of product placed within specific time intervals and are reflected in rates of production (such as dollars per month). Both the budget and schedule are intensely involved in these processes and consequently, both the budgeting and scheduling updating processes must be kept current and reflect what is actually happening on the job. In order for the production projections to be both valid and current, it is important that the procedural processes and steps which are prerequisite to the update of production projection be completed in a proper and orderly sequence.

### Production Projections and Schedules

The most important initial requirement is the update of the production schedule. The production schedule must be updated using information from the turnaround document, which provides the actual status of work being performed in the field environment. Any adjustments to the logic of the schedule as a result of changes in the work which must be made should be made at the time the data entry processes are completed and prior to the recalculation of the schedule. When the time performance ratios indicate consistently longer duration for activities in the project and when this is reflected in the plan, the durations of the activities yet to be completed should be adjusted to properly reflect the extended durations. Conversely, when the time performance ratios are consistently running less than the planned durations, the discrepancy should also be reflected for the durations of the remaining activities.

Each monthly schedule update which is prerequisite to the production projections should be considered a major schedule review. A major schedule review involves, among other things, the complete review of the logic for the remaining activities in the project based on the performance of activities already completed. The schedule projection to completion serves as a distribution of the remaining production values on a time scale to properly reflect the billing period and rate of placement of product.

## Production Projections and Cost to Completion

The second major item to consider for the production projections is the projection of cost to completion. These processes have been described in detail. To the extent necessary, budget maintenance should be performed to reflect any changes in logic or in the description of activities included in the logic network and schedule. The projection of cost to completion is a formal analytical procedure which provides information on the overall cost performance of the project. Keep in mind, however, that the dollar values assigned for the production curve and consequently included in the production projections are the dollar values included in the budget plus the overhead and fee.

The production projection deals with budget amounts as earned amounts when the activities are completed. The single issue with respect to the dollar value of production is the rate of placement. The rate of placement is determined by the distribution of the dollar value of each activity along the time scale. These values are summarized, usually by monthly intervals, for production projection purposes.

## Production Projections and Contract Sum to Completion

The third major update is the projection of contract sum to completion. This, unlike the projection of cost to completion, has a direct effect on the values assigned for projecting production, especially when there have been significant changes to the contract affecting the contract sum. When significant changes in the contract sum occur during the projection period, these should be reflected in adjustments to both the budget and schedule as added activities to allow the distribution of the earned value associated with the performance of each in the production projection.

## Production Projections and the Cost Curve

The fourth major update is the creation of the cost curve based on the information generated for the production curve. The cost curve reestablishes the planned data for doing the cost performance ratio calculations.

With the foregoing items completed and updated, it is now possible to generate the production distribution by summarizing the planned or budgeted value of each activity falling within each production planning interval (usually monthly). Of course, for those production planning intervals already completed, the actual performance data should be included. Projections are then developed for the current period (or month) and the remaining periods through the completion of the project. What is now available is a complete distribution of expected production values and expected costs for each production period throughout the remaining life of the project. This data provides a clear indication of expected production, expected costs, and net income for each month of each project. A composite summary may be prepared of all projects within a work environment as has been previously described.

## Differences Between Production and Cash Projections

The preparation of the six months projection may be quickly and conveniently prepared by using the single line distribution previously described (Exhibit 28-1). One job is shown on each of the summary lines included in the six months projection. Each includes a summary of job related receipts and disbursements and net income from jobs. An adjustment is made to net cash for overhead expenses. Net income after overhead expenses is also shown. Net income balances are shown when starting balances have been added. The insertion of future work to start within the projection period at a 0.9 probability or higher provides the capability of analyzing the effect of new work on production and net income. It is interesting to note that a variety of summary levels may be included in the analysis and a wide variety of information generated from the production projection. (The processes described here deal with production projections without the usual allowances for retainages, discussed earlier in the chapter.)

It is also possible to do cash projections based on the cash income curve and the cash requirements curve. Where it is the desire to do the projections on a cash basis, the cash income curve is generated as well as the cash requirements curve. These are both summarized on monthly intervals and these data are used in the tabulation in place of the production data. An example of a production projection based on cash rather than on production is shown in Exhibit 13-9.

There are several critical differences between production projections and cash projections. The most important of these are the significant differences in the amounts calculated for a particular period and the critical time phase differences between the two. Not only will the cash produced as a result of production for a specific production period be significantly less than its related production value, but the receipt of cash for that production period will fall significantly outside of the production period. Fortunately, the dynamics of cash requirements tend to shift with cash income, which to some extent lessens the time phase and dollar difference disparities.

The production projection provides data on the real (actual) product to be placed within the time frame of the current production period. Cash projections, however, present data on cash income, cash requirements, and net cash resulting from real production. Cash income as reflected by the cash income curve will always be below and to the right of the production curve, reflecting the time and retainage differences between the two. Cost and cash requirements are similarly related.

These differences are primarily related to terms and conditions of payment and payment of retainages. The summary levels for each job, for all jobs, for general overhead, net monthly income and overall business operations remain essentially the same. As has been previously explained, in an ongoing business operation with a steady business volume, the issue of retainages tends to balance out with jobs being completed and jobs just starting to a point where the differences between the two projections may not be very great. It may, therefore, be primarily a matter of convenience and management discretion as to which method may be used if only one can be routinely prepared. Where the cash method is used, however, the analysis is primarily a cash projection rather than a production projection.

In an ongoing dynamic business environment, both production projections and cash projections are of vital interest to management. Where practical and to the extent that clerical or computer resources are available for completing the projection, it is strongly recommended that both projections be carried as part of the management information system. Both production and cash projections should be routinely prepared on a production period basis since each presents significantly different but equally vital and important information. The method of making the projections are fully developed and illustrated in Chapter 13.

## Minimum Requirements for Production Projections

If there is a particular advantage of production projection updates, it is that they are less formal, less structured and, to that extent, perhaps less dependent on other systems integration requirements. The production projection updates, in fact, can be empirically derived with a minimum of information: adjusted current contract sum, the last historic month performance, an estimate of overhead and fee, and expected completion for projects in work. With the use of the production curve, standard "S" curve, or other data, a production distribution of expected income and receipts can be prepared. These empirical methods may also be used and are perhaps more appropriate for new projects to be started within the planning horizon when budgets and schedules are not complete.

In the context of integrated cost and schedule control procedures, however, production projections and related updates of these projections should be integrated with other informational systems which will routinely provide all of the information required. There are a number of prerequisites to using data and information from other work environments and there is a natural sequential relationship between these. The flow diagrams presented in Chapter 15 give some indication of these items and their flow. Each of these should be the natural consequence of the monthly close of the job.

## Update of Other Contemporary Data

The items which should be updated prior to preparing the monthly production projections and updates based on actual performance include the budget, the schedule, the cost projected to completion, the contract sum projected to completion, the production values, and the Schedule of Values. Also prepared as part of the month-end administrative close of the project will be the cash income and cash requirements data. This information, however, is not directly used in the production projection updates.

The prerequisites to updating the budget include a complete administrative review of all formal changes occurring in the project during the previous month affecting the roster of activities such as change orders. Work authorizations would be a part of this category. These routinely require budget maintenance and should be properly reflected by the addition of activities into the budget. The claims activity within each work package also should be updated to reflect all changes proposed by the subcontractor or principal vendor involved, or occasioned by claims which have been asserted. Standard budget control procedures require that proposed changes and claims be accounted for as budgeted items so that proper job cost accounting may be maintained for the project in general.

The next item to be updated prerequisite to updating production projections is the production schedule. The update of the production schedule is particularly important and follows the update of the budget since activities may be added or deleted as a result of budget maintenance. The importance of the schedule update is that it provides the only information on the distribution of production values over the remaining life of the project for contracts in work and over the entire life of the project for proposed contracts.

The cost projected to completion analysis of the project should also be completed. This analysis will determine the cost to be incurred and also determine production values, both of which will be distributed using the updated schedule information. A disparity could occur between the updated budget, which may include claims and proposed changes at their full value, and cost projected to completion, which may include those same claims and proposed changes reduced by a profitability factor. The more conservative picture would be provided with the updated budget rather than cost projected to completion, but the more likely result is cost projected to completion since not all claims will become changes in the contract.

The contract sum should also be updated by the processes which have been described. That is, the same conditions pertain to the contract sum, which would be adjusted or modified by claims, profitability analysis, those conditions pertaining to the difference between the budget and cost projected to completion. In the case of cost projected to completion, the more conservative presentation is made by using the full value of claims. In the case of the contract sum projected to completion, however, the most conservative picture is presented by not recognizing any changes to the work which have not matured to full change order status.

Production values may be updated by the foregoing processes using (1) the contract sum plus changes plus the probability evaluation of changes and proposed claims or (2) the formal contract sum plus formal changes. Management would have to determine which presentation would most likely serve its purposes.

Where the Schedule of Values is involved in requisitioning, it should also be updated. Keep in mind, however, its primary use is with cash income projections.

The routine monthly updating of budget, schedule, production values, Schedule of Values, and the resultant projection of cost to completion and contract sum to completion will routinely provide all of the information required for the monthly production projection update. Keep in mind that these updates are begun at the activity level within work packages and are summarized for each job. The individual job summaries then become single line entries for the production projection update. The remaining summation procedures may be followed as these have been developed and described.

## PRODUCTION PROJECTION UPDATES AS A SALES AND MARKETING TOOL

One of the major applications of the monthly production projection update is in sales and marketing. Most organizations in construction will have predetermined production goals and objectives. These production goals and objectives will be supported by sales and marketing strategies in order to maintain or exceed the planned levels of production or to achieve increases in production where these are part of the business development strategy. The production projection update provides a clear indication of the organizational probability of maintaining planned levels of production within the planning horizon. When the production projection update indicates a significant change in production levels, appropriate sales and marketing strategies may be designed and implemented to cope with these changes.

When the production projection indicates a significant increase in production which may tend toward exceeding organizational capacity and resources, a reduced, or more selective, sales and marketing effort may be appropriate. When the production projection indicates a significant reduction in production within the planning horizon and the decline is *not* due to internal inefficiencies, contrary adjustments in sales and marketing strategies may be required, including, perhaps, addressing more accessible but less profitable work in the open marketplace such as that obtained through competitive bidding. Trends in production consistently lower than goals and objectives are almost always a clear indication that the marketplace and/or client base must be expanded and production projections adjusted to meet the new realities.

### Production Forecasting and Organizational Development

Another significant use of production forecasting concerns organizational growth and development. When production objectives indicate an increase in production, the regular update of production projections will provide information on the extent to which these objectives are being achieved. As confirmation on actual performance and the related production projections is received, organizational growth and development may be implemented in ways appropriate to organizational requirements while being carried out in the most cost effective way.

Since a component of the production projections and the monthly updates of these also projects net income, some measure of confidence may be developed that, along with organizational growth and development, will occur the appropriate growth of profitability. Such is available *only* through production projections. Not even cash analyses and projections provide this kind of information since cash projections are distorted or skewed to the right as a result of terms and conditions of payment and retainages. The production projection in this regard is a superior business development tool to the cash projection.

## SUMMARY

Production projection update is an important ongoing management information process. When maintained as a part of integrated cost and schedule control, production projections provide a wide variety of information to management which is not otherwise available.

The processes and procedures for preparing production projections are discussed in detail in Chapter 13. The methods of updating these projections and the prerequisites to these updates, however, require special consideration. Included in these processes are additional capabilities of adding future production to the analysis to measure its long term effect. Beyond this, an additional capability is also provided. When a new project is under consideration, its effect on production may be measured by adding it separately to the current production projection. This information can be used in developing appropriate sales, marketing, and bid strategies.

Production projection updates depend on information concerning schedules, cost to completion, and contract sum to completion, and are directly related to both production and the cash curves. While it is possible to use empirical methods for developing production projections, maintaining these projection tools within integrated cost and schedule control will provide more valid data for the projections.

There are significant differences between production projections and cash projections. The primary differences involve the *amount* to be paid in cash and *when* the cash payment will be made. Production projections and cash projections provide different kinds of information, both are recommended and encouraged. Production projections have additional value when these are used to measure production *trends* on an ongoing basis against production *goals.* These, however, are not to be confused with production performance ratios, which are historic measures of performance and compare the actual production in place for a specific time interval with the planned production for that same time interval. Since increasing rates of production tend to consume cash and decreasing rates of production tend to increase cash, the trends indicated by production projections are of great interest to management from cash, liquidity, and capitalization points of view.

# 29
# EXCEPTION REPORTING

The very nature of the construction processes requires continued attention to a wide variety of details. These informational details surround the several hundred activities which make up a project and are handled in the context of both budget and schedule. They probably also involve the myriad of details associated with the design of the project, the preparation and approval of shop drawings, and other technical data required for the fabrication of the specialized materials which make up the project. In addition, checking and rechecking of dimensions, placement of product, and other qualitative controls in the field environment require a considerable amount of attention. In a multiproject work environment where one project manager may be handling several projects, it is very difficult for the project manager to give equal attention to all details relating to the performance of the project. The project manager, however, must retain overall control of the project and be in a position to positively influence adverse performance trends as quickly as these occur.

Since all activities requiring attention are not of equal importance, and activities which are performing satisfactorily require less attention (and project management action) than those not performing properly, some management information system must be devised to distinguish between information items and action items to improve the efficiency and productivity of the project manager.

## MANAGEMENT BY EXCEPTION

In recent years, fortunately, a management information process has developed which specifically addresses the need for this kind of information and divides the informational output into two categories; those items performing satisfactorily and those requiring corrective action. This management process is called *management by exception.*

## The Two Premises of Management by Exception

Management by exception works on two basic premises. The first is that for a properly set up and managed construction project, 95 activities out of every 100 will be performed within the limits and constraints of the statement of expectations. This means that activities falling into this category require a simple confirmation of performance and no other attention. These are the items which are unlikely to be the source of significant problems with respect to cost and schedule control.

The remaining 5 activities out of 100 are those which are likely to cause the problems and which require timely project management action to correct adverse performance. The informational processes on performance must have the capability of sorting out the 5 items or so out of every 100 which require special attention.

The second basic premise in management by exception is that those items likely to cause problems must receive immediate, effective, and vigorous action. This kind of action, particularly in the early stages of the project, will significantly reduce the need for similar action as the project moves toward completion. This premise involves the psychological issues previously discussed wherein the reputation of a job relating to the efficiency of its administrative control is established very early in a project. The degree of tolerance permitted in performance is very quickly recognized by the team players. When tolerance of poor performance becomes the standard, the overall quality of the job inevitably deteriorates very rapidly.

## Preparation for Management by Exception
## Informational Procedures

The statement of expectations, which has been developed throughout this reference, provides the basic framework for informational reporting systems which will, properly applied, greatly facilitate the flow of information. The statement of expectations clearly sets out cost and time performance parameters which are acceptable for each activity and, as a consequence, clearly provides a basis for measuring the performance of all activities and for flagging those for which the performance is not acceptable.

## SCHEDULE EXCEPTIONS AND REPORTING

Several prerequisites must be established to enable management by exception reporting procedures to work effectively. Management by exception with respect to time is, perhaps, easier to identify. Clearly, any time performance which utilizes or runs the risk of utilizing all of the slack or float available for an activity should be flagged as an exception. A much more stringent requirement is to flag any item which does not start on time, which is not completed on time, or which occupies a longer duration then planned for its completion.

The later criterion is, perhaps, too stringent for practical purposes. The dynamic nature of field construction suggests that, except for critical activities, utilization of slack and float might be a better criterion for exception reporting. Of course, any slippage in a critical activity has an immediate impact upon the schedule of the project and should be flagged for exception action the moment it becomes visible.

## The Turnaround Document

The turnaround document becomes a particularly useful vehicle for communicating the exception status of activities in terms of time. Where the start date actually is later than the start date planned, the items should be clearly flagged by field supervision for project management attention; this is the earliest warning of a schedule deviation. The second level of reporting which should involve an exception flag is where a percentage duration is at variance with the percentage of time utilized. For example, if 25% of an activity has been completed but 50% of the alloted time utilized, an immediate exception flag should be generated via the turnaround document as the reporting vehicle. The same procedures are applicable to the completion of the activity also and should use the turnaround document.

Overall, the weekly and/or monthly update of schedule should serve also as a basis for flagging exceptions in overall performance. When particular milestone dates extend beyond the planned time earliest for that milestone, an exception should result. Carefully monitoring time performance through the use of the turnaround document and exception reporting procedures will significantly reduce the number of items requiring attention and give earliest visibility to those which do require attention while they are yet correctable. In this connection, it is recommended that the project superintendent, who first is informed of poor time performance, assume the lead responsibility for flagging time exceptions and for presenting these to management for action.

## The Turnaround Document and Exception Reporting

When the turnaround document is regularly being used for production management in the field environment and is the primary data capture tool for job site performance, exception reporting in the three time related categories may be easily accomplished. Exhibit 29-1 provides a sample turnaround document with partial data entries on time performance.

## Exhibit 29-1. Turnaround Document with Exception Status Indicated

```
        JUNE  6, 1985              GENERAL CONSTRUCTION CO.,' INC.              MONDAY 11:20 AM
                              Actual Schedule sorted by Durations
PS075    JOB 9010 CONSTRUCTION MANAGEMENT SAMPLE      1 FLRS   75000 SQ FT              ESTIMATE  PAGE 04
         Daily Production Activity Check List & Time Report                            Report No :    1
         Period Covered   0 - 3/ 1/85 to    88 -  7/ 1/85           Work Day: 71       Report Date: 06/07/85
=========================================================================================================

     Activity        Description           Comment Early Start Completion Dur. Actual Actual Percent  Req  Pay  Hold Close
    Vendor Number/ Name                                                        Start  Compl Complete Quant Adjust Pay  Out
                                                                               A d j u s t  R e a s o n / C o m m e n t s

    3320 (9    1) CAST-IN-PLACE FLOOR SLAB              06/14/85 07/05/85   15_____:_____:_____:_____:_____:_____:_____


    3335 (9   1) SITE CAST-IN-PLACE CONCRETE           06/27/85 07/18/85   15_____:_____:_____:_____:_____:_____:_____
    437 3365-1

    4200 (9    1) UNIT MASONRY              ALL MASONR *04/19/85  06/07/85  35_0520✪¦_____:_____95_:_____:_____:_____:_____
    273 2356-1 GOOD MASONRY, INC.     P.O.# 5185

    4200 (9    2) UNIT MASONRY              MASY FILL  04/19/85  06/07/85  21_04✪¦_05/30_:_100_:_____:_____:_____:_____
    273 2356-1 GOOD MASONRY, INC.     P.O.# 5187

    5120 (9    1) STRUCTURAL STEEL          ALL STRUCT 06/21/85  07/12/85  10_____:_____:_____:_____:_____:_____:_____
    997 7211-1 STURDY STEEL CO., INC.   P.O.# 5193

    7150 (9    1) DAMPPROOFING                         05/19/85  05/20/85   1_05/19_:_05/20_:_100_:_____:_____:_____:_____


    7206 (9    1) CAULKING                  CNTRL JNTS *05/10/85 05/27/85 * 13_05/10_:_05/22_:_100_:_____:_____:_____:_____
    273 2356-1 WATERPROOFERS, INC.   P.O.# 5206

    8100 (1    1) METAL DOORS & FRAMES      INSTALL    *05/10/85 05/10/85   0_05/10_:__✪__:____45:_____:_____:_____:_____
                EMP #    EMP NAME    HRS :EMP #    EMP NAME    HRS :EMP #    EMP NAME    HRS :EMP #    EMP NAME    HRS
    MONDAY   :                         :                         :                         :
    TUESDAY  :                         :                         :                         :
    WEDNESDAY:                         :                         :                         :
    THURSDAY :                         :                         :                         :
    FRIDAY   :                         :                         :                         :
    SATURDAY :                         :                         :                         :

    8100 (2    1) METAL DOORS & FRAMES      MATERIAL   *05/10 85 05/10/85 * 0_05/10_:_05/10_:_100_:_____:_____:_____:_____


    15400 (9   1) PLUMBING SYSTEMS          UNDER SLAB *03/29/85 04/12/85 * 4_04✪✪_:_04/11_:_100_:_____:_____:_____:_____
    587 0013-1 GREAT CONNECTIONS, LTD.   P.O.# 5212

    16111 (9   1) CONDUITS                  UNDER SLAB *04/01/85 04/04/85 * 1_04/01_:_04/04_:_100_:_____:_____:_____:_____
    384 1053-1 POWER PLUS, INC.      P.O.# 5216

    16400 (9   1) ELECTRICAL SERV.& DISTRIBUTION       06/01/85 06/08/85   5_✪__:_____:_____:_____:_____:_____:_____
    384 1053-1 POWER PLUS, INC.      P.O.# 5216

    81110 (4   1) Builders Risk Insurance   BY OWNER   06/01/85 06/01/85   0_✪__:__✪__:_____:_____:_____:_____:_____
    ---------------------------------------------------------------------------------------------------
                                                                          CONFIDENTIAL
    ------------------------------------------------------------------------------------------------------
    WEATHER:_____ :_REMARKS:_____
    TEMP__:_____:_____
    FACTORS_CAUSING_DELAYS:_____:_____
    ------------------------------------------------------:_BY:_____DATE:____
```

While using the turnaround document for production management in the field environment, field supervision will also use the turnaround document to assign work to the field work force. When an activity is to start, it is first identified on the turnaround document. The actual assignment of work is made from the turnaround document and the actual start date is recorded in the space provided. When the actual start is later than the planned start data, a star or other designation is inserted in the same space with the actual start date. This notes an exception status relating to the start of that item.

When the immediate supervision of the work force reports that the item is completed, it is again located on the turnaround document by field supervision and the actual completion date is inserted. Again, the actual completion is compared to the planned completion; where actual completion is later, a star or other highlighting is inserted in the actual completion date block with the actual completion date. Upon actual completion, the actual duration is calculated by counting the number of days required to complete the project from actual start to actual completion. This duration is then compared to the planned duration; if the planned duration is less than the actual duration, a third star is inserted in the actual column.

## Three Level Flagging of Exception Status

While there may be a variety of ways of flagging exceptional status on the turnaround document concerning time, three level flagging provides the best instant visual reference and attention getter for project management. One star means a late start with a potential opportunity by application of resource at an accelerated rate to maintain the planned completion date. The exact extent of the delay can be determined by comparing the planned start and actual start, and the time remaining to meet the planned completion date. On the basis of this information, corrective measures may be designed to maintain the schedule. When an activity fails to complete on time, the second star appears although at this time the opportunity for any corrective action on that activity has been lost. The third star indicates that an activity not only started late and completed late but it took longer than the planned duration.

The value of three star flagging is not so much related to the activity in question since all opportunities for corrective work will have been lost once a second star appears. Its value, lies in highlighting all of the activities in a single work package which are the responsibility of a specific subcontractor, principal vendor, or supplier. It will become quickly apparent whether the performance of that resource is generally at fault or whether poor performance on a single item may be simply an isolated case.

## Schedule Float and Production Management

The illustrations above do not take into account the slack or float which may be available to this activity as determined by the critical path analysis. Earlier recommendations have suggested that in the conversion of the critical path logic network to time scale each activity be placed in time scale based on its early start plus duration. The result of this conversion to time scale is that the time scale network schedule represents the most optimistic schedule. When schedules are updated on a weekly basis including new scheduling calculations, the slack or float will automatically be taken into account and the new turnaround document will reflect the new status which includes adjustments for slack and float.

Assigning responsibility for flagging scheduling exceptions to field supervision is consistent with the concept of capturing data on performance as close to the site and time of the work as is possible. Clearly, the first visibility to potential schedule problems at a supervisory or management level will be with field supervision. Exception flagging on schedule may be addressed promptly and efficiently when this earliest visibility becomes a standard practice.

More sophisticated applications in the computer environment will present to management a separate exception report. This exception report may include, among other things, the estimated effect of time delays on the ultimate schedule. Whether these procedures are manual or automated, they are equally effective provided the exceptions flags are attended to and acted upon.

The exception procedures with respect to schedule also provide constant feedback on production. For example, if activities are being started ahead of the planned early start and completed ahead of the planned early completion, it is reasonably safe to assume that these activities also result in a production performance ratio greater than one, indicating that product is being placed at a rate faster than originally planned.

## COST EXCEPTIONS AND REPORTING

Several elements involved in cost exception reporting are worth noting. The first of these is that significantly fewer items may be involved in reviewing a project for cost exceptions than for time exceptions. For example, in most general contractor environments, a substantial portion of the work will be subcontracted to subcontractors or let by purchase order to principal vendors on a lump sum basis. The lump sum purchases in subcontracted activities fix the cost at a point in time far in advance of the actual performance. The amount to be paid to the subcontractors and principal vendors and, consequently, the cost when it is incurred will be controlled by these lump sum amounts.

The point of caution is that when purchasing or subcontracting has not been all inclusive or claims have been incurred, these may affect the cost and hence the cost performance ratio. The methods of dealing with claims in the calculation of cost performance ratios in the projection of cost to completion have been dealt with earlier. All unresolved claims where money has been spent represent cost exceptions.

Exception reporting concerning costs should relate primarily to the claims issue and the exceptions properly noted for action. The flagging of an exception with respect to a subcontractor or vendor claim should immediately initiate those administrative actions, described in detail, which include pursuing such claims to a final conclusion as quickly as possible. An exception relating to a subcontractor or vendor claim should immediately cause the item to be placed on the master X list, the basis for the claim determined, and appropriate proposals for change or other action taken.

## Cost Exceptions, Control, and Purchasing

The general concept of the control of cost through the purchasing process has been discussed earlier. The ability to influence cost through the purchasing process continues to be reduced as the project proceeds. There is usually very little opportunity to reduce the cost of an activity or work package once a lump sum subcontract has been awarded or a lump sum purchase issued to a principal vendor. The cost performance ratios for these items are then fixed and the time or opportunity for positively influencing the cost will have passed. When these items are purchased over budget, cost exceptions will result. Such activities will be the subject of a cost exception but with no opportunity for corrective action.

## High Cost Exception Risk Items

Most likely to result in cost exceptions are activities performed on a unit price basis or as a function of time. These include one's own labor, one's own equipment, materials bought from suppliers on a unit price basis, and subcontractor equipment acquired on a rental basis. These must be closely monitored using accurate data capture procedures and certain calculations to reveal cost exceptions.

## Allowed Variances

Before the calculation of cost exceptions, some range of tolerance should be established. A range of tolerance for cost variance insures that minor or

**Exhibit 29-2.** Establishing Need for Dual Variance Criterion

| ACTIVITY | BUDGET | COST | CPR | VARIANCE OVER/UNDER | EXCEPTION |
|---|---|---|---|---|---|
| A | $     100 | $     150 | 1.500 | $    50 | * |
| B | 100,000 | 104,500 | 1.045 | $4,500 | |

insignificant variances do not constantly result in cost exceptions. In the absence of any other criteria, a cost variance of plus or minus 5% is likely appropriate since normal cost reporting data will probably contain reporting variances in that range under most circumstances. Each manager should, however, set the acceptable range of variances.

Variance may also be set as a dollar amount. As a matter of practicality, both a percentage variance and a dollar value variance may be used as a dual criterion for a cost exception report. Exhibit 29-2 presents two examples to demonstrate the value of a dual variance criterion. A variance of plus or minus 5.0% is allowed.

In activity A, a percent variance activity at 5.0% over budget results in a cost exception accounting for a $50 cost overrun. Activity B with a CPR of 1.045 is only 4.5% over budget and even though the cash impact is 90 times greater than A, it does not show as a cost exception. If a maximum cash variance is also included, say $250 (or any amount selected by management), an exception results ($4,500 is greater than $250).

## Cost Performance Ratios

The simplest way to identify cost exceptions in the categories just described is through the cost performance ratio along with a maximum cash variance. The cost performance ratio is determined by the procedures previously set forth and is used as a basis for cost exception. A cost exception occurs at any time the cost performance ratio falls outside the range of 0.95 through 1.05 inclusive or the dollar variance is exceeded. Note that both positive (over) and negative (under) variances are calculated since either may be a significant indicator of a cost exception or a calculation error.

The procedure for making the necessary calculations are similar to those detailed in Chapter 27 and illustrated in Exhibits 27-9 through 27-16. Group I, II and III are the most likely costs to be affected although any may be applicable. The primary difference between calculating the cost performance ratio for cost projections to completion and for exception reporting is that exception reporting is related to an early specific point in the performance of the activity rather than to the ultimate cost outcome. In this sense, exception reporting is historic with respect to performance but contemporary with respect to future corrective actions. Of course, a CPR above 1.05 or below 0.95 at completion would also indicate a cost exception but with no opportunity for corrective action.

The use of a dollar amount as a second exception constraint requires an additional simple calculation:

$$\text{Actual cost} - \text{planned cost} = \text{variance}$$

$$\text{Example:} \quad \$1,200 - \$1000 = \$200 \text{ (over budget)}$$

Cost performance ratios can also take into account variances with respect to required quantities, unit prices, and required time for performance. The time performance exceptions should result from schedule exception reporting but may not be visible through the scheduling process. It is recommended, therefore, that cost projection to completion processes be used also to produce cost exception reports since both consider cost performance ratios, time performance ratios, and other significant variables. Cost exceptions may then be flagged for project management and field supervision action along with schedule exceptions.

## SUMMARY

Integrated cost and schedule control requires that jobs be prepared for production with valid budgeting and realistic scheduling. Since these two processes are primarily communications tools on expectations, the project plan, schedule, and budget must be fully implemented using all of the processes and procedures which have been recommended for these purposes.

One of the major factors involved in the control of the schedule and the budget is the relatively large number of details to be controlled in relationship to the people resource available for that task. While companies' historic performance records may vary to some degree, experience indicates that most of the activities in a properly planned, scheduled, and budgeted project will perform as expected within reasonable tolerances. There is a group of activities, however (usually about 5%) which, in spite of best efforts at planning, scheduling, budgeting, and production management, will not perform within acceptable tolerances.

The design of management by exception procedures allows routine confirmation of satisfactory performance on the bulk of activities while isolating or flagging those not performing satisfactorily for management attention and action.

A recommended process is schedule exception reporting directly through the well established turnaround document. Three star exception flagging includes (1) delay start, (2) delay completion, and (3) extended duration initiated by field supervision and provides the earliest visibility to time performance exceptions while these exceptions are correctable. Consistently poor performance within a particular work package will become apparent and corrective action at that level can be initiated. Some sophisticated computer systems provide a separate exception report.

Cost exceptions and the methods of reporting cost exceptions are somewhat more complex. The procedures previously developed for projecting cost to completion, however, provide the analytical and informational basis for establishing cost exceptions. It is recommended that two measures of variance be used in conjunction with each other for cost exception reporting. The first of these is a percent variance, say plus or minus 5% of budgeted cost. This variance alone is adequate for inexpensive, low volume items. The second restraint is a dollar variance, say $250, preventing large dollar variances on a high volume or costly item which otherwise would occur when the percent variance, even if minuscule, is the sole restraining factor.

In final summary, exception procedures allow performance on all activities falling within acceptable parameters to flow through the management information system with little or no attention on the part of field supervision or project management, while flagging both time and cost exceptions where performance falls outside established parameters. Exception reporting should significantly reduce the informational review loads of project management, field supervision, and all support staff as well. The major objective of exception reporting is to create early visibility to potential problems in a time frame which allows project management, field supervision, and the support staff to identify potential problems as they occur, to ascertain the causes for these problems, and to attack them vigorously in terms of cost and time while there is opportunity to bring them under adequate control.

# 30
# COST CONTROL

Cost control may be defined as the ability to influence positively the cost outcome of a project by modifying negative performance trends. From the discussions throughout this reference, it should now be clear that cost control goes far beyond the simple monitoring of performance and, where possible, adjusting the basis for performance in some way. These actions are only a small part of the larger issue of cost control. In a typical construction project, cost control should be of paramount concern from the earliest involvement to address the construction needs of an owner through the completion of the last punch list item.

## THE ESTIMATE AND COST CONTROL

The cost control issues begin to unfold with the earliest conceptual estimates and cost engineering work. These issues continue through the monitoring, cost engineering, and estimating phases of the project. By the time the project has reached the bidding or contract evaluation stage, many of the opportunities for cost control have already passed.

In competitively bid projects, where there are few opportunities for cost engineering work as a primary cost control tool, the estimating function becomes the primary opportunity for cost control. What is required is accuracy and validity of estimates and a thorough competitive and comparative analysis of pricing from subcontractors, principal vendors, and suppliers as a prerequisite to using such pricing in the estimate. Estimates should be put together in sufficient detail and with sufficient validity and accuracy to minimize the cost control risks involved. The assumption should always be that, when an estimate is prepared and the bid offered, the real likelihood is that the bid or offer will be accepted and therefore the project must be buildable for the amount of the bid.

## THE PROJECT STRATEGY OF COST CONTROL

The estimate or bid as a cost control tool presents opportunities to control the cost of the project, essential to the estimating process. What is required as a prerequisite to the actual estimating is some initial project strategy, which will affect significant factors relating to the cost of the project. For example, a major component of cost is the general requirements cost, which is more or less a direct function of the duration of the project. It is not possible to estimate accurately the time related general requirements cost without first clearly establishing (1) an expected duration of the project and (2) the intensity of project management and supervisory effort which will be required in order to produce the project in view of its difficulties and complexities. The interrelationships of the activities required to produce the project are also significant, along with construction strategy, in having a direct effect on the overall duration of the project as well.

## PRODUCTION, PRODUCTIVITY, AND COST CONTROL

A further major factor in the cost of the work is the methods of production to be utilized. Maximizing production, that is, maintaining the maximum rate of placement of product which the physical constraints of the project allow certainly influences cost control. Note that maximizing the rate of production is a issue different from productivity. Production deals with *methods* of placement including tools and equipment to be utilized; the kinds of work force to be applied; the close coordination relating to availability of materials, tools, and equipment to minimize delays; and similar items. Productivity, on the other hand, deals with improving the *efficiency* of the selected method of production. The major opportunities for cost control lie not in the incremental issues of productivity and efficiency as much as in the selection of the best production methods for cost reduction.

No sensible approach to estimating a project can be dealt with apart from the issue of production methods. Production methods are handled in one of two different ways each time an estimate is prepared. In the first case, an assumption is made with respect to production methods. The way in which an organization customarily produces product is assumed to be applicable with respect to the current project. The second method, however, challenges the validity of "This is the way we always did it." It offers opportunities for applying to a project different production methods which can significantly enhance the rate of placement, resulting in lower production costs, a more competitive bid, and greater profitability.

## ESTIMATE VALIDITY, SOURCES OF INFORMATION, AND COST CONTROL

Another aspect of estimating for negotiating and for making competitive bids, which goes beyond the accuracy of the estimate, is the completeness of the estimate. Every effort and process should be utilized to make certain that *all* items required for the project are included in the estimate.

There are two general areas in which missed items affect the ability to control costs. The first of these is the miscellaneous items not clearly covered in pricing from subcontractors, principal vendors, or suppliers. These "orphans" can in total account for a rather substantial cost gap when they are not fully recognized and built into the estimate. Perhaps the most convincing argument for creating a roster of activities and for preparing a parametric estimate based on the roster of activities independent of sub prices is the likelihood of missing these miscellaneous items when the only pricing information is from subcontractors or vendors. When this kind of estimating is done independently of sub pricing, a clear activity checklist for comparative analysis is generated.

With this information in hand, a comparative analysis among sub prices and against parametric estimating can immediately be performed. In comparative analysis, values may be assigned to any items not included in the subcontractor, principal vendor, and supplier prices to create a value for the complete work package. As a general rule, the complete work package price for a particular group of activities is equal to the marketplace pricing from subcontractors and vendors plus a value assigned for all items excluded from marketplace pricing. When a detailed comparative analysis is not made and such items are missed in the estimate as a result, cost control is severely compromised at the very outset.

## COST CONTROL AND MISSED ITEMS

The second general concern beyond miscellaneous items is always the risk of missing a major work component or of using pricing for an item on the assumption that the item complies with the requirements of specifications when such is not the case. Either kind of omission usually results from a failure to meticulously cross-check all construction and bid documents. There may be a tendency within each organization to rely on one group of the contract or bid documents as a primary source for such technical information. This group may be the drawings or it may be the specifications.

There may, however, be other technical data or bid information which impacts on the cost of the work. For example, an item may be clearly required by the bid drawings but missed in the specifications; the converse may also be true. There may be items not covered in either the drawings or the specifications but included in some subordinate test data or information such as test borings. In addition, the bid documents will frequently be the only source of information on items affecting the cost of the project covered in neither drawings nor specifications, for example, performance and payment bonding requirements and insurance requirements.

In building the roster of activities from which estimating will be done, it is essential that these be cross-referenced to all the available data to eliminate the probability of missing either the "orphan" kind of items or major elements of the project. Another caution stems from using any fixed or limited job cost chart of accounts as the exclusive checklist for inclusions of items in the job. There may be, in fact, items in the job not clearly indicated in such a chart of accounts.

Prior to commencing estimates, there should be a careful and meticulous review of the drawings, the specifications, all of the technical data, and the bid documents, all as a prerequisite to setting up the roster of activities and as a cross-check with the roster of activities while it is being prepared. Upon receipt of any addendum, the effect on the scope of the work should be considered immediately and the roster of activities adjusted to include the addendum.

To recapitulate, no cost control method can fully or successfully cope with cost gaps resulting from miscellaneous items not included or from significant items which may be missed or misinterpreted as a failure to thoroughly examine the technical documents. The ability of the estimating process to contribute significantly to cost control decreases as the estimating process itself proceeds and reaches a zero point when the price is offered or the bid is filed.

## SUBCONTRACTING AND PURCHASING AS A MAJOR COST CONTROL TOOL

Perhaps the most significant factor in cost control once the project has been awarded is the subcontracting and purchasing of the project. Prerequisite to subcontracting and purchasing must be the completion of the budget and the schedule. The budgeting processes, which have been described in preceding chapters, involve a postbid pre-purchasing detailed review of the project. At this point, any last missed items and any miscellaneous items should be discovered. It is essential that the focus be on the discovery of these items, if any, and that these be included in the budget even when the estimate will not have provided budgetable dollars for the items.* Since the cost of work and the schedule for its performance are inseparably bound together, it is also essential that the schedule be in place before purchasing commences.

---

* It is much better to realistically reduce budgeted overhead and fee to cover *all* required items for cost control and to know this before the job proceeds.

## Purchasing Criteria

The recommended criteria for purchasing involve a four step evaluation of the technical data and a five step negotiating process. The four step evaluation is as follows:

1. Not less than three competitive prices.
2. A comparative evaluation for complete scope of work within the work package.
3. Sub or vendor pricing plus miscellaneous items at or below budget for the work package.
4. Independent evaluation of the fair value of the work.

**Competition Defined.** The definition of competitive prices is prices taken for the identical work package based on the same technical information in the same marketplace and at the same time. The competitive prices are both time and scope related. Given the volatile nature of the construction marketplace, prices taken a month apart may vary significantly solely on the basis of market changes during that time interval.

**Scope.** The comparative aspects require that the same definition of scope be provided to all bidders. They also require that a detailed comparison of the competitive pricing be made and this comparison extended to include the complete work package. This methodology will significantly reduce, if not eliminate, the possibility of missed items. It will also facilitate comparisons to the budget.

**Control Purchases at Lowest Level.** Significant in cost control is making certain that all awarding of subcontracts and letting of purchase orders are made at dollar amounts at or below budget. It is a truism but worth recognizing that "If the cost is controlled at the level of each work package, then the cost will be controlled for the job in its entirety." Making certain that each work package is purchased at or below budget is crucial to purchase and cost control.

Careful consideration should be given to the miscellaneous items, which fall into two categories. The first of these is the miscellaneous item normally excluded by a particular trade in a particular marketplace. In this case, the opportunity is to recognize that the work will be required even though it is excluded from a particular set of sub pricing and to make provision for its purchase or the performance of the work by others.

The second involves the inclusion of items in certain trade groupings which may be performed more effectively by others. While it is desirable to make the purchases with respect to work packages as all-inclusive as possible, significant opportunities for cost control involve an examination of the kinds of activities to be purchased, or performed separately. For example, it is not at all unusual for an all-inclusive pricing from a mechanical subcontractor to include rather large amounts of concrete construction for equipment foundations and pads. Since this work may be subcontracted in turn by the subcontractor, it may be best to exclude the item from the purchase of that work package and to make provisions for it to be performed in a less costly way by people more expert in that kind of work.

Further critical issue with respect to subcontracting and purchasing as a major cost control tool is the buy/sell position of the general contractor previously discussed. Crucial is using precisely the same technical data for describing what is to be bought as for what has been sold previously to the owner.

**Fair Value and Cost Control.**   The fourth prerequisite to purchasing involves an independent evaluation of the fair value of the work package to be bought. This simply means standing back from the detail and looking at the whole work package in the context of historic cost, experience, and any other factors which may be appropriate, while raising the question, "Does the proposed subcontract or purchase order amount clearly reflect the dollar value for the work package?"

This process is based on historic experience and will, on occasion, reveal significant market-related cost factors. The item may have a market value which competitive pricing has set significantly higher than its fair value. When this occurs, some alternate negotiating strategy may be required. It may be necessary to appeal to the long term working relationships with the subcontractor or principal vendor to bring the pricing in line with fair value rather than its inflated market value.

## TIME AS A SIGNIFICANT VARIABLE IN COST CONTROL

In two ways time is a significant factor in cost control. The first involves the portions of job costs which are time related as distinguished from those which are quantity related. Most general requirements items, for example, are functions of time. Project management, supervision, temporary facilities, utilities such as field office and storage trailers, and similar items are included in the time related general requirements items.

If a project is budgeted for 12 months completion, each month of time represents 8.33% of the time related general requirements cost. While the principal effort should be directed to realistically planning the schedule and budgeting these items, it should also be directed to controlling the time of the project as a whole. If it is possible through careful monitoring of the project to reduce its schedule by one month, an 8.33 percent cost reduction in the time related items can be effected. Conversely, one month delay in completion will increase the cost of these items by 8.33%.

Controlling time itself is the most effective tool for controlling the cost of time related items.

## The Time/Cost Influence Relationship

There is yet another way in which time is related to cost control. As we have seen, the maximum ability to influence the cost of a project commences during the bidding and negotiating phase. By the time the project reaches the contract phase, a significant portion of the opportunities to influence costs has already been expended. But earlier, as the project is moving through the planning and scheduling phase to the contract phase, significant opportunities to influence cost remain available relating to production (alternate methods of placing project) as distinguished from productivity (incremental improvements by a particular method).

Perhaps the greatest opportunities (or risks) to influence the cost of a project prior to the commencement of work in the field lies in the subcontracting and purchasing phase. These opportunities come in the form of the careful negotiating and awarding of subcontracts and purchase orders to principal vendors as closely to the fair value as is possible. A closely related opportunity involves clear attention to the descriptions of the scope of the work and exactly matching what has been sold to an owner with what is being purchased from a subcontractor or principal vendor.

## Negotiating the Subcontract

By the time the subcontract and purchasing phase of a project is completed, the ability to influence the cost of the project (that is to reduce costs below budgeted values or to improve costs by reducing same) tends to be negative. That is, all of the positive opportunities to reduce costs have been expended and there remains only the negative effects of failure to control cost in the field environment. Failure to properly negotiate the scope of work and subcontract can only result in claims from subcontractors and principal vendors and the consequent risk of increasing cost with a negative influence on cost control. Since subcontracting and purchasing is so critical to cost control and to the ability to influence costs, the six major steps in negotiating a subcontract or purchase order for a principal vendor (Chapter 15) are again worth reviewing. These are:

1. Review the historic performance of the subcontractor or vendor.
2. Review scope of work in detail by reference to contract documents.
3. Review the time schedule for the project and the resources required for the performance of the work.
4. Review the conditions of the work in the field environment where the work will be performed.
5. Review the terms and conditions under which the contracted work will be administered.
6. After (and only after) all issues relating to items 1, 2, 3, 4, and 5 have been negotiated and agreed upon, negotiate the subcontract amount or purchase order sum.

Resisting all pressures to negotiate the subcontract amount or purchase order sum until all other matters have been agreed upon is a crucial but important negotiating strategy since the first five items will surely have a direct influence on the cost of the work.

It is important to keep in mind that one of the major opportunities for cost control of the project is during the negotiating processes with the subcontractors and principal vendors. Great care, therefore, should be taken to provide complete accurate information to the subcontractor and vendor base for pricing. This step must be followed by a careful review of the scope of work and other details which have been discussed. Completing the negotiations and executing of the subcontract and purchase documents well in advance of the need for performance in the field environment is, to repeat one of the last major opportunities to positively affect the cost of the project.

Expediting is a particularly critical portion of cost control which relates primarily to minimizing or eliminating all possible negative effects connected with the delivery of important material. Expediting during the construction phase of the project is especially important since delays will cost time, time will cost money, and the expenditure of money will reduce the control of cost. The issue in expediting is controlling negative cost influences rather than enhancing positive influences.

## PRODUCTION AS A COST CONTROL TOOL

Beyond establishing and maintaining a detailed and accurate schedule for the project, maintaining the expected production or rate of placement throughout the project is also an important cost control tool. When significant and unforeseen delays are anticipated, alternate strategies must be examined and implemented to maintain the level of production set in the original plan. Unexpected delays do not necessarily mean an extension of the time schedule if alternate strategies allow for maintaining the planned rate of placement for a product. The production performance ratio is the best indicator of maintaining production and must be utilized in conjunction with the time performance ratio to get a clear indication of project performance.

## COST CONTROL FOR LABOR AND EQUIPMENT ITEMS

One's own labor, one's own equipment, and equipment rentals differ significantly from subcontractor and principal vendor items in that there is no guaranteed cost performance associated with these items. The cost control of these items is, therefore, somewhat more difficult because careful monitoring of such items must be maintained and immediate action taken when the cost performance ratios are not within acceptable levels. The cost performance ratio is the best indicator of these items' performance whether the monitoring and calculations are done on a completed activity basis or on a percent completed basis.

Significant factors not only include the amount of time required to perform the work, which is a measure of efficiency or productivity, but also mobilization and demobilization for the performance of the work. Historic studies on productivity indicate that the significant factors in low productivity are inadequate supervision, inadequate information about the work to be performed, the absence of specific tools required, the absence of materials required, and the like. The ancillary management and supervisory tasks for direct labor and equipment rental items turn out to be a more significant key to controlling the cost than the actual performance of the physical labor or the work by the rented equipment.

## TIME, COST CONTROL INFLUENCE CURVE

Historic studies indicate that the ability to influence the cost of a project significantly diminishes as the project moves from implementation toward completion, as has been stressed in this chapter. Exhibit 15-2, repeated here as Exhibit 30-1, shows a curve indicating the cost control influence on the time and phasing of a project.

Note that with the progression of time and the completion of each phase or milestone of the project the ability to influence or control the cost of the project moves from positive to negative. This simply means that it is not the items which have been properly handled, properly controlled, and properly administered which will have a negative influence on the cost, but those which have not been so handled. For example, all of the items included in the properly negotiated and executed subcontracts are less likely to create cost control problems than the one item which may have been overlooked in one of the subcontracts. Any serious effort for cost control will be directed

**Exhibit 30-1. Time/Cost Control/Influence Curve**

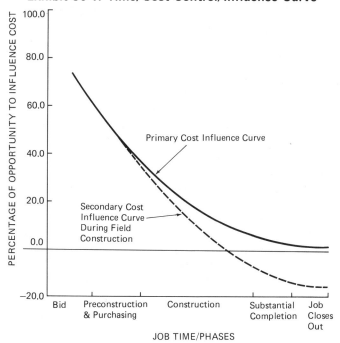

primarily to the proper estimating, budgeting, planning, scheduling, purchasing, expediting, mobilization and implementation of the project rather than to correcting adverse trends as they develop. Positive action up front is a much more effective cost control tool than later corrective action, though the latter will most likely be required and should be anticipated.

## THE COST CONTROL MONITORING AND REVIEW CYCLE

There is a repetitive cycle relating to effective cost control while a project is in progress:

- Strategy or expected performance;
- Implementation;
- Actual cost;
- Comparison to planned;
- Evaluation of performance data;
- Corrective action (new strategy);
- Reimplementation.

This cycle repeats throughout the job and is especially important when adverse trends develop. It is as applicable to time or schedule control problems as it is to cost control problems.

## COST PERFORMANCE RATIOS FOR JOB TO JOB COMPARISONS

Cost performance ratios have an additional, significant value. Since the cost performance ratio is calculated by comparing actual performance to planned performance, the cost performance ratio by individual activity, work package, trade classification, and entire job provides a valid comparison for job to job performance. Job to job performance comparisons are especially valuable in evaluating the performance of subcontractors on different jobs.

Job to job comparisons have other uses as well. Performance of project management and supervision on a job to job basis is of interest. Also of interest is the comparison of the performance of project management and supervision on a particular group of jobs compared to the performance of other project management and supervision on another, similar group of jobs. As integrated cost and schedule control procedures are developed within an organization and as the flow of high quality contemporary information becomes routine, many additional uses of the information will be found.

## SUMMARY

The purpose of this chapter on cost control is not so much to introduce new and novel ideas concerning cost control processes but to summarize the effects and influences of different work environments on overall cost control. The processes of integrated cost and schedule control have their primary value in creating an organized system for planning, scheduling, and budgeting which in turn provide an informational framework for cost control purposes. These processes, procedures, and planning activities do not, however, present some magical formula by which cost control is accomplished. The basic principle of proper planning of work and proper working of the plan still remains the most effective weapon in the cost control arsenal.

The cost control processes must begin with the first visibility to a project. They must continue through the preparation of the estimate, on through the development of the project strategy, and ultimately through the full development of the game plan and the selection of the team members who will be responsible for producing the project.

Issues of production, of productivity, are crucial in cost control. Production deals with selecting the best methods for accomplishing the work at hand. Productivity deals with performing that work most efficiently.

Integrated cost and schedule control procedures provide an abundance of data and information which may be used for cost control purposes. The effectiveness of cost control depends however on how well the information and data available is utilized.

The nature of construction estimating raises the real possibility that items may be overlooked in the estimate. Significant to cost control is that these items be identified when the estimate is converted to a budget, that they be properly built into the budgeting and scheduling processes so that their effects may be considered as a part of the overall project strategy. This, by the way, is one of the more compelling arguments for a formal procedure which reviews the estimates in the process of converting the estimate into a production budget.

In most construction environments, it is hardly possible to overstate the importance and value of purchasing as a cost control procedure. Time and effort spent in the proper and thorough negotiation of all subcontractor and principal vendor agreements will pay substantial dividends not only in terms of cost control but also in terms of the level of effort required by project management and field supervision which may otherwise be required to deal with claims and disputes arising from poor subcontract negotiations. In purchasing subcontractor and principal vendor items, control must be exerted at the activity level since this provides the foundation for complete integrated cost and schedule control and, of course, maintaining the appropriate buy/sell relationships in the negotiations.

Time is a significant variable in cost control. The cost of many activities in a construction project is a direct function of time. To the extent that time is controlled, the cost of time related activities is also controlled. Indeed, reducing time for performance represents one of the few opportunities for cost reductions once a project is underway in the field.

To set all of these in context, we note, construction cost control assumes the ability to influence positively the cost of the work. It is true that as time progresses the ability to influence cost positively (that is, to reduce costs) shrinks until project is bought out and is placed in construction. At this point in time, opportunities for positively influencing costs are nearly exhausted. What remains is the penalty of negative cost influences by failure to control costs. The conclusion to be drawn is that cost control is more directly influenced by the proper planning, scheduling, budgeting and implementing of those plans than by detailed monitoring and reviewing of cost in actual performance. This is not to say that the latter is not important. Simply, the payback is far greater when integrated cost and schedule control procedures are being utilized in support of competent project management and field supervision.

# 31
# THE SCHEDULE AND OTHER
# INFORMATIONAL PROCEDURES

Historically, the construction industry has always had an intense interest in cost control. While the management sciences associated with cost control have gone through some rather dramatic changes over the past 40 years, the basic issues of being able to estimate a job prior to its construction, to create a budget, and to maintain or control the cost within the budget have always been central to profitability in the industry.

Especially from the discussions in Chapter 30, it should be apparent that cost control and schedule control are so closely related that it is difficult to deal with either separately. This closeness is especially true in terms of the effect of schedule control on cost. Effective schedule control will significantly enhance the opportunities for cost control. The inability to control the schedule will significantly depreciate the ability to control costs, especially so on those items which are time related. In fact, it is not realistically possible to control costs without controlling the schedule.

Even as the budget is a cost budget, the schedule for the project is essentially a time budget. The project schedule represents, as a result of the planning processes prerequisite to the schedule, the strategy not only for the time frame in which the project is to be completed but also for the relationships between the activities making up the project through their times for performance.

More than any other of the performance models, the schedule represents a significant communications tool concerning expected performance of the project. All members of the project team should have an intense interest in the time schedule for performance. The architect, owner, and lenders are particularly involved with such issues as cash demands created by the schedule for performance, when the product may be delivered for useful occupancy, and like schedule-control items. The subcontractors and principal vendors need the schedule to measure resource demands for the performance of their work and, of course, the general contractor needs the schedule as a production management tool and also as a primary communications tool. What is required prior to the implementation of the schedule is a detailed review and understanding of the schedule by the project team and an intense commitment to the schedule. Everyone must cooperate to control the schedule.

## MANAGING THE SCHEDULE

The general contractor has the primary responsibility for managing the resources at the job site. The schedule provides information on when activities are expected to commence and when they are expected to be completed. It also provides information on when materials are required to be delivered to the job so that these may be placed when the schedule requires their placement. While not extensively emphasized in this reference, the administrative activities and processes which are necessary to expedite the delivery of materials nonetheless play an important part in overall schedule control.

A separate administrative and purchasing schedule should be prepared for each project based on the production schedule milestones for delivery of materials. Separate monitoring of the administrative schedule will confirm that these processes are being carried out on a timely basis consistent with job site requirements. Such seemingly routine tasks as the preparation of shop drawings by a vendor, submission of these drawings through the contractor to the architect for approval, the review and approval processes, and the return of these to the vendor for fabrication of materials can have a significant impact if not carried out properly. This is one area which is similar to cost control, for which the influences beyond a certain point tend to be negative. The relationship shown in the cost/time ability to influence the cost curve (Exhibit 15-2) presented in Chapter 15 apply to the administrative details which are a function of time as well.

The review cycle applies to schedule control as it does to cost control. The strategy needs to be clearly articulated and communicated; it also needs to be implemented. Data on actual performance needs to be captured. Comparisons in the form of time performance ratios need to be made and, simultaneously, production performance ratios also need to be calculated. Both time performance and production performance ratios are required to get a clear indication on job performance and the probability of a job completing within the allotted time. The comparisons need to be evaluated and a determination made concerning the necessary corrective action. Corrective action is, in effect, a new strategy to address specific problems that occur. The new strategy then needs to be implemented, and the cycle repeats itself.

As in the case of cost performance, time performance is greatly enhanced by using management by exception procedures. This process is discussed in detail in Chapter 29, on exception reporting; it allows those activities performing within the time periods to flow through without direct management attention. Only those items whose performance is falling outside time expectations require project management attention. Exception reporting processes greatly facilitate identifying items requiring attention without spending unnecessary time and effort on items being performed properly; that is, exception reporting allows direct focus on problem areas.

When the information cycle is sufficiently short, alternate strategies for correcting defects in time performance can be designed and implemented. The resultant processing cycle includes developing a strategy for expected performance, implementation of the strategy, data capture on actual performance, comparisons of actual to planned, evaluation of the data, designing corrective action (new strategy), and reimplementation. This cycle continues until the particular item is either completed or no longer appears on the exception report.

The cycle of review thus becomes a critical element in schedule control. The definition used in connection with cost control (the ability to influence outcome) is similarly applicable to schedule control. Schedule control assumes the ability to influence the outcome of the schedule. Mere information on actual performance is of very little value unless compared to expected performance and directed toward correcting adverse trends in performance.

## THE SCHEDULE UPDATE AS A SCHEDULE CONTROL TOOL

There are three kinds of schedule updates. The first of these is a routine update of the schedule based on performance at or near the performance expected, with no significant deviations and no significant modifications to the schedule. This kind of schedule update is the easiest to deal with and involves primarily the recalculation of the schedule to completion based on actual performance to date.

A performance related major update of the schedule, on the other hand, occurs when there have been significant changes resulting in either a deceleration or an acceleration in the actual schedule compared to the planned schedule. A major performance update can also result from significant changes in the work. The performance related update is especially important in connection with the schedule control procedures since it provides an updated statement of expectations taking into account the significant changes.

The third kind of update is logic related. It occurs when significant changes in job strategy have developed or when major modifications to the construction program necessitate a reconstruction of the logic in the schedule for completion. To maintain the validity of the planned schedule, it is important to update or reconstruct the logic from the current date to completion when significant change occur. This allows the actual performance data to be captured in ways closely related to the production control schedule. The comparisons and the generation of the time performance ratio and production performance ratio will then be more closely related and the performance ratios more meaningful.

## MAINTAINING PRODUCTION AS A SCHEDULE CONTROL TOOL

The importance of maintaining production as a cost control tool has been discussed in some detail. Similarly, maintaining production is an important schedule control tool. Production control is, perhaps, even more significant in schedule control than it is in cost control. As we have seen from earlier discussions, the production curve for a project which has been properly planned, scheduled, and coordinated takes the shape of a typical "S" curve as shown in Exhibit 31-1.

This means that the project begins slowly and accelerates to a particular point in the project called an inflection point, at which point it begins to decelerate and slow down moving toward the completion of the project.

When even minor delays occur in the schedule for the project, an unplanned inflection point may occur early in the project. This deceleration of production at an unplanned location must be reversed with accelerating production if the project is to be completed within the original time schedule. This new

**Exhibit 31-1. Typical Production "S" Curve with Point of Inflection**

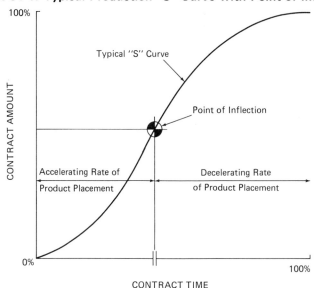

**Exhibit 31-2. Delayed Production Curve with Multiple Points of Inflection**

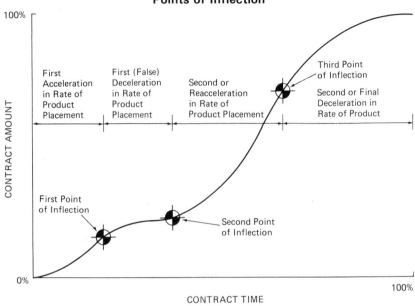

acceleration will result in a third inflection point when the project begins to decelerate into final completion. These effects are shown in Exhibit 31-2.

The very nature of the activities in a project are such that the major or bulk production items occur in the center range of the time of the project. The initial activities, while important, are less cost intensive. The same is true with regard to the activities toward the end or completion of the job.

When delays are encountered, production is maintained by rearranging the schedule for the project. The total dollar value of product placed within a time period is essentially the same even though it may be made up of a totally different grouping of activities from those originally planned. The fact that different activities are involved in production is of no consequence in maintaining the production curve provided a logically developed schedule has been prepared allowing the orderly flow of work toward the completion of the project.

## THE TIME PERFORMANCE RATIO AS AN INDICATOR OF SCHEDULE CONTROL

The methods of calculating the time performance ratio are discussed in Chapter 25; each gives a different view of the dynamics of the project as related to schedule control. Performance ratios (which can be easily and quickly calculated when the project is properly planned and scheduled and when data is captured on actual performance) offer perhaps the most effective way of getting the information for designing corrective action when it is required. The time performance ratio must be interpreted along with the production performance ratio to get a clear indication of overall job performance. When certain elements or phases of the work are running behind schedule but an alternate strategy for the project has been developed and implemented (as evidenced by maintaining production), the likelihood of effective schedule control is greatly enhanced.

## THE IMPORTANCE OF A GOOD START

Perhaps the most significant element of schedule control occurs in the process of launching the project. A properly started job tuned to maximum production from the first day with deviations in the schedule and production being vigorously attacked the moment they occur is the best assurance of schedule control. Second only to a proper job start is maintaining maximum production for the project through both substantial and final completion. There is a strong tendency in the industry to produce 90% of the product in 60% of the time and the remaining 40% of the time to complete the remaining 10% of the project. This tendency can only be overcome by proper planning and scheduling, by the proper implementation of the plan and schedule, by carefully managed periodic updates of the schedule, and by an informational system sufficiently responsive to provide current information on performance while correction is possible.

### Most Likely Delays

While a project may be delayed by weather or as a result of certain processing requirements of public agencies, the most likely source of delay is failure to attend to routine administrative detail. The cure is sufficient lead time to allow an orderly flow of administrative processing which results in material being at the job site when required. Under normal circumstances, the project does not usually lose time in extended or protracted blocks of months or even weeks. Individual days are lost as a result of a failure to get an approval of the shop drawings, failure to award a subcontract or purchase order on a timely basis, a day's delay in an inspection, a day or two delay in receiving material at the job site, and similar items.

Typically, a project will experience a whole series of short delays of a day or two. It is therefore especially important for such delays to be vigorously attacked and solved as quickly as possible, especially during the initial stages of the project when its pace is being established. A delay of a single day may seem of little consequence on a project which may be a year in duration. The accumulative effect, however, of single days can impact substantially on the overall duration of the project and must not be permitted. Careful monitoring of the schedule for field production is essential. Careful monitoring of all the administrative processes, which put product at the job site when it is required, is equally important in schedule control.

## THE CRITICAL PATH METHOD OF ANALYSIS

Historically, the construction industry has been intensely interested in production. The importance of rates of production and controlling schedules have been significantly heightened in recent years due to the increasing effect that time has on several major indirect cost components, such as the cost of construction money during the construction process and the offsetting value of income once the project is completed. As a result of this heightened interest, new techniques for managing the time of a project have been developed, the most significant being the logic network analytical techniques.

For construction practice, the logic network technique most applicable is the critical path method of analysis since it deals with activities to be performed and the relationship between activities. Both the critical path method (CPM), activity oriented, and the program evaluation review technique (PERT), event oriented, have their origins about 1958 in several major government sponsored programs for which controlling the time schedule and, indirectly, costs were major factors.

The critical path method is now recognized as a useful tool for the construction industry. Earlier planning and scheduling methods for construction such as the bar graph or Gantt charts made no clear division between the planning and the scheduling processes. Planning how the project was to be put together was interwoven with the scheduling process. These methods tend to focus on trade groupings in a rather crude way rather than on the activities which are required to be performed to produce the project.

### Separate Project Planning

One of the significant major contributions, then, to the construction industry provided by the critical path method is the clear separation of the project planning process from the scheduling process. Planning using the critical path method allows the diagrammatic layout of a project by concentrating on the activities required to complete the project, and their inter-relationships of these activities, without consideration of the project's schedule. Once the plan for the project is completed, as articulated in the logic network diagram, a separate consideration of the resources required and already available to perform each activity can be made independently. A further analysis, referred to as *workforce leveling,* can also be made to deal with potential conflicts or overload demands for the same resource to perform multiple activities which may occur in the same time frame. These analytical techniques allow for the determination of the duration expected for each activity independent of that activity's relationship to other activities in a logic sense.

## Time Boundaries

Establishing the planned time boundaries for all events and all activities within the logic network diagram resolves itself into several sets of simple arithmetic computations. These computations are capable of accurately determining the time earliest and time latest for every event in the logic network diagram and establishing early and late start and early and late finish for all activities.

## Identifying Critical Activities

Critical path analysis has the capability of clearly identifying the particular chain or sequence of activities which, within the context of the logic network diagram, controls how long the project will take from start to completion. These activities are referred to as the *critical activities,* from which the name *critical path method* is derived. Critical activities may not be the most important by other measures but they do control project time.

## The Critical Path Method as a Communications Tool

A later discovery with respect to the critical path method is its extraordinary value as a communications tool. Previous attempts to graphically present schedule information through Gantt or bar charts or other methods were effective in communicating the planner's intentions only in the broadest sense. The critical path method, however (when the list of activities are generated at the level of detail—and control—required for implementation of the project) has the capability, in the hands of a skilled planner, to completely eliminate ambiguity with regard to the planner's intentions.

With the recognition of the potential weakness in the description of the scope of each activity within the logic network diagram, every technically competent critical path method practitioner will come to precisely the same conclusions about the *relationships* between specific activities shown in the critical path logic network diagram. Further, each skilled practitioner will arrive at exactly the same arithmetic results in time boundary calculations.

Though each planner may have a different view of logic relationships, particularly in those areas in which the controlling logic is planner's choice rather than physical constraint, there can be no misinterpretation of the logic actually shown in the logic network.

A recognition of this simple fact about the analytical result or output from the critical path method clearly nominates the method as an extraordinary communications channel extending far beyond simple planning and scheduling techniques. The capability to articulate a complete project strategy showing the relationships between activities to be performed in an unambiguous way is a tool which has been desperately needed by the construction industry for many years. This capability opens horizons of systems integration which would have been difficult if not impossible to achieve by any other process.

## THE NEGATIVE ASPECTS OF THE CRITICAL PATH METHODOLOGY

The critical path methodology for construction has not received universal acceptance. In fact, its potential utilization as an effective ongoing production control tool and informational system has not been fully realized. The extent to which it continues to be utilized is more a matter of owner or specification requirements than the fact that it has been found to be effective in production management. The negative reputation acquired by the critical path method in construction is rather unfortunate since it does represent a rather substantial advance in the science of planning and scheduling techniques and is an extraordinary communications tool.

### CPM Does Not Solve Problems

There are, perhaps, two major reasons the critical path method for production management in construction has lost favor from its initial acceptance. The first of these stems from expectations far beyond the capabilities of any planning and scheduling method. The expectations were such that if a project were planned and scheduled using the critical path method, the automatic result would be a project properly controlled in the field environment in terms of production. The unreasonableness of this expectation is clearly seen when applied to any other construction scheduling method; any planning and scheduling method of itself, apart from implementation, management, monitoring, and corrective action, will *not* result in a controlled production schedule and in this regard, CPM is no different.

## CPM as an Informational System

The critical path method is primarily an *informational* system. It has extraordinary capabilities to provide a clear articulation of a statement of expectations. Apart from careful management and administration, the statement of expectations will not automatically result in a well controlled project no matter what method used in its preparation.

These problems are an integral part of the construction process. The critical path method is (perhaps surprisingly) not a *problem solving* method. It is, in fact, an information generating method which provides the problem solver (project management and field supervision) with sufficient current valid information on the cause of a problem so that creative solutions can be designed to solve it. In short, these informational processes provide the *basis* for problem solving.

Noting the basis for delays (cause and effect and other items) does not of itself provide for a solution. When the critical path method is clearly recognized as an analytical and information generating method and project management and field supervision assume the responsibility for problem solving, *then* the information and analytical capabilities of the method are quite extraordinary.

## CPM Provides More Information Than Is Useful for Production Management

The second characteristic of the critical path method which has led to its disfavor within the construction industry is, unfortunately, directly related to the analytical power of the method. The critical path method not only identifies the critical activities but also the near critical activities, the amount of float or slack available for each activity which is not critical, the time earliest and time latest for every event throughout the schedule, and a total of eight different time boundaries for every single activity. These informational capabilities are quite impressive.

However, from a field production point of view not all of the information generated by the analytical method is useful. The classic (older) computer outputs from computer critical path analyses have the capability of literally burying field supervision and project management in data. Much of this data, however, is not at all useful for production control and management. In fact, the plethora of data is detrimental to production management since the sheer abundance of data requires much time and attention which is not directed toward enhancing production.

**What Is Helpful in Production Management.** Of the eight activity time boundaries and the two event time boundaries provided by critical path analysis, surprisingly *only one* is directly useful in production management—in planning for maximum production, *the early start* ($E_S$) *for each activity.* Early start is most significant to project management. The second characteristic necessary is the planned duration for each activity. While other factors may be used in measuring performance and in generating the exception reporting capability, the time scale network for production control should be clearly based upon early start plus duration.

Indeed, this reference prefers early start plus duration as the basis for production management. In the field production environment, it is unnecessary to deal with the other time boundaries either for activities or events or to deal with the issue of slack and float. Slack and float represent resources owned by project management to deal with contingencies involving the overall production schedule. Early start plus duration, however, will provide a valid production schedule (and most optimistic as well) for the project and allows for the flexibility of resource leveling when needed.

Considerable practice and discipline is required to set aside or discard a substantial portion of the critical path analytical output. Success in production management, however, is greatly enhanced when this discipline is achieved.

## THE BASIC FRAMEWORK FOR SYSTEM INTEGRATION

Much research and effort has been expended in recent years to determine an appropriate method for the integration of both cost and schedule control. The results of these investigations have, until recently, been singularly less than successful. The usual result is an extraordinary and very complex system which cannot be effectively managed in a manual environment and which requires substantial computer resource in a computer environment.

Several basic discoveries recently have led to the keys to systems integration. One is the uniquenesses of the critical path method in its movement away from trade orientation of estimates and schedules to an activity orientation. The planning processes using CPM are directed to creating the list of activities for a project which must be completed if the project is to be completed. The list of activities and their relationships to each other are used in the critical path method to define the way in which the project is to be constructed. It is therefore apparent that production management in the field environment should be carried out on the same basis. The complete scheduling of the project using the critical path output involves calendar dating the time boundaries of events and activities to provide practical, implementable schedule information.

## Integration of All Informational Systems

A concomitant discovery was that there are no impediments to estimating by activities rather than by trade groupings. When an estimate is prepared by activities and converted to a budget by activities, and a common list of activities exists for both budget and schedule, the clear basis for schedule and cost control integration is at hand. The data reported from the field environment on activity start, duration, and completion provides sufficient information on performance in the field environment. When the budget, sub-contracts, purchase orders, and workforce assignments are integrated into the same informational framework, field confirmation of actual performance automatically provides actual performance data for every other informational process.

Based on the performance models discussed earlier in detail, knowledge that an activity is completed in the field environment provides the following cost related information with respect to that activity:

1. The amount earned stated in budget terms.
2. The amount earned stated in production terms.
3. The scheduled value for requisitioning purposes.
4. The cash value to be received in accounts receivable terms (A/R).
5. The cost incurred to produce the work in job cost terms (J/C).
6. The cash value to be paid to subcontractors and vendors in accounts payable terms (A/P).
7. Cash income anticipated.
8. Cash requirements.

Specific information on all the foregoing items may be generated by the simple knowledge of when an activity has been started, what its current percentage of completion may be, and when it is completed. This level of systems integration is quite extraordinary. Using the activity as the lowest common unit of work for informational descriptions is, without a doubt, the principal key to systems integration.

The activity is crucial to the critical path method for planning and scheduling. Using the activity as the lowest common unit of work requires a basic modification in the estimating technique. In reality, estimating by activity with summary capabilities at the trade level is not significantly different from estimating by trade groupings. Estimating by the separate activity, however, provides an estimated cost based on the simple quantity × unit price formula for every single activity throughout the job; this capability lends itself directly to a budgeted value for every activity. Establishing the quantity, unit price, and budget value for every activity opens horizons for directly linking all activities in all subcontracts and purchase orders to the roster of activities with values assigned to each activity for cost control and payment purposes. This establishes integration of accounts payable with budget and schedule and provides significant cash flow and cash management opportunities.

## COMPUTER APPLICATIONS

Integrated cost and schedule control tends to increase the level of detailed information and data which require processing and handling. While it is entirely possible to deal with integrated cost and schedule control by the manual processes described in this reference, these opportunities are significantly enhanced with the use of computer data processing, particularly appropriate since the critical path method was generated and validated out of computer technology. Integrated cost and schedule control, which does not require recasting of data to address the requirements of information and data for the individual construction work environments, is, in fact, a unique application.

The future of the construction industry and the ever heightening economic impacts on project management and production management will require integrated cost and schedule control to deal with all of the complexities of field production and business management as well.

## SUMMARY

The relationship between schedule control and other informational processes in integrated cost and schedule control is especially important. The design of integrated cost and schedule control processes is such that the capture of data on actual performance in the field environment provides the basis for all other informational processes. The management and control schedule is, therefore, of particular importance from an informational point of view.

The process of managing the schedule involves the normal review cycle, previously discussed, which detects negative trends early so that corrective action may be taken. Considering the dynamic nature of construction in the field environment, regular schedule updating as frequently as weekly to relate the planned future on site performance to actual historic performance is crucial. In addition to these routine updates of schedule, major updates may be required when there are significant changes in the level of production at the job site. On occasion it may be necessary to update the schedule logically to reflect broad changes in the job or other significant factors.

One of the most effective schedule controls is maintaining production. It is necessary to modify the application of resources in the field environment to prevent unplanned reductions in the rate of on-site production. Therefore, both the production performance ratio and the time performance ratio are good indicators of how effectively the schedule is being controlled and how well field production is being managed.

As a schedule control tool, the critical path method has both advantages and disadvantages. Its primary advantage is its capability of separating project planning from project scheduling and of developing a complete project strategy based solely upon activities to be performed. The critical path method represents an extraordinary communications tool for integrated cost and schedule control. There are, however, negative aspects of the critical path methodology. The first is to recognize that CPM does not solve problems but generates information. The second is to recognize that not all information generated by CPM analysis is directly useful in schedule control and, consequently, in terms of production control.

The basic framework for systems integration is provided by the way in which the project is planned and scheduled. The key reporting vehicle on actual performance in the field is the turnaround document, which feeds all the other contemporary informational systems. These contemporary informational systems allow current valid information to be extracted on job performance for management purposes far in advance of the more formal job cost accounting systems.

Integrated cost and schedule control processes involve the handling of large quantities of raw data and the manipulation of this data into useful management information. The clerical support requirements for these informational processes increase with job size and complexity and with the number of jobs being handled simultaneously. It is obvious that integrated cost and schedule control processes lend themselves in a rather extraordinary way to computer applications. While computer applications will provide additional useful capability, the processes and procedures for integrated cost and schedule control which have been presented can be handled and will work in a manual environment.

# Appendix A
# Preconstruction Job Planning

Three levels of planning are required to prepare a project for integrated cost and schedule control. Each level is described below in the context of how it is introduced and used in this reference.

Level 1. The Project Strategy
    1. Logic—Using the logic network, describe diagrammatically HOW a job will be built.
    2. Schedule—Using resource analysis, describe the resources to be used, set the duration of each activity, and calculate the time boundaries to establish WHEN.

Level 2. The Statement of Expectations
    3. Cost—Add to the descriptions of work from 1 and 2: cost and budget values; and to HOW and WHEN: the additional information on HOW MUCH.

Level 3. The Game Plan
    4. The Team—Add to the information from 1, 2, and 3: the specific assignment of responsibility for performing each activity; that is, WHO will perform the work.

The game plan is complete only when all four questions have been answered for each activity: HOW, WHEN, HOW MUCH, and WHO?

# Appendix B
## Summary Levels of Detail for Project Informational Processes

Eight summary levels are used within the reference. These are listed in order from the lowest unit of work to the project level.

| NUMBER | DESCRIPTION | DEFINITION |
|---|---|---|
| 1 | Activity | Lowest (indivisible) unit of work. |
| 2 | Task | Any group of several related activities. |
| 3 | Work package | Any group of activities to be assigned to one subcontractor, principal vendor, supplier, or one craft group. |
| 4 | Trade group | Any group of activities relating to one technical specialty (such as the Standard CSI 16 Divisions). |
| 5 | Functional group | Any group of activities of which the functional components are usually used for parametric estimating (such as the Standard UCI 12 Divisions). |
| 6 | Production unit | Any group of activities making up one phase of a job. |
| 7* | Job | Any group of activities and/or production units making up a phase of a project. |
| 8* | Project | Any group of activities, production units, and/or jobs having a specific origin and objective, the completion of which is the same as the completion of the project. |

* "Job" and "project" are used interchangeably in this reference.

# Appendix C
# Cost Types and Their Uses

Cost types used in this reference are a convenient way of numerically identifying the kinds of resources to be used to perform each activity in the project. Cost types are assigned to each activity for planning purposes before the actual resource to be used is known. This procedure allows the planning processes to continue and for all performance models to be prepared prior to the buying of the job. The table is divided into "own" and "sub" resource groups; each cost type is described.

| COST TYPE | DESCRIPTION | DEFINITION |
|---|---|---|
| | **Own** | |
| 1 | Own labor | The employed workforce making up one's own labor. |
| 2 | Own material | Material from one's own stock or purchased from supplier's stock. |
| 3 | Own equipment | One's own equipment. |
| 4 | Own, other | Other cost not directly fitting into 1–3. |
| | **Sub** | |
| 5 | Sub labor | Labor other than employed workforce. |
| 6 | Sub material | Material worked to specific job requirements and furnished by a subcontractor or principal vendor. |
| 7 | Sub equipment | Equipment other than owned. |
| 8 | Sub, other | Other cost not directly fitting into 5–7. |
| 9 | Sub lump sum | Any combination of 5–8 awarded on a lump sum basis where cost or schedule control does not require cost breakdown into 5–8. |

# Appendix D
# Performance Ratio Comparisons

Contemporary cost and schedule control requires current, valid performance information while any indicated poor performance is correctable. The performance ratios developed by the author in the context of formal planning, scheduling and budgeting, and matching data capture on actual performance, quickly provide the information required and—just as important—when it is required.

Actual performance is always compared to planned performance:

$$\text{Performance ratio} = \frac{\text{actual performance}}{\text{planned performance}}$$

While performance ratios may be calculated on any parameter, the most useful are the *cost performance ratio* (CPR), *time performance ratio* (TPR), and *production performance ratio* (PPR), shown in the table below.

As a general rule, nominal performance will result in a performance ratio of 1.00; good performance, less than 1.00; and poor performance, more than 1.00. The production performance ratio (PPR) is a notable exception.

| | PERFORMANCE VARIANCES | | |
| --- | --- | --- | --- |
| | POOR | EXPECTED | GOOD |
| Cost performance ratio: | | | |
| CPR = actual cost/planned cost | CPR greater than 1.00 | 1.00 | CPR less than 1.00 |
| Time performance ratio: | | | |
| TPR = actual time/planned time | TPR greater than 1.00 | 1.00 | TPR less than 1.00 |
| Production performance ratio: | | | |
| PPR = actual production/ planned production | PPR less than 1.00 | 1.00 | PPR greater than 1.00 |

# INDEX

# INDEX